Designing with Community Participation

COMMUNITY DEVELOPMENT SERIES

RICHARD P. DOBER, AIP, Series Editor

Bechtel/Enclosing Behavior
Becker/Housing Messages
Bednar/Barrier-Free Environments
Bell, Randall, and Roeder/Urban Environments and Human Behavior: An Annotated Bibliography
Branch/Planning Urban Environment
Branch/Urban Planning Theory
Coates/Alternative Learning Environments
Conway/Human Response to Tall Buildings
Finsterbusch and Wolf/Methodology of Social Impact Assessment
Golany/Strategy for New Community Development in the United States
Hesselgren/Man's Perception of Man-Made Environment: An Architectural Theory
Hester/Neighborhood Space: User Needs and Design Responsibility
Koncelik/Designing the Open Nursing Home
Küller/Architectural Psychology: Proceedings of the Lund Conference
Lang, Burnette, Moleski, and Vachon/Designing for Human Behavior: Architecture and the Behavioral Sciences
LaPatra/Applying the Systems Approach to Urban Development
Lawton, Newcomer, and Byerts/Community Planning for an Aging Society: Designing Services and Facilities
Michelson/Behavioral Research Methods in Environmental Design
Moore and Golledge/Environmental Knowing: Theories, Research, and Methods
Perks and Robinson/Canadian Planning Practices
Plēsums/Townframe: Environments for Adaptive Housing
Pollowy/The Urban Nest
Popko/Transitions: A Photographic Documentary of Squatter Settlements
Preiser/Facility Programming: Methods and Applications
Procos/Mixed Land Use: From Revival to Innovation
Rondinelli/Planning Development Projects
Sanoff/Methods of Architectural Programming
Sanoff/Designing with Community Participation
Sleeman and Rockwell/Instructional Media and Techonology: A Professional's Resource
Stearns and Montag/The Urban Ecosystem: A Holistic Approach
Unesco/Planning Buildings and Facilities for Higher Education
Way/Terrain Analysis: A Guide to Site Selection Using Aerial Photographic Interpretation, 2nd edition
Zube, Brush, and Fabos/Landscape Assessment: Values, Perceptions, and Resources

Designing with Community Participation

Edited by

Henry Sanoff

North Carolina State University
at Raleigh

CDS/40

DOWDEN, HUTCHINSON & ROSS, INC.
Stroudsburg, Pennsylvania

McGRAW-HILL BOOK COMPANY
New York St. Louis San Francisco Auckland
Bogotá Düsseldorf Johannesburg London
Madrid Mexico Montreal New Delhi Panama
Paris São Paulo Singapore Sydney
Tokyo Toronto

To Joan, Ari, and Zoe

DESIGNING WITH COMMUNITY PARTICIPATION

1234567890 HD KP 78321098

Library of Congress Cataloging in Publication Data

Main entry under title:
Designing with community participation.
(Community development series ; v. 40)
Bibliography: p.
Includes index.
1. City planning–Citizen participation–Addresses, essays, lectures. 2. Community development, Urban–Addresses, essays, lectures. 3. Cities and towns–Civic improvement–Addresses, essays, lectures.
I. Sanoff, Henry.
HT166.D387 1978 309.2'62 78-3723
ISBN: 0-07-054692-4

Contents

Series Editor's Foreword

Professionals concerned with the built environment in recent years have increasingly embraced the principle of client-centered design. They expect owner and user to help determine what is to be constructed, to help find the most beneficial ways to join together programmatic elements into a functionally workable scheme, and to have some influence on those decisions that give color, texture, and other visual attributes to the eventually constructed space.

While client involvement is welcomed by many planners and designers, the actual carrying out of client-centered design processes too often remains mere rhetoric. Under the pressures of budget, time, and organizational issues (and occasionally from inexperience), what began with good intentions is abandoned for more traditional problem-solving approaches.

In this perspective, Henry Sanoff's new book *Designing with Community Participation* makes a substantial contribution not only to the theory of client-centered design, but also to its practice. Multiple examples of successful projects, including ideas, methods, concepts, and start-up strategies, are given that await only intelligent adaptation. Like any worthwhile contribution of this kind, the proposals have been kitchen-tested by Sanoff and his associates, which means they can be used as a reliable point of departure for as many variations on a theme as available creative energy and thought permit. The illustrations that show how to organize and present design information insure that this book will be a handy reference work for all practioners.

The Community Development Series is intended to facilitate the exchange of information, expert advice, and experience among professionals concerned with the built environment. CDS publications include state-of-the-art books, work manuals, handbooks, and other forms of information transfer. They are offered to planners, architects, social scientists, landscape architects, engineers, and those in related disciplines who will benefit from having such knowledge in a readily convenient form.

Themes central to CDS books include: systematic searching out of patterns, relationships, and behavioral settings as a prelude to design; a high regard for the interdependence of communities at all scales of human habitation; ecological ethics; and an interest in not just finding appropriate design solutions, but in establishing ways and means for having those solutions implemented. *Designing with Community Participation* helps meet an important CDS publishing objective—better tools for enlightened professionals.

Richard P. Dober, AIP

Preface

The increased sense of social responsibility of architects and planners took root in the mid 1960s when a new feeling of community consciousness prevailed over many low income urban neighborhoods. One result was the establishment of community design centers whose purpose was to offer design and planning services to enable the poor to define and implement their own planning goals.

During the last decade the militancy towards social change has abated yet there is still the predominant feeling that keeps the eighty community design centers operational throughout the nation. This persistence has shown us evidence of the absorption of many of the CDC goals into design offices, public planning agencies, and community development legislation.

This volume represents second generation thinking about community participation and is dedicated to design center people whose impact on the design and planning world has not adequately been recognized.

Henry Sanoff

Designing with Community Participation

Introduction

This book is paradoxically about designing and planning without the presence of designers and planners. It stems from the recognition of the scarcity of a design resource that is often unavailable and inaccessible for a wide range of community needs. Nevertheless, decisions are often made by a few that affect many; decisions are made because a crisis situation demands immediate action or decisions are not made and there is no change in the environment.

The results of almost a century of research in the field of medicine have produced volumes of precautionary aids, guidebooks for parents, and a vast array of options for maintaining good health. A lack of any sustained research efforts in the design field has not permitted consumer guides to develop, although there seems to be an almost alarming need for this type of aid, despite some controversy about the meaning of good design.

This book attempts to fill the void of providing participatory methods for solving design problems. There is no attempt to offer quick solutions or magical remedies to recurring classes of problems. Quite the contrary. It is my belief that many environmental problems requiring technical guidance can best be solved through the active participation of those affected by the design decision.

Interest in user needs or user participation is not rooted in romanticism about human involvement but rather in the recognition that users have a particular expertise different than, but equally important to, that of the designer. This expertise then needs to be integrated into the process that concerns itself with environmental change. Although designers and planners have gracefully embraced this aspect of human involvement in decision making, the question of the designer's accessibility to many community groups is still left unanswered.

The material substance of this book represents an exploratory probe into what might be described as design assistance techniques. My intention for organizing this book is to transfer or distill (for general application) those processes unique to designers. The content represents an accumulated knowledge and experience base that has been documented for the use of architects, planners, and people desiring professional assistance. The procedures described herein may not eliminate the need for technical assistance; in fact, they may suggest a greater reliance on a professional acumen. Whether the concern is for creating outdoor play space or exploring options for a central business district, the focus is on citizen involvement through structured decision-making processes.

Most of the material included in this volume was not necessarily prepared for this purpose. My intention was to locate off-the-shelf methods; a social technology that has been tested in the field, that could benefit a diverse audience of users.

Each strategy described in this book represents a departure from traditional practice; the designer has substituted cumulative experience, procedures for designing and simulated conditions, in lieu of the direct contact normally experienced with client-user groups. The designer has facilitated the user group's ability to reach decisions pertinent to the environment through a communicable procedure. In some instances facilitation takes the form of an awareness that will permit citizens to be critical of environmental alternatives. Design assistance in this case is oriented toward an educational approach whose purpose is sensitizing groups to certain environmental issues.

Facilitation can also include systematic procedures whereby nondesign trained people can organize themselves for the purpose of creating a planned action. This populist orientation is by no means an attempt to discard traditional design practice. Yet it seems that increased concerns for environmental quality and appearance have not manifested themselves into the environment that we have contact with daily.

Clearly, alternative approaches are necessary to remedy the increasing visual blight that we are experiencing. Neither aesthetic manifestos nor public policy has served to thwart effectively the creeping monotony of the environment.

Public awareness must be approached through public participation. The reward for participation is power. People will join together if it is clear that change can and will occur. Participation can function if it is active, directed, and a sense of achievement is experienced by those who get involved.

Our cultural roots stem from the principle of sharing and participation. The romanticism we hold for precivilized house and city forms requires us to develop new insights into the methods by which these architectural legacies were created. The directness of communication and unyielding clarity of purpose distinguish between architecture by the anonymous and by our contemporary architectural efforts.

The additive, often incremental growth of precultural communities, is characteristic of a natural process of change. As social order became more prominent, people became further removed from direct control of their welfare. Although the functioning of any social group requires a distribution of skills, there is a consequential loss of individual power associated with social development.

In order to regain a human concern for environmental quality, more attention should be directed towards methods of participation without the encumbrances of technological obstacles. Currently employed methods of user participation disenfranchise the user because the methods of communication have not changed to accommodate a nondesign oriented population.

The contributors to this volume recognize that any form of participation requires a re-examination of traditional design procedures to insure that participation becomes more than affirmation of the designers' intentions.

Participation in design has come to be construed as an acceptable concept. However, in most instances, serious demands and responsibilities are placed upon participants. Where economic constraints are severe enough to exclude design assistance, citizen groups may voluntarily organize themselves for action. Although social and church groups as well as professional societies have assumed an orientation towards accomplishing social goals and do participate in community projects, the technical complexity of creating a playground for a day care center or revitalizing a downtown business district requires the availability of design assistance. In addition to the technical complexity of the issues, there is also the need to incorporate sound design principles in the developmental process. When there is a lack of guidance through a design or planning process, citizen groups often disassemble, respond only to crisis situations, and subsequently cannot achieve the broad goals that originally united the group. Quite often community volunteers cannot draw upon personal experiences for resolving environmental conflicts and select courses of action that create unforeseen, deleterious consequences. Therefore, the management of participatory efforts directed at environmental change is viewed as central to a productive process.

Structuring group efforts can take many forms; at least as many as correspond to the diverse array of environmental issues that confront communities daily. The basic ingredient in the participatory process is individual learning through an increased awareness of the problem. Learning occurs best when the process is clear, communicable, and open; encouraging dialogue, debate, and negotiation. As more participants learn about environmental issues and impacts, the decisions they make will have an abounding effect on the quality of the environment.

People, however, have different learning styles and can participate in many diverse modes. Learning about other group members' value differences and beliefs can contribute to a constructive problem solving session provided that the issues are clear and the solutions are within the realm of implementation. This approach usually requires a large expenditure of time and the involvement of people directly affected by the outcome. As the issues become more ambiguous, beyond the expertise of the participants or without a perceptible schema for implementation, group interest wanes and productivity disintegrates.

In its most usual form, participation is conceived of as citizen reaction to the public presentation of community plans. Recent legislative requirements include supportive statements of environmental impact or costs and benefits of the proposal. This type of forum for exchange is referred to as a public hearing and is usually mandated by law for public expenditures.

Participation, however, has many other modes of involvement ranging from increasing personal awareness to simulated experiences or group decision making. Simulation methods are primarily educationally oriented since their purpose is to prepare people to act. The methods are based on abstractions of complex processes achieved by compressing time and capturing the interest of participants by employing "parlor game" strategies.

Gaming simulations engage participants in the decision roles normally included in the "real life" process. Participation in a game process assumes individuals will confront their own community problems with an increased awareness of the issues at large, the process, and the appropriate decision strategies. This approach is utilized to sensitize and ultimately influence individual behavioral changes.

Another form of participation is the direct involvement of community residents in a systematically organized decision process. This process can be professionally designed for a specific community issue or a general planning process can be utilized for typical classes of problems. Success in this approach is often associated with the quality of guidance through the decision procedure. Leadership is necessary to insure that all the participants contribute to the fullest of their ability. Similarly, the procedure should be tailored to the willingness level of those required to work together, yet not obligate involvement beyond the participant's competence. Overzealous attempts at total involvement of community residents in all stages of the design process may lead to early withdrawal, particularly if progress toward implementation is not evident.

When a high level of interaction is desired among members of a community that share a common purpose, a workshop can be planned whereby participants learn from each other as they explore issues. This type of structured experience can focus the learning of the participants and permit them to engage effectively in other forms of group decision processes. A design workshop can be viewed as a catalyst whose objective is to increase the participant's level of awareness to particular community problems. This belief stems from the growing understanding that intense and concentrated learning through workshops can provide decision makers with tools for effective planning.

The workshop, as a learning structure, evolved from the need to keep decision makers current amid the increasing amount of information that is central to the issue under investigation. Experiencing the intensity of human interaction in a highly directed event has generated widespread interest and support for the viability of this technique in many fields of interest. Games and group decision processes in conjunction with design workshops provide the gamut for effective citizen participation in environmental matters.

The thrust of this book is in the transfer of power from the designer to the user through a social technology. This material should be viewed in a larger context of people organizing for local environmental efforts where accomplishment entails a wide range of activities. The process of winning requires a knowledge of the support constituency, the opposition, individual leaders, goals, and points of intervention in the decision process. Designing for community participation prepares citizens for involvement in technical issues and an awareness of the consequences of environmental decisions.

The basis for a participatory democracy lies in a volunteer society whereby citizens can work in partnership with public and private efforts to accommodate human and environmental needs. Citizen participation can no longer be used as a last resort, but must be seen as a vital resource in implementing a quality environment.

HOW TO USE THIS BOOK

Consider the techniques presented in this book as a resource to be adapted to your particular situation. In some instances this material could be directly suitable for your needs, such as the section on Environmental Awareness Techniques. If you are planning a workshop, seminar, or presentation, the visual material in this section may be pertinent and should be used.

In other instances an approach suggested in this book may be adapted and substituted with your particular circumstances. Perhaps the idea of a planning ballot may be appropriate to your situation so you can substitute your planning options for those proposed by the Yerba Buena team.

The material in this book is for the use of lay citizens and designers interested in expanding the possibilities of citizen participation. Although the copyright controls reprinting, this material can and should be used, modified, or adapted to analogous situations. Where our material will be included in a published report, the publisher should be contacted for permission to release this material.

Communities involved in projects that require technical assistance in the preparation of visual materials should refer to local professionals or service institutions for aid. There are community resources such as design centers, schools of architecture, technical institutes, as well as professional societies such as the American Institute of Architects, American Institute of Planners, or The American Society of Interior Designers, that can direct community groups to the appropriate type of assistance.

Design centers, where available, generally provide free technical assistance to groups or agencies that cannot afford professional fees. In many communities, other similar arrangements can be made with local public service institutions or socially oriented professionals.

Part I

Models of Design Assistance

This section represents particular design assistance methods generated by the Community Development Group at the School of Design, North Carolina State University. Our role as a public service oriented program often requires assistance to community groups identifying a wide range of interests and capabilities. Most often, communities that request assistance cannot afford professional designers or planners and frequently do not have access to professional assistance. This inaccessibility renders many communities powerless in responding to proposals that are not representative of the best interests of its citizenry.

Our approach to community needs identified in this section is related to their recurring interest in this country and abroad. Each form of design assistance is based on a set of sequential procedures requiring lay participants to interact, decide, and finally implement the solution. It is the process, then, that is transferable to other community groups expressing assistance in areas such as children's environments, historic preservation, or central business district planning. The solutions, included in each approach, illustrate the nature of the information required. They would, however, be subject to modification as the approach is applied to other communities.

Design assistance strategies contain an organizational and environmental component. Clearly the size of the project influences the organizational control maintained by the community interest group. In the case of children's environments, the design and organizational control are well within the realm of the interest group. When the project size increases to an historic or central business district, other interest groups may provide input into the process, thus shifting the balance of control in less predictable circumstances. This variation in project size modifies the type of design assistance that is feasible from a sequential plan of specific operations to a set of alternative strategies that are modified every time a decision is made.

The methods of design assistance in this section concentrate on recurring types of community problems that include the need for various degrees of participation. The role of participants changes considerably when the problem is to build a playground, program a children's center, or develop a plan for an historic district. Ideally, participation, control, and implementation are corner stones of our democracy. Time and urgency, however, seem to modify and limit community participation.

The creation of places for children to play illustrates the possible roles for participants, ranging from organization to fund raising, planning, design, and implementation. Often however, implementation in the form of self-help construction, particularly on larger projects, may be constrained by lack of available skills or union requirements.

A particularly effective role for community members is that of programming for the needs of occupants of community buildings. Recognizing that the user is the principal ingredient in the building process, participation can occur through a systematically organized procedure that would include all people affected by the design. A system was developed for graphically transforming verbal statements into visual notations that are clearly communicable to all participants. This approach towards systematically inventorying and recording user requirements can be used in the programming of various community facilities including those for education, health, public service, and the like.

In situations where changes in the environment occur over long periods of time, such as in the planning for an historic district, a downtown redevelopment, or similar community projects, the participation process should be equally responsive. The developmental planning approach described in this section is based on the need to disaggregate problems into small manageable components and provide alternative solutions for each component. To illustrate this process, an historic district in a small town was considered as a collection of issues and geographic areas, each requiring design solutions. The unique feature of this approach is that community residents select the most appropriate alternatives for each of the areas and issues in the historic district. Since the process was developed for a planning period of ten years, a workbook was designed for participants to record decisions as they were made, to refer continuously to previous decisions, and to anticipate the consequences of present decisions.

1

PLANNING OUTDOOR PLAY:

A manual organized to provide design assistance to community groups.

by HENRY SANOFF and GREG CENTENO with the assistance of Sam Caltigirone, Jackie James, Kim Sumner, David Tester and Bodil Vaupel.

The guidelines were prepared by the Community Development Group at the School of Design, North Carolina State University, as a result of many experiences in the design of outdoor play.

So, you want to... build a playground!

We think it's a terrific idea! There is the benefit of saving money, the opportunity for creativity and self-expression, group interaction and, best of all- a custom designed playground!

As a beginning, we need to examine a few sobering facts:

- Playground planning is serious business
- Do-it-yourself playgrounds require a plentiful labor supply.
- Playgrounds should encourage learning as well as provide recreation.
- Organization is the key to successful playground planning.

Organization is the basic theme of this booklet. P.O.P. was created especially for groups working on a limited budget who are in need of professional assistance for designing a playground. We encourage the learning-through-play approach, and we firmly believe in user participation in the design process.

What follows is a step-by-step sequence for planning outdoor play, from the original inspiration to the final workday. The information is taken from the collective experience of various groups (church, school, community) and is flexible enough to allow for any changes you may wish to make for the uniqueness of your own situation.

2

3

We might assume at this point that a core person or group has generated the idea of having a playground. What is needed next is a well defined structuring of the group. You may wish to increase the size of your group to include the maximum number of persons who will be affected by the playground (for example, an entire community or neighborhood).

If this is the case, call a meeting: **5**

Ways to announce a meeting:

A church group could announce it in the bulletin.

A day care center could send notes home with each child.

A telephone chain or network lets news travel fast.

Radio or newspaper announcements of news.

Discussion

Let the discussion proceed informally, encouraging everyone to share their thoughts. A good group leader will be able to identify the critical aspects of an issue under discussion and present them clearly to the group. Decisions may be made by a formal vote or a compromise may be negotiated.

During the discussion, these general guidelines should be followed:

A person commenting on another's idea should state two things he likes about the idea adding his comments. For example, Mr. Smith says, "I appreciate Mr. Jones' enthusiasm for this project and I also would like to have animals in our playground but I think that building a duck pond on this small site is impractical."

Do not permit one person to dominate the discussion even though that person may have design or construction experience.

6 Defer important decisions until all the facts have been collected.

Goals for the first meeting:

Getting Acquainted. (List names, addresses, and phone numbers including names and ages of children in the family; distribute list at the next meeting) Wear name tags; provide time for mingling.

Presentation of Idea (a member of the core group should do this)

Description of Tasks (committee duties)

Selection of Organizers (coordinator, recorder, committee chairpersons)

Soliciting of Committee Members

Listing of Resource Persons (provide sign-up sheets for carpentry skills, construction experience, tools and equipment, materials, trucks and heavy equipment)

General Discussion about Playground Idea (if time and energy permit, you may want to break into small groups for discussion and regroup at the end to review ideas).

Plan the Next Meeting (set a date, place and time for the second meeting; schedule committee reports for then) **7**

committees

You will want to involve many people in the planning and construction of your playground. The work will go much faster when duties are divided among several people. We suggest three committees, a coordinator and recorder.

Recorder

It is assumed that someone at this first meeting is to act as recorder, jotting down the ideas of the group. This person could also be responsible for the sign-up sheets and the directory of group members.

8

Design committee

The design committee has the responsibility to prepare the presentation of the playground. Their responsibilities include:

Visiting similar playgrounds located near the community and analyzing how they work and identifying problems.
Gathering information from sources such as those listed in the bibliography.
Possibly visiting local university's or technical institute's architecture department to seek design assistance.
Obtaining information on the community needs.

The committee is responsible for informing the group about all that it has found.

Coordinator

Shares duties and records. Acts as a group leader to keep meetings running smoothly. The coordinator should be elected by the group at the first general meeting.

9

Communication committee

Its responsibilities include:

Informing people about progress.
Advising the group about future meetings.
Scheduling workers for workdays and advising them on what materials to bring.
Recruiting neighborhood volunteers.

Materials committee

This committee is responsible for:
Funding the site.
Locating the materials.
Having materials delivered and stored on schedule.

Finding and having available in time all the heavy equipment and tools that will be needed for the construction.

Organizing the construction on workdays.
Keeping to the budget.
Scheduling safety advisors on workdays.

10

behind the scenes

Between the first and second general meetings, the committees' move into ACTION. Organization is the "name of the game" in completing the necessary behind the scenes tasks. Each committee should use the recommendations that follow to organize their efforts. When possible, try to divide the workload equally between committee members.

11

Design committee

The design committee meets, schedules trips to collect information and plays the P.O.P. game.

planning outdoor play

pop

Planning Outdoor Play (POP) is a gaming strategy developed to facilitate group participation in playground planning. It can be helpful in sensitizing your group to the many purposes of outdoor play.

The game process is based on a sequence of negotiation stages that facilitate dialogue about issues critical to playground planning. POP is a prop intended to aid players in reaching consensual agreement through negotiation.

Members of this committee should also read the sections on planning, ideas, surfaces, safety and equipment before submitting their first report to the general meeting.

12

Planning

A playground is for children, so they should be involved in the design process. It is their environment, a place of their own and should include physical and mental experiences that encourage individual development through social interaction, sharing, imagination, and muscle development. A playground should encourage learning and extend the school into the outdoors. The area should also permit choice and provide diversity. It is a child's world and the children will try to do everything, so there must be a sufficient and challenging number of activities in a restricted space. Interconnecting equipment with integrated activities and separate areas for active and quiet play is essential. The individual areas and equipment should be designed to a child's scale to aid in the realization that it is a child's environment.

13

Safety

The safety of the playground is important, but the area must still be challenging. If the playground is constructed properly with a great deal of diversity and the use of sturdy structures, it will be safer than what the children might be doing otherwise.

Activities of different kinds should be separated by natural barriers or pathways. Invisible dangers, such as high slides that the younger children do not understand, should be avoided. Instead of a high slide, which becomes dangerous as the children become bored, a low, wide slide that is safe for passing and going down by twos is recommended. The ground cover underneath equipment should be soft and at a depth of 1'. If a child does fall, it should not be upon a sharp piece of equipment. Anything over 18" high should have a railing, and the safe height for climbing is about twice that of the child.

If enough diversity is offered and enough changes made the children will not get bored with the area and begin to perform unsafe stunts. For structural safety see the section on construction.

If the child enjoys the playground and has good experiences there, both physical and mental, then he will be safe.

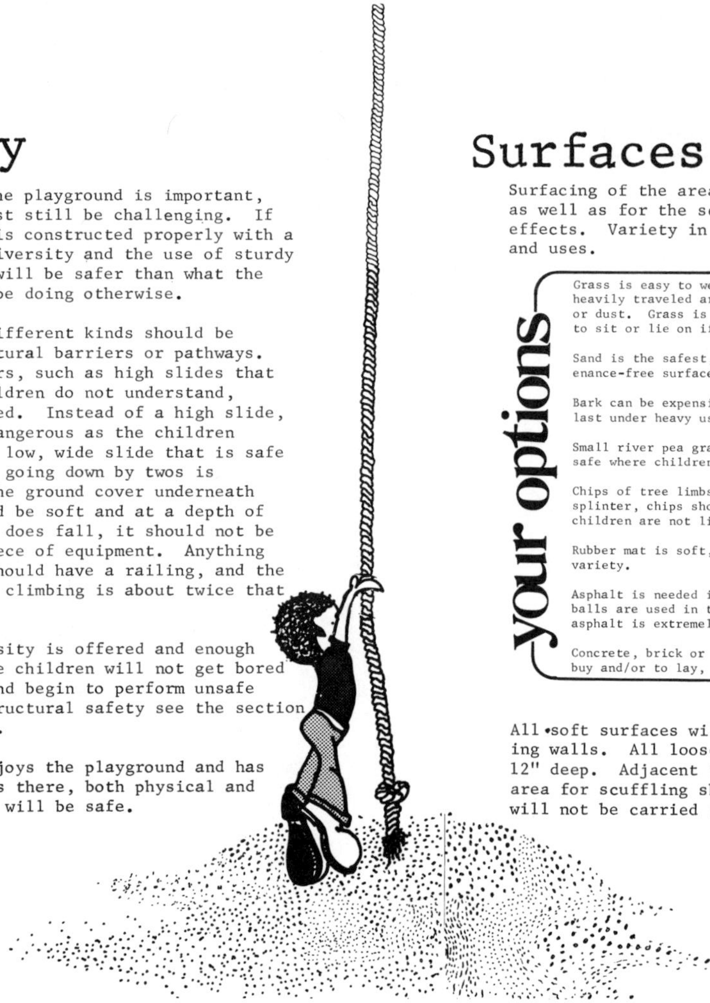

14

Surfaces

Surfacing of the area is important for safety as well as for the sensory and aesthetic effects. Variety in surfaces marks off zones and uses.

your options

Grass is easy to wear out, and if used in heavily traveled areas, will develop into mud or dust. Grass is a soft, comfortable surface to sit or lie on if it can be maintained.

Sand is the safest, cheapest, and most maintenance-free surface for areas of heavy use.

Bark can be expensive, and like grass, does not last under heavy use and will turn to dust.

Small river pea gravel is smooth but is not safe where children will be jumping or falling.

Chips of tree limbs may be fine, but since they splinter, chips should be used in areas where children are not likely to be jumping.

Rubber mat is soft, expensive and has little variety.

Asphalt is needed if bicycles, tricycles, or balls are used in the area, but too much asphalt is extremely dull.

Concrete, brick or stone may be expensive to buy and/or to lay, but can be very attractive.

All soft surfaces will need some form of retaining walls. All loose materials should be placed 12" deep. Adjacent to doors there should be an area for scuffling shoes so that sand and mud will not be carried into buildings.

15

If there is a traffic route across the area, it needs to be rerouted. The children should not cross a drive or parking lot in order to enter a building.

A fence will probably be necessary to separate the children and the street. Retaining walls to contain soft surfacing may also be required.

ideas worth considering

There are other factors and other areas to be included in the design of a playground.

Weather must be considered in the playground plan. There should be shelter from the wind and the sun either naturally or by screening devices. Hills form good protection from the wind, and trees offer shade.

Hills and trees should also be a part of the basic design of the play area. This space should not be built upon and should be left for quiet activities and imaginative play.

Water is a fun but expensive area to add to the playground. Drains and water sources must be available, and it does require more maintenance. Water areas should be considered, whether a wading pool or an elaborate sprinkler system.

16

Equipment options

We have listed several pieces of equipment with brief explanations about how they can be built. Your committee should review these and consider whether or not your group would be able to use them. Concerns as the cost of materials, complexity of design, loading and hauling, limited space, etc. should be discussed.

17

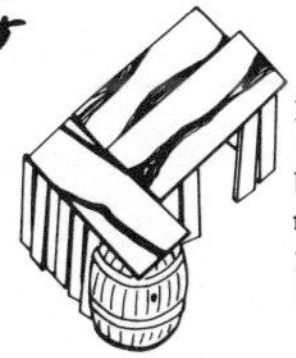

BARREL HOUSE - Construction is with loose materials and should be done by children. Pieces for this equipment and for the junk section of the playground should be light enough for one or two children to handle.

TIRE TRAIN - Tires should be placed solidly in sand.

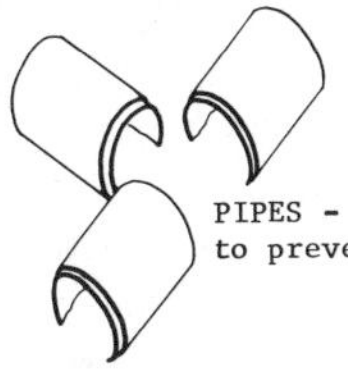

PIPES - Pipes should be deep enough in the sand to prevent any rolling from occuring.

TIRE TUNNEL - Place tires in sand. Make the tunnel short enough for an adult to reach a child inside. Place tires close together.

TIRE HOUSE - Place the tires solidly into the ground at right angles to each other.

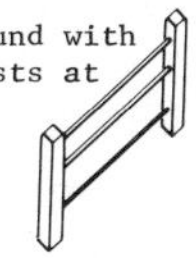

FENCE - Two 4" x 4" set 3' in the ground with three 1" pipes connecting the wood posts at 6", 1½', and 2' heights.

TREE - Check for rotten areas. Trim the small branches off closely.

18

COOKING - Cut out both ends of an oil drum and punch holes in the side for air. Place the oil drum 1' into sand. Place a grill over the top.

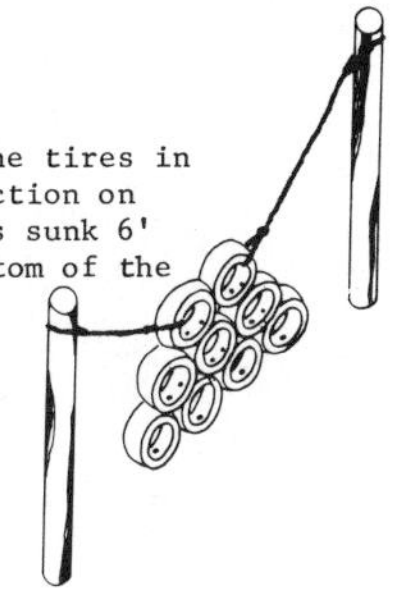

TIRE CLIMBER - On the ground connect the tires in a pattern for maximum support. See section on connections. Hang from two phone poles sunk 6' in the ground. Drill holes in the bottom of the tires for drainage of water and dirt.

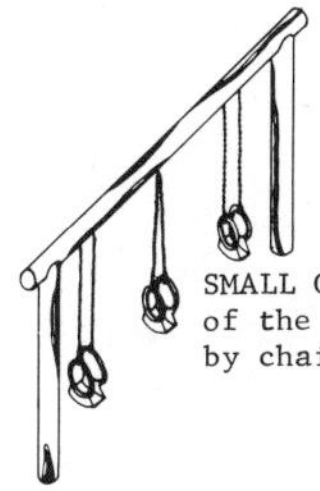

SMALL CHILD's TIRE SWING - Cut out 3/4 of the tread and sidewall and attach by chain and rope to supports.

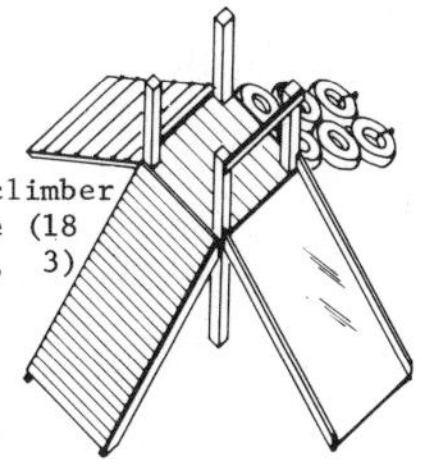

WIDE SLIDE - Wood platform with 1) tire climber secured at top and bottom, 2) metal slide (18 guage stainless steel on plywood backing), 3) wood slide, and 4) cleated wood climber.

SEE-SAW - Use rubber tires for bumpers on the ground. Insert 1¼" pipe through the supports and the long piece. Secure the ends of pipe with caps.

19

PLATFORM WITH ROPE RAILING - Wood platform is made from the side of a cable reel. Pole supports should be placed in a triangular pattern smaller than the diameter of the platform. Connect the top of supports with 2" x 8" 's. Saw a 18" circle of plywood then nail six 2" x 4" 's spaced equally around the circle. Lift into position over platform. Space 2" x 4" 's equally and nail to platform. Tie ropes to make railing. Secure ladder to ground and to platform. Platform should be 5' to 6' high with a 10' clearance. The space underneath may be partially blocked to produce private play areas.

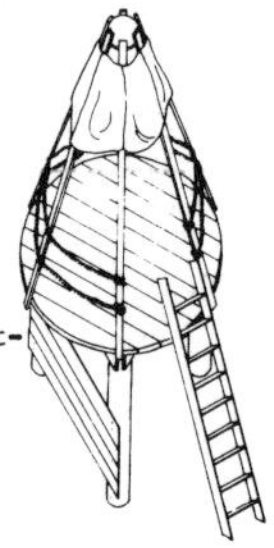

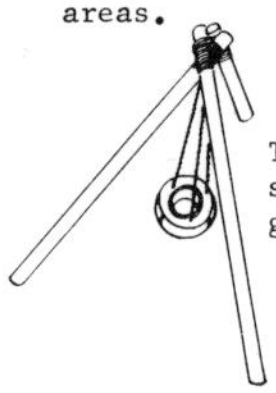

TRIPOD TIRE SWING - Use the tripod lashing as shown Lash three 15' poles on ground, then set them 4' into ground.

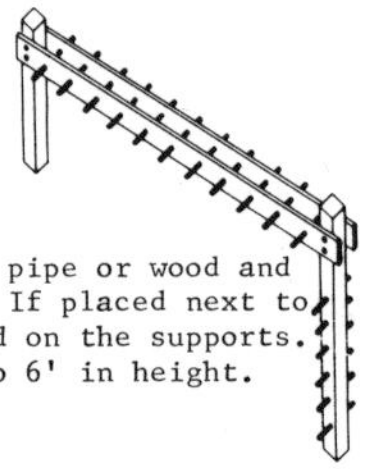

MONKEY BAR - The bars may be pipe or wood and should be 12" inches apart. If placed next to a platform no bars are needed on the supports. The equipment should be 4' to 6' in height.

CAR - Secure four tires in ground connected by a wood platform with a steering wheel.

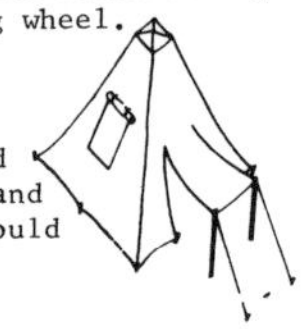

TENT - If a tent is available it is a good experience for the children to put it up and disassemble it. A storage area nearby should be provided.

20

SAND BOX - Use 2" x 12" stained or painted lumber to form an 8 ft. square sand area with a 2 ft. wide table across one end.

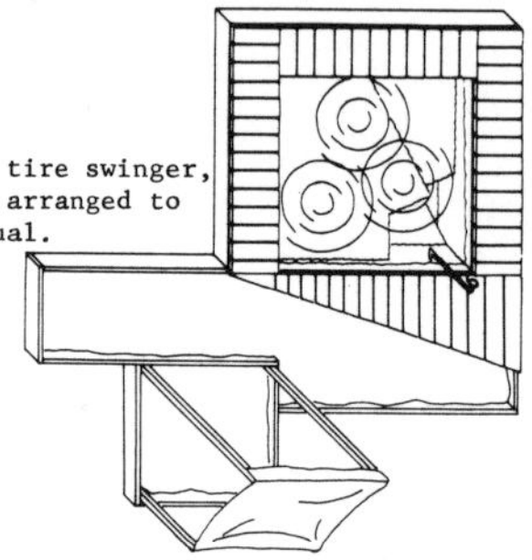

PLAY STRUCTURE - Combination of tire swinger, platform and slide, and sandbox arranged to keep children's movement continual.

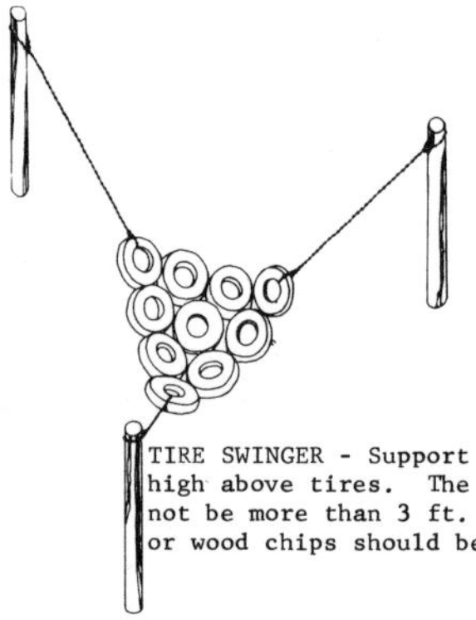

TIRE SWINGER - Support cables should be suspended high above tires. The bottom of swings should not be more than 3 ft. above the ground. Sand or wood chips should be underneath the swinger.

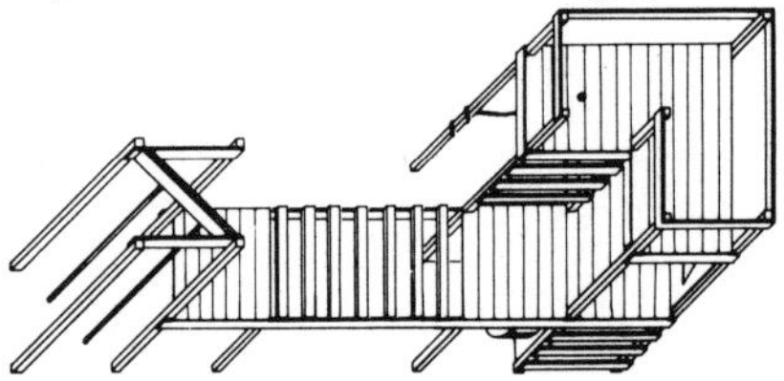

PLAY STRUCTURE - Combination of platform, stair and swinger to create a linear pattern of movement.

21

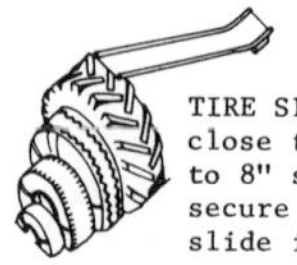

TIRE SLIDE - Set four different sizes of tires close together into the ground leaving a 6" to 8" step between each. Cut into sidewall and secure slide inside tire with bolts. Secure slide into sand at the base.

REVOLVING WOOD SWING - Place two telephone poles 6' underground leaving approximately 8' above the ground. Bolt two 2" x 10" onto the poles. Place a 10" pipe through the center of the 2" x 10" for support of the platform. Secure the ends of the pipe with a cap. Put rope over pipe and secure it to the platform or tire in three places. If chain or cable is used a swivel must be included.
Do not hang platform more than 3' above the ground.

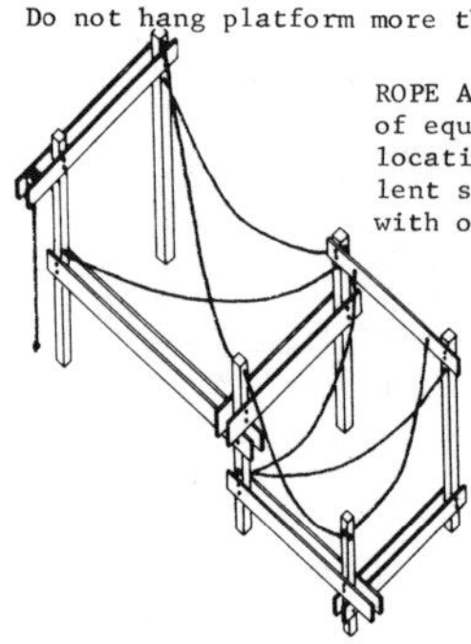

ROPE AND WOOD STRUCTURE - This piece of equipment may be designed at the location and may serve as an excellent segment to integrate or link with other pieces of equipment.

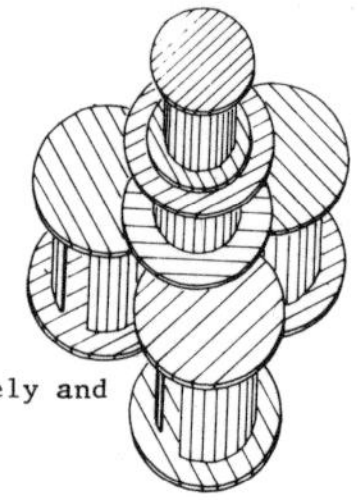

SPOOL STRUCTURE - Stack spools securely and then bolt.

22

BALANCE BAR - Cut a notch 1½" wide and 4" deep into the ends of the two smaller pieces. Set the two different length pieces of telephone pole vertically into the ground at 2' depth. Place a 2" x 6" into notches and secure with nails. Instead of notches the board may be nailed to the side or the top of the two smaller pieces at a height of 6" to 2' off the ground.

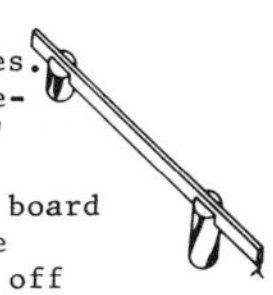

RAILROAD TIES - 4" x 4" lumber set 6" apart with two 4" x 4" nailed or bolted on top perpendicular to the others to resemble a railroad track.

STUMPS - They should act as a barrier or path between different kinds of activities. Place them at varying heights.

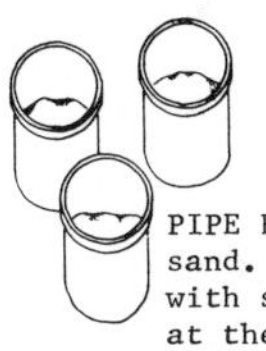

PIPE HIDEOUT - Secure pipes 1' to 1½' deep in sand. Fill the inside with gravel and then with sand to leave a clearance of 1½' to 2' at the top making escape easier.

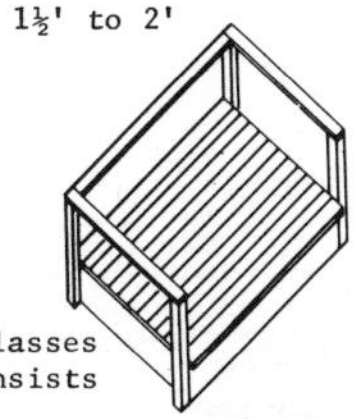

STAGE - This piece may be used for small classes as well as for drama presentations. It consists of a 6' x 8' platform with a railing

Useful References for Designing and Building Playgrounds

- M. Paul Friedberg, "Handcrafted Playgrounds", "Designs You Can Build Yourself", N. Y., Vintage Books. 1975.
- Jeremy Joan Hewes, "Build Your Own Playground", San Francisco, Houghton Mifflan, 1974.
- Richard Dattner, "Design for Play". N.Y., Van Nostran Reinhold, 1969.
- Martin Helick. "Elements of Preschool Playgrounds". Pittsburgh, Pennsylvania. Regent Graphic Service, 1973.
- Paul Hogan. "Playgrounds are for Free: The Utilization of Used and Surplus Material in Playground Construction. Cambridge, Massachusetts. MIT Press, 1974.

Materials committee

The Materials committee looks into site choices, locates and lists materials and resources and prepares a report.

Members of this committee should also read the sections that follow on selecting a site, the money puzzle, resources and sources of materials and equipment.

DON'T FORGET TO CHECK ON A BUILDING PERMIT!!!

The money puzzle

Money may be required for renting larger equipment and purchasing some materials. There are several methods available; federal, state, local funds, and private donations are sources for additional support. The Parks and Recreation Department can provide assistance to the group about available public funds. Fund raising projects in the community promote interest and children can be especially helpful in these endeavors. Sales, door to door, collections, and barbeques, etc can involve them as well. Some people will donate to the cause in order to keep children out of their yards or shops.

Selecting a site

Finding a site for a community playground is sometimes difficult. The children know the neighborhood and can help in the selection of an appropriate site.

In choosing a site you may want to consider that:

Children should have safe accesss to the playground and be free from motor vehicle hazards.
The site should not be in an area where children's noise will upset neighbors.
Existing structures can help shield parts of the playground from noise, wind and sun; existing structures can be used to create interesting spaces in the playground.
Children like variety; a better site offers several types of ground cover.

The insurance and ownership of the area should be investigated to decide whether the area will be open to the public. The owner has final authority whether public or private.

Resources

The labor that builds the playground should appropriately be that of the group members. However, the group may want to seek additional labor. Some volunteer organizations that may be willing to donate their time and resources are:

Boy Scouts; Girl Scouts.
University fraternities.
Military reserve units.
Men's or Women's Clubs.

The group will also require a person with construction experience, preferably from outside the group. This person will:

- Review the plans for feasibility.
- Demonstrate the correct use of tools and safety procedures at the beginning of each workday.
- Supervise construction, give technical advice when necessary and insure that safe working procedures are followed.

Obtaining free or inexpensive materials and the use of necessary equipment for building will require time and energy. All the leads suggested by group members need to be confirmed and other sources must be solicited. To find private sources in your area, make a thorough review of the telephone directory's yellow pages. List any firm, its address and phone number, that may have materials or equipment you may need. Also, note the products or services of the firm. Public agencies and utility companies may be found directly in the phone book or through information sources in city, county, or state governments.

Solicitation should be done by telephone or in person especially when there are a few possible sources for a particular item. It is important to involve as many members of your committee as possible in finding ideas for materials and in locating them. It is effective to find someone who knows someone who has something... effective scrounging of materials is best done by personal contact. If one of your committee members' is a friend of the hardware store owner and thinks he might donate materials, have the member ask him.

When soliciting in person, persistence pays off. In a large organization, find out who might be in charge of answering your request. Try to deal with people in decision making positions. When you encounter someone who doesn't know the answer to your question, always ask to be referred to someone else.

Indicate your willingness to meet the firm halfway by picking up materials or providing own driver for bulldozer, for example.

If a donation is offered work out the details of quality, quantity, delivery, and storage. Send a letter of confirmation restating the details of your agreement and thanking the person as well as the organization he represents.

Letters of solicitation should cover the agenda listed above. Any donation should also be answered with appropriate letters of acknowledgement.

☆ When you have reached the key person, these are the points to remember:

☆ Write down your name and your position.

☆ Give your name and briefly explain what your group is doing.

☆ Ask for donations of specific items (prepare list beforehand).

☆ Ask if the firm can supply other used or surplus materials.

29

Keep a file of contributions and available resources on index cards. On each card list:

name of group member who solicited this contribution and date.
firm -- name, address, phone number.
name of person responsible.
material (quantity and quality) or equipment.
delivery or pick-up arrangements.
storage on or off-site.

★ ★ ★

Here is an example of an index card. You may want to write key words in the upper right corner to facilitate later processing and sorting.

utility poles
auger--NOTIFY

Main Town Electric
Robert Smith, Public Relations
123 Industrial Drive

12 20' utility poles and use of auger truck to sink poles - poles to be delivered to site at company convenience prior to June 20. Has been sent map to site. Call him back week before workday to remind him to arrange auger truck.

30

Sources of materials and equipment

Tires: tire dealers, recapping services, junk yards, trucking contractors
Cable Spools: utility companies, industrial hardware dealers
Telephone Poles: utility companies
Peeler Cores: lumber yards, plywood manufacturers
Chain, Cable, Rope: government surplus
Concrete Culvert Pipe: heavy construction contractors, concrete pipe manufacturers, department of public works, concrete pipe manufacturers
Steel Drums: scrap metal yard
Railroad Ties: utility companies, salvage & wrecking companies
Hardware: utility companies, industrial hardware dealers
Scrap Steel Pipe: scrap metal yards, junk yards
Scrap Lumber & Plywood: lumber yards, building sites, plywood manufacturers
Dead Trees & Logs: parks & recreation department, tree service companies, building companies
Boulders: parks & recreation department, heavy construction contractors
Used Lumber: salvage and wrecking companies, local government
Old Cars, Trucks, Boats: government surplus, junk yard, marinas, auto used parts & body shops
Old Ladders: department of public works, junk yard
Sand, Cement, Concrete: sand, gravel, & concrete suppliers
Paving Materials: department of public works, building contractors, heavy construction contractors
Fill Dirt: building & construction contractors
Cloth, Canvas, Parachutes: government & military surplus
Plants, Shrubs, Landscape Materials: park & recreation department, tree service companies, nurseries
Used Carpets: carpet & remodeling companies
Hammocks & Cargo Net: gravel & concrete suppliers, military surplus
Wood Packing Crates: industrial hardware dealers, furniture & appliance dealers

31

Communications committee

The Communications committee compiles a directory of interested individuals, prepares copies for distribution, and hands these directories out at the next general meeting. This committee also:

- keeps tabs on the progress made by the other two committees.
- writes up and distributes progress reports.
- looks into the possibility of media (newspaper, church bulletins, T.V., etc.) announcement of the project.
- and announces the next general meet-

32

here's the plan

(At the 2nd General Meeting)

- Reports from Design Committee and Materials Committee
- Discussion and Vote on Reports

At this point the group should realistically assess its ability to implement the project.

If group members are doubtful about their abilities, or if the general feeling is that more participation or more information is needed, another meeting may be necessary before a decision is made to continue or abandon the project.

33

If, however, the group is ready, the Design Committee should complete the playground design, fitting it to the site and materials available. The HELP game could be useful here.

help

handmade equipment layout planning

When the planning concepts in POP are understood by your group, a second planning game called Handmade Equipment Layout Planning (HELP) should be introduced to permit your equipment choices to be incorporated into a specific site.

This game process is to help your group organize the desired activities and equipment into an arrangement that fits your site and which meets your requirements for safety, variety and learning.

The Committee should determine how the group wants the final plan to be presented- at a third general meeting, or posted at a convenient place for viewing by members (school bulletin board, for example).

34

Communications

The Communications Committee needs to establish workday schedules, based on input from the total group, about most convenient dates and times.

Convenience is the most important factor that will contribute to the success of the planning phase. Things to consider are:

- REFRESHMENTS: Seek volunteers to bring or provide light refreshments during or after the meeting.
- BABYSITTING: Child care should be provided for the younger children since older children will be participating. Parents may have to share expenses for an outside sitter.
- TRANSPORTATION: If transportation will be provided, indicate the phone number on all announcements so that interested people may call in advance and plan to attend.

Everything should be made as convenient as possible, to insure the continued cooperation and participation of all the people involved. If planned in advance, meetings can run efficiently, giving everyone a feeling of accomplishment.

35

When your project is well under way, you may want to issue a press release to the local newspaper. Copies of the published article can be used in the solicitation of materials and equipment. Here are some major points to be included in a press release:

36

1.) Name the leaders of the group and how they may be contacted.

2.) Identify the group and the organization (if any) with which it is associated.

3.) State exactly what you are doing and how it benefits your group, organization or community. State where the playground is to be located.

4.) Describe what you are doing to raise money (the press release may advertise your money-raising project).

5.) Indicate when your playground is to be built.

6.) Indicate the nature and sources of contributions already received from outside the group and invite further contributions.

Options

A press release is most appropriate when the project is a community playground open to the general public. The article should emphasize community participation in the project and the benefits to be realized such as cleaning up a littered lot, giving children a place to play, and fostering community spirit.

A private organization may wish to use a press release to advertise a fund-raising project. The article should state that the organization is non-profit and should emphasize its role in the community.

A sample press release is included here.

37

38 Sample press release

publicity

The Main Town Cooperative Children's Center will hold a Pancake Jamboree on Saturday, May 31 in order to raise money for building a playground. The Jamboree will be held at the Main Town Shopping Center parking lot from 8 am til 3 pm and a variety of pancakes will be served. All tickets are $1.50; children under 6 eat free.

It is children under 6 who will benefit the most from the planned playground. The Cooperative Children's Center is a non-profit organization which serves over 70 families in the area, involving 83 children from ages 2-6. The playground will be located behind Holy Church, which houses the Center.

Designed by a group of involved parents and teachers of the Center, as well as church members, the playground will utilize donations from several city groups. Main Town Electric will contribute a dozen poles and the use of their auger truck to sink the poles for a tree house. National Tires Service will provide over a dozen large "off-the-road" tires for the site. The group continues to seek additional contributions.

According to John Fawley, chairman of the Construction Committee, the group plans to begin construction the last weekend in June.

Volunteer workers are invited.

39 all together now

We are now ready to prepare the site for construction of our playground. Organization and participation are the key words. Job assignments and coordination of site activities are extremely important and warrent careful planning.

40 Site Preparation

Site preparation is one of the most expensive parts of building a playground. The area may need to have its ground cover removed if it is hard topped. The space may need grading for adequate drainage, but natural hills may be left if they will be incorporated into the design. Hills add variety, and, if they are already landscaped, the erosion will be much less.

Ask one or more excavating and grading contractors or drainage contractors to visit the site. Describe what is desired and get their opinions for what needs to be done and an estimate of the cost. If grading is necessary, your local or state Parks and Recreation Department may do the work for you. Contracting companies, or even military reserve groups may donate their services.

If some type of drainage needs to be installed to prevent erosion and puddles, it is suggested to slope the ground towards the street or toward dry wells. Dry wells can easily be installed by digging a hole three to four feet across and eighteen inches deep. Fill it with gravel and cover it with a grill or the dirt itself when in soft surfaces.

Construction

Before each workday begins, the planning group should decide exactly what is to be built. All materials must be gathered beforehand and at the site in time for the work to begin on schedule. There should be an ample number of workers to lift, carry, and hold heavy materials. Each worker should know what to bring to the site each day and have a checklist to insure that the necessary tools are assembled. For instance, a good checklist might include, donated tools, food and drink, and special work clothing.

It is preferable to have one supervisor that is not from the group who is skilled in the construction field. Safety advisors should also be assigned to keep an eye on all work in progress. Instruction should be given on the use of necessary power tools.

It is recommended that lunch be eaten on the site, so that the work group can discuss what has been done, and what must be completed by the end of the day.

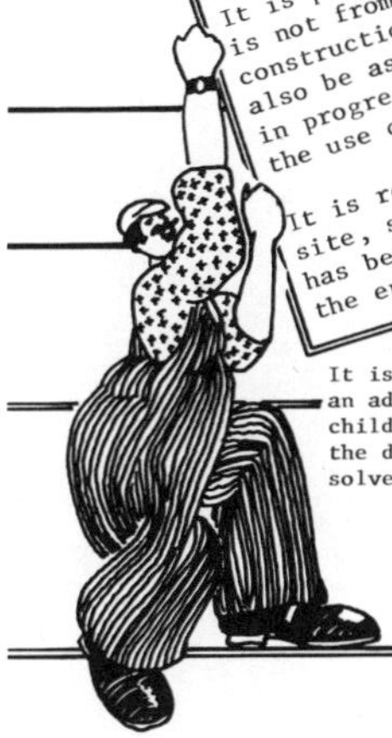

It is advisable to require each child to have an adult sponsor on workdays. Nametags for the children should be issued at the beginning of the day. Children can be given problems to solve for themselves.

41

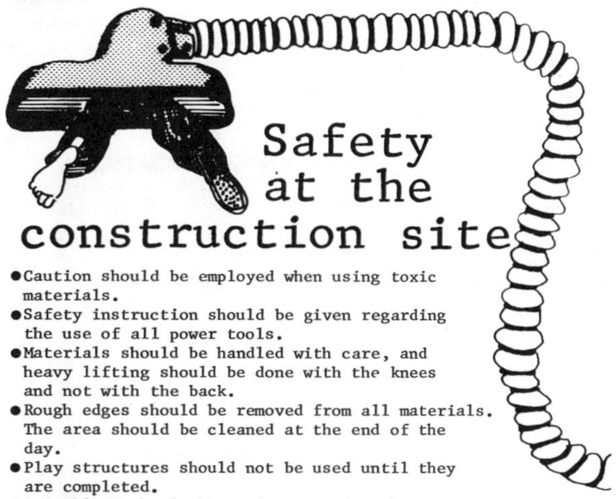

Safety at the construction site

- Caution should be employed when using toxic materials.
- Safety instruction should be given regarding the use of all power tools.
- Materials should be handled with care, and heavy lifting should be done with the knees and not with the back.
- Rough edges should be removed from all materials. The area should be cleaned at the end of the day.
- Play structures should not be used until they are completed.
- Suitable work clothes and protective shoes should be worn.
- Hair and loose clothing should be securely fastened.
- Grounded heavy duty electrical cord should be used.
- Power tools should not be used on wet ground.
- Wood should be supported correctly when sawing.
- Protective goggles and gloves should be worn where necessary.
- Tools should be kept sharp at all times.
- Work at an even pace and do not rush.

42

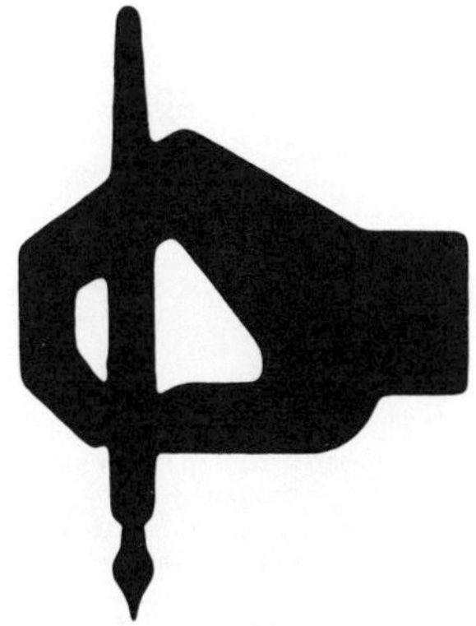

finishing touches

Maintain a steady workday schedule until the project is completed. The Communications Committee should closely monitor the success of response and try a variety of days and times until a suitable level of participation is reached.

43

Some HINTS for sparking morale and levels of participation:

- Allow for social activity on the workday, such as a potluck picnic.
- Mix-up workdays- have a kid's day with projects they can easily handle.
- If the group is very large, try subdividing into smaller workday teams. Or, take names ending in A thru J one day, another alphabetical section the next.
- Do not allow workdays to fall on the same day of the week too often, so that participants can have some flexible free time.

If you have come this far, chances are that you've done it all! Why not make an occasion of opening day-

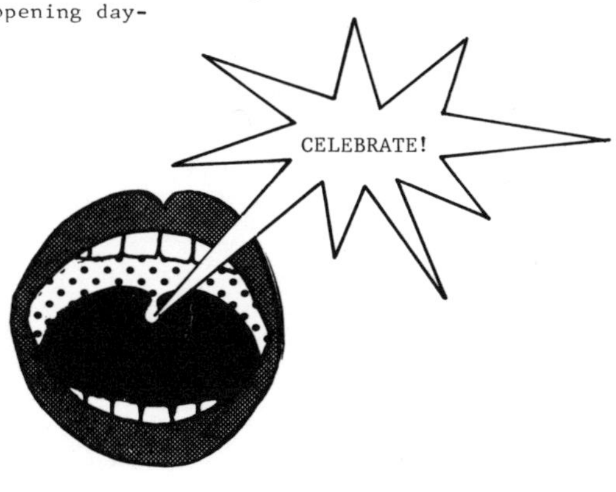

44

Community Outdoor Play

Keep a check list for future maintenance checks and record what was found to be wrong and how it was corrected.

- ☐ Check to see if the grounds are clear of trash. Mow and trim plantings.
- ☐ Do not use toxic insect or weed sprays.
- ☐ Check to see if equipment is still sturdy.
- ☐ Check the wood and sand the rough edges.
- ☐ Check the metal for rust.
- ☐ Check for worn rope.

45

planning outdoor play

pop

Planning Outdoor play (POP) is a method of facilitating the design of children's outdoor play and the selection of appropriate equipment.

POP consists of four (4) sets of card categories:

OBJECTIVES (yellow cards)- the purpose of outdoor play
ACTIVITIES (orange cards)- what children do outdoors
ZONES (pink cards)- areas of related activities
SETTINGS (see POP sheets 4 & 5) equipment used by children

NOTE: YOU WILL NEED TO SUPPLY YOUR OWN CARDS, USING THE LISTED CATEGORIES ON POP sheet 7.

pop

The game is to be played by one to four people (teachers, assistants, parents, etc.). To begin, each player selects four of the most important OBJECTIVE cards. After each player has chosen cards, the individual OBJECTIVE cards are pooled. Through negotiation, the group must choose from these, no more than four OBJECTIVE cards. Players are urged to forcefully support their individual choices, especially if other members did not make the same choice, until they persuade or are persuaded by others that an OBJECTIVE should be included in the final set. Use the game record sheet to report the final decision.

Next, as a group, examine each OBJECTIVE and select no more than three ACTIVITY cards that best satisfy each OBJECTIVE. Record all the selections.

Now, as a group, rearrange all the ACTIVITY cards (twelve) so they appear to fit into an appropriate ZONE. It is important to note that not all of the ZONE cards need to be used. Record all selections.

Finally, select the appropriate SETTING that corresponds with each of the ACTIVITY choices and record the number of the SETTING. The set of rules is only a guide for gaining insight into the planning process. Players should feel free to change the rules in order to accommodate more specific needs.

pop 1

ACTIVITIES

mixing	painting
feeling & handling	cooking
hitting	pouring
crawling	stretching
splashing	sliding
role playing	constructing
dressing up	planting
molding	swinging
contact with animals	balancing
digging	vehicular motion
climbing	throwing & catching

OBJECTIVES

concept formation	social development
cooperation	initiative
problem solving	communication
emotional development	self expression
self confidence	positive self image
language development	perceptual development
sensory discrimination	sensory development
self motivation	

pop 2

RECORD SHEET

OBJECTIVES

a ____________________

b ____________________

c ____________________

d ____________________

ACTIVITIES

a	b
1 ________	1 ________
2 ________	2 ________
3 ________	3 ________

c	d
1 ________	1 ________
2 ________	2 ________
3 ________	3 ________

pop 3

Equipment Options

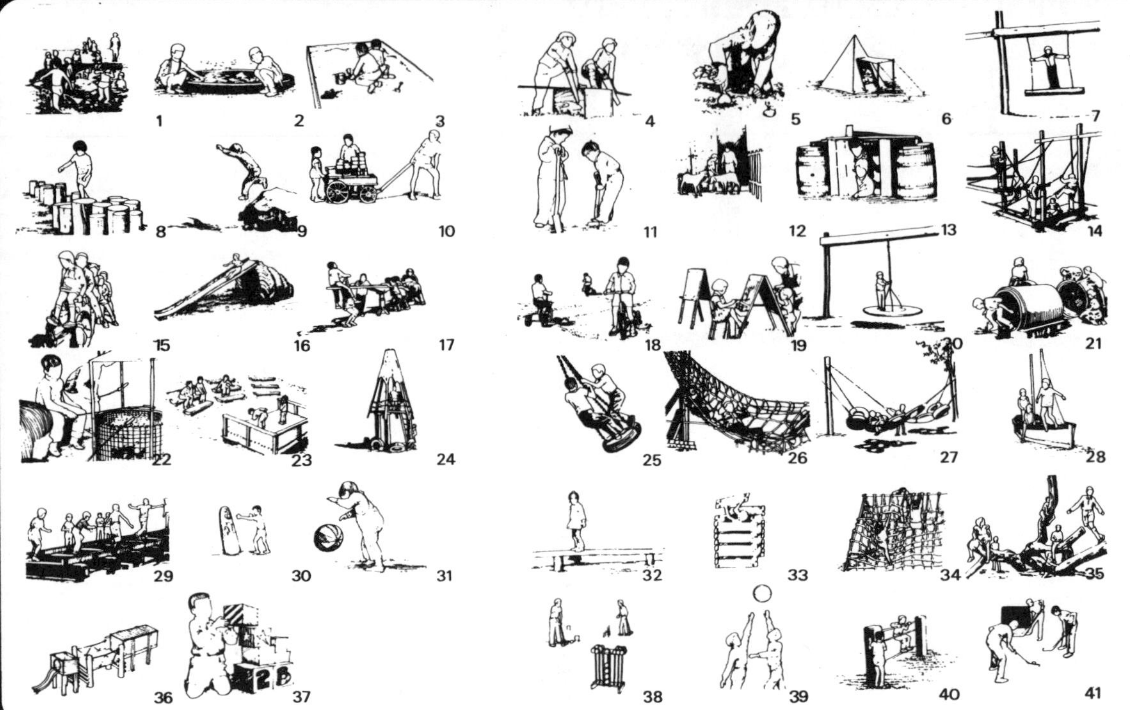

pop 4

pop 5

ZONES	ACTIVITIES	SETTINGS
Creative Play Zone		
Imaginative Play Zone		
Dramatic Play Zone		
Nature Zone		
Adventure Play Zone		
Private Play Zone		
Manipulative Coordination Zone		
Large Muscle Development Zone		
Open Area Play Zone		

pop 6

To save time, the cards should be prepared prior to the meeting at which you plan to play the game.

To make the cards, we suggest you use:

3 sheets of construction paper, 1 yellow, 1 orange, 1 pink
scissors and a felt tip pen

Cut the paper into one inch strips; cut the strips in half. Use the strips as playing cards. You will need 22 orange, 15 yellow, and 9 pink cards.

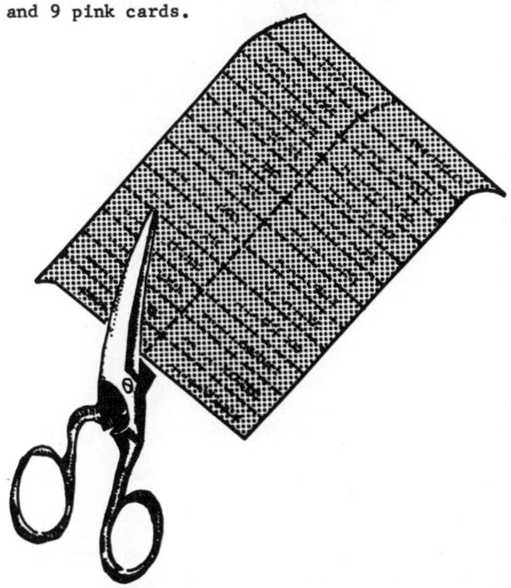

pop 7

Handmade Equipment Layout Planning

H.E.L.P. is a useful method for organizing playground layout ideas. It can aid you in discovering various problems that may arise before you actually build and locate the equipment.

The game includes two stages: (1) Preparation and (2) Planning.

PREPARATION

You will need: the graph sheet (included in the H.E.L.P. package) scissors, tracing paper (at least 17in. by 12in.) tape, ruler, pencils, a flat drawing surface, equipment cards (also in the H.E.L.P. package) and the P.O.P. game sheet results.

Measure the length and width of your playground site and record this information on the graph sheet. Each square on the graph represents a 4 foot by 4 foot area. To match your site measurements to the graph squares, divide the length and width by 4. Fo example, if your lot is 100 ft. wide and 60ft. long:

100 ÷ 4 = 25 squares wide
60 ÷ 4 = 15 squares long

Draw the outline of the site on the graph sheet.

Next, tape the graph sheet to the table top in the following manner:

PLANNING

Divide the playground site into the zones you have chosen from the P.O.P. game zones list:

- PRIVATE PLAY ZONE- children use small protected areas for individual or quiet activities.
- DRAMATIC PLAY ZONE- children exercise their imagination to create roles.
- ADVENTURE PLAY ZONE- children spontaneously build and rebuild their environment.
- MANIPULATIVE COORDINATION ZONE- children develop coordination skills frequently with repetitive motion.
- OPEN AREA PLAY ZONE- children use large spaces for group games and individual activities that may require hard or soft surfaces.
- CREATIVE PLAY ZONE- children combine materials to make a different object.
- LARGE MUSCLE DEVELOPMENT ZONE- children overcome physical and mental obstacles, exercising all possible muscles.
- NATURE ZONE- children interact with natural objects.
- IMAGINATIVE PLAY ZONE- children exercise imagination and limited muscular effort, but no may be produced.

Decide what equipment need be placed in each zone before estimating its size. Then, locate your zones to capitalize on characteristics unique to your site-e.g.-shaded areas, existing surfacing, slopes, trees,etc.

Cut the equipment cards along the solid lines, choosing the pieces that match (or are similar to) the settings you selected in the P.O.P. game.

Match the equipment cards to the zones and arrange them on the graph sheet.

When moving equipment around on the graph sheet, four things need to be remembered:

1.) There is a top and a bottom to each drawing on the cards provided. Keep the direction of all your equipment standard.
2.) Always keep the edges of your equipment cards aligned to the grid.
3.) You can turn the equipment cards over, to reverse reverse their direction.
4.) If you want to decrease the distance between pieces of equipment, feel free to overlap cards

For example, here's a layout scheme for an adventure play zone.

Once you have an arrangement that you feel is satisfactory, use the Guidelines Checklist provided to evaluate your layout scheme.

If you are satisfied that your arrangement meets the checklist guidelines, place a sheet of tracing paper over the graph sheet and cards, tape it down, then trace the site plus equipment.

(HINT: Use a fine tip felt pen to outline the final tracing for presentation).

You should try several different zone and equipment arrangements, following the same procedure each time, until you have several tracings to compare.

Your committee should then choose one of the tracings as the final plan to show to the entire group at the next general meeting.

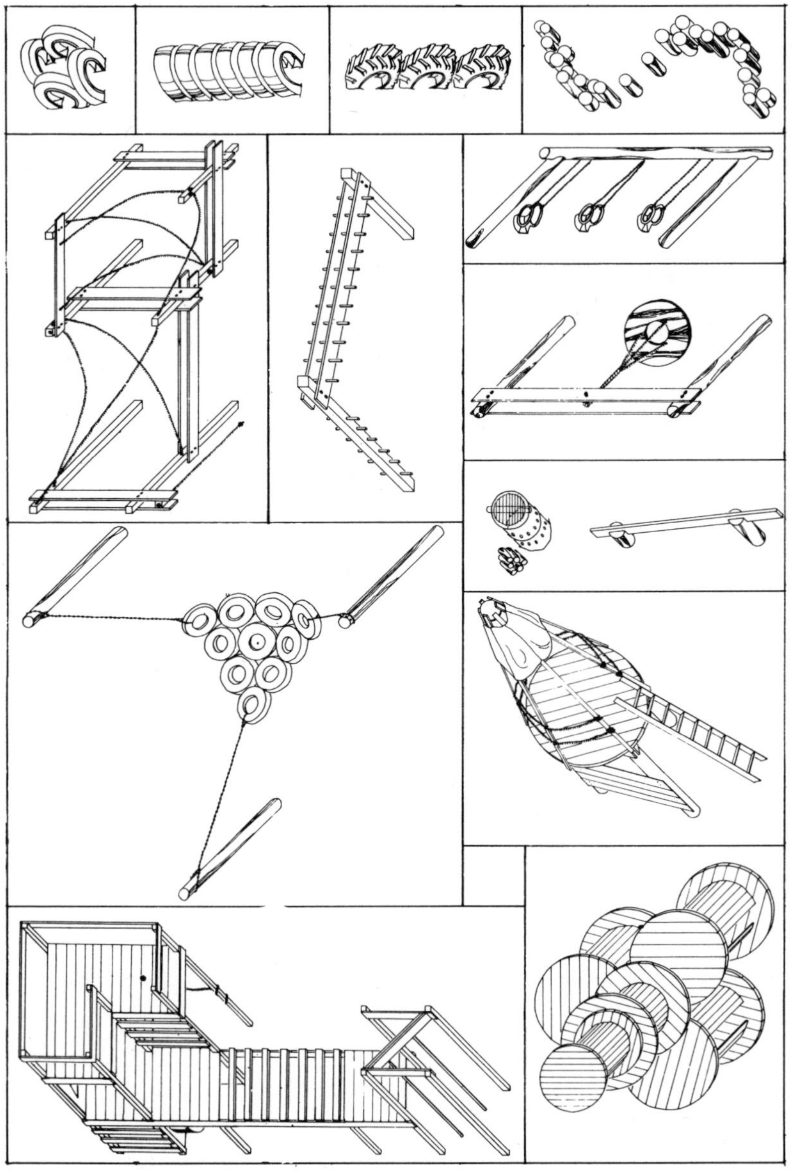

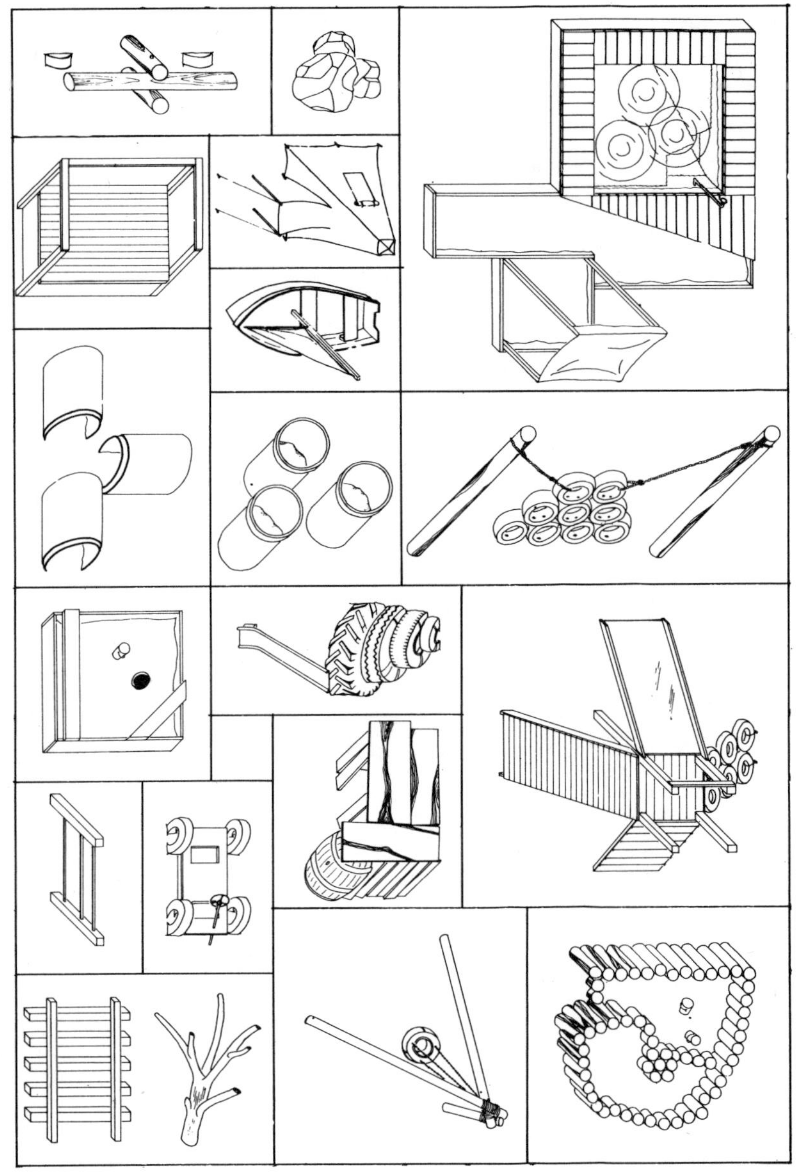

Guidelines Checklist

☐☐☐☐
If there are swings, allow space for moving in front and behind without children getting hit.

☐☐☐☐
If there are good, sturdy trees and suitable grassy hills, try to include them in your design plan as play places.

☐☐☐☐
If there are tall structures, use soft surfaces underneath in case of falls (Refer to the section about surfaces and safety).

☐☐☐☐
Be sure to include places for litter and for small toy storage, and for storing wheeled toys. It is better to have several small closely spaced play areas than one or two large zones.

☐☐☐☐
If you have riding toys (trikes, wagons, etc.) leave an open smooth surface where it will not interfere with other activities. The same is true for ball playing - leave enough room for bouncing, throwing and rolling away from windows, streets, or places where the ball may interfere with other children playing in sand, etc.

☐☐☐☐
If you need garden space - put it in a well protected area. Planted areas need borders for added protection.

☐☐☐☐
If there are slides, make sure there are at least 4 feet of space around the base, for waiting a turn or just running.

☐☐☐☐
Younger children should have separate play areas for safety, but they do not need to be completely separated from older children's play areas.

☐☐☐☐
If children of different ages and sizes will be using the playground, make certain that you have allowed for the differences in skill levels. Each age group should have a place where they feel comfortable and play safely.

☐☐☐☐
The most popular playgrounds have variety and are located in pleasant areas (sunny in some places, shady in others). Children prefer playgrounds that have familiar types of equipment (swings, slides, jungle gyms); but it also good to have a space for adventure play where they can build their own structures out of tires and light weight boards.

2

Programming Environments for Young Children

Henry Sanoff
Director

Susan Adams
Gregory Centeno
Ronald Wells

Community Development Group
School of Design
North Carolina State University at Raleigh
Raleigh, North Carolina 27607

and

Stephanie Fanjul
Director
Amity Again Inc.
1300 Brookside Avenue
Raleigh, North Carolina 27604

If schooling is to complement the natural way children learn, it is only appropriate to understand the impact that the child's environment may have on the learning process. Our objective is to describe a process for creating a responsive environment. The intent is to stress the interrelatedness of the goals of children's developmental programs and the physical environment in which they are housed.

Traditionally, school planners have based their decisions on their own individual values reflecting their own cultural, educational, and economic biases, and social class characteristics. Often people that plan and design school buildings are unrelated to the people that use the facilities. There is, however, a growing belief that the physical environment does profoundly influence psychological states and social behavior. It is important to understand this interaction between the physical environment and people's behavior in order to create settings that can enhance one's intentions.

Neil Postman and Charles Weingartner in *The School Book*[1] address the issue of the free school movement vs the traditional school concept. Postman and Weingartner state that the structure of a conventional school requires that everyone be there when school is ready, that everyone learn what the school teaches, when it teaches it, and where it teaches it—all together, at the same time, and in the same place. From 1965–1970, critics of the traditional school method felt that the schools were not devoted to the interests of the individual child, to helping him/her become an autonomous, creative, inquiring person who has the will and intelligence to determine his/her own destiny but instead that the schools functioned as a service agency to the dominant bureaucracies of our society. Ivan Illich in his book *De-Schooling Society* (1971)[2] states that "the trouble with school is the basic structure. Schools discriminate. Schools compel. Schools judge. Schools mandate. Schools certify. Schools punish."

The Free School movement emerged as an alternative route to the public schools that many people felt were beyond reform. Critics wanted to overhaul the procedures governing the relationships among students, teachers, and administrators and parents.

Postman and Weingartner[3] ask the question: what role can school realistically play in the education of our youth? They answer this important question with their ideas of what school should become in the future. They see future schools becoming a medium of communication through community and resource based programs. Schools would become an environment for the development of healthy egos. School would be designed to enhance mental health, not restore it.

John Dewey stated that the purpose of studying a subject was to learn how to think. Postman and Weingartner[4] believe that in order to facilitate learning, schools must be designed systematically to be supportive and rewarding, rather than disparaging and punitive. They must be designed to provide children with consistent support and challenge in order to reaffirm each child's curiosity and natural love of learning. In order to create an environment that is responsive to the desires and aspirations of children, it is necessary for the designers, teachers, and students to articulate clearly their intentions for learning and understand the process of learning.

A CASE STUDY

An alternative school in Raleigh, North Carolina, whose purpose was to develop a building program that could enhance the educational purposes of the school, was the subject of our study. Amity Again is a small school composed of two teachers, fifteen children, an active parent-resource, and a network of participants who expand and extend learning beyond the school setting. The families involved represent a mixture of life styles and incomes. What binds them together is the belief that public schools are too competitive and repressive, rather than liberating, and do not involve parents or children in decisions affecting their welfare. The teachers at Amity recognize the need to teach the basic educational skills, but also place an emphasis on the child's psychological growth toward developing a good self-image. This philosophy is based on the assumption that the children at Amity learn through discovery and exploration and that their school complements the natural way children learn. The children's feelings and ideas deserve consideration and their teacher acts as an initiator and guide primarily working with small groups and individuals at their own pace and in their own style. Finally, the children should find a measure of success in each and every day. In order to accomplish these purposes, Amity teachers encourage child-initiated use of community resources, internal resources, and educational materials.

The differences in the way Amity functions as a school compared to the traditional classroom concept are: At Amity, emotions are dealt with directly. There is more individual teacher's time available per child. The older and younger children work together to solve problems. The teachers relate to the group as participants rather than observers. The children are not divided by grades, but rather, by self-selected interest groups.

Emotional growing is not something that is only observed at Amity Again, it's participatory. For example, when conflict occurs between teacher and child, their roles are reversed to re-enact the conflict with the other children becoming a listening audience. The group, as a whole, discusses the problems and comes up with several productive and aware insights. The expectation, which the children accept, is to help solve the conflict constructively. One half-hour meeting can yield an emotional interaction; a role reversal; a group problem-solving session; a verbalization of feelings; and an opportunity, for the school as a group, to share thinking and feelings, and in turn come closer together.

When we consider the way children learn, it helps to recognize the main factors involved in each activity. The role of the teacher and child are two of the most important ingredients in the learning environment. Available to the teachers and the children are the roles of initiator, director, and evaluator. The role choice is made according to the child's needs, specifically considering how the activity could be designed to be challenging.

These are the approaches to activity selection that take place at Amity where the focus is always on the interaction between the child and teacher.

Teacher Initiated Activities

Teachers initiate the activity by proposing a theme to an individual child or group. Various modes of presentation may be employed such as film, tapes, or books. The children choose their way of working within the theme construct, limited only by the teacher's objectives and expectations. The teachers direct the children because of their interests or a specific academic, social, or physical need they may have. They also use this method to challenge a child to attempt a new or different task within the theme framework.

The teachers select themes such as dinosaurs or transportation that can be extended into other areas of learning. They outline the activities, specify the timetable for the child to master the tasks, and suggest a format for the final product. The children's progress is continuously monitored to insure that they are learning the concepts initially stated.

The teachers usually evaluate the results of this type of activity although children or parents may be involved.

Teacher Initiated, Child Directed Activities

Teachers propose a concept or skill that requires mastery to the child. Together, they outline the procedures necessary for the completion of the task. The discussions include suggestions for appropriate methods and materials as well as the child's feelings about the task; whether any emotional blocks exist that interfere with learning and how they can support each child to do his or her best. The children proceed with their tasks and the teachers facilitate them as they require assistance. They participate in the evaluation of the children's efforts and occasionally involve parents.

Parents or volunteers can also participate in activities with children. The teachers might initiate the idea and direct the volunteer to assume the role of the teacher or the children could be encouraged to engage in follow-up exercises by themselves. Everyone had a good experience when a parent came to school to set up an aquarium with the children. Later the children organized a feeding schedule and sketched the different fish.

Child Initiated Activities

The natural curiosity of the child is also an appropriate initiator of activities. A child may discover something unusually interesting and spontaneously enlist others in the exploration. On one occasion crayfish caught in the nearby creek stimulated other children to participate in a sand building project.

Teachers may become an integral part of the adventure or propose to expand the activity by assigning tasks or encouraging the children to seek more information. Teachers use this approach to identify children's interests and direct them to a skill or concept they should acquire. Their concern is to attend to the progress and participate in the evaluation with each child.

Children can also initiate and direct their own activities, either individually or in groups. This child-centered approach begins when the child makes a discovery, explores it, organizes the activities, and completes the task. Teachers enter into the activity peripherally; through facilitating the child's needs to accomplish the task.

There are many activities that the teachers do not become involved with at all. These activities are organized, directed, completed, and evaluated by the children. Many large group games are predominantly managed by the children.

Observing Children's Behavior

Although there are numerous interactional events occurring daily, some form of systematic recording was deemed necessary to generate information regarding space usage.

In the study *Activity-Setting and Social Interaction: A Field Study,* Paul Gump and Brian Sutton-Smith use on-the-spot recordings of ten minutes of each boy's behavior in each setting.[5] To understand how space is actually utilized, it is necessary to identify a procedure, and to record reliably and unobtrusively behavioral occurrences, who performed them, where, and when. Gump and Sutton-Smith address the issue: How do activities limit, provoke, or coerce the expression of children's needs and problems? The hypothesis tested was that the amount and kind of social interaction is significantly affected by the variation of activity settings.

The aim of this phase of our activities was to describe what children do in the environment through the unobtrusive monitoring of their behavior in its setting, as well as casual observations. (See figures 2-1 and 2-2.)

To simplify the process of recording behavioral information, data sheets were constructed to facilitate the observation process. Three observers recorded the children's activities for two consecutive days. Children were observed in various activities and settings for periods ranging from five to sixty minutes. The variables recorded included the activity description, materials used, ages and number of children involved, and the initiator of the activity. Since all the activities occurred in and around a seven-room house that was modified to accommodate basic learning centers, the plan drawing on the data sheet provided a useful reference for each activity that was observed. From the systematic analysis of children's behavior, discussions with teachers, children, and parents, particular requirements were established for each of the learning centers and will be described in the programming section.

Briefly, however, the daily schedule at Amity is designed to be flexible for the teachers and children. The early morning is devoted to the child's individual interests such as art, reading, or games. The children and teachers then come together, as a group, to discuss school business and plans for the day.

During the following two hours the children are divided into groups to work in mathematics and language arts. They work within their level of competence in their group where explorations are designed to achieve success.

After lunch the children become involved in individualized projects. In addition to teacher initiated activities, the children are encouraged to explore their own interests. Their projects range from library research and insect collecting to sand box constructions and outside group games.

Programming the Activities

In order to develop a prescriptive program that incorporates the observational information as well as other types of information, a method was utilized to describe the events of the school in terms of activities and activity areas. We define an activity area as a section of the learning environment described by specific materials and physical boundaries. (See figure 2-3.)

In an article "An Ideal Environment for Learning," Dwayne E. Gardner indicates the importance of space usage within a learning environment. He states that: "the organization of the space, the placement within of centers of interest, dictates the flow of the learning activities" and "sufficient areas are program-organized to accommodate the learning activities in a functional manner."[6]

Eight primary activities were identified in specific settings within the building. They were mathematics (manipulative play), language arts, art, food preparation, sand and water play, block play, dramatic play, and science. From the literature in the field and conversations with teachers and children, we identified three more activity centers that could be recognized spatially. They were environmental education, large group, and social science. In conjunction with the development of spatially defined activity centers, there was an expressed need for individual work space.

For the proper development of the activity centers, secondary or molecular activities were generated. These are the actions that would occur within the primary activity. For example, weighing and measuring were molecular activities frequently observed in the mathematics center. When spatially arranging an activity center, it is important to identify all of the molecular activities that might occur as well as to identify the degree of simultaneity of occurrences of each activity. (See figures 2-4 and 2-5.)

When the centers and their constituent activities are identified, a data sheet (figure 2-6) is constructed to itemize the requirements for the successful performance of the center. The design requirements are statements pertaining to the provision of certain

(Text continues on page 38.)

Location O. T.: Date: Observer:

amity again OBSERVATION SHEET

General Info:

ACTIVITY ______________________

PARTICIPANTS:

AGE ______________________

NUMBER ______________________

STORAGE REQ. ______________________

OBJECTS USED ______________________

Classification

Type	High	Medium	Low
Gross Motor			
Manipulative			
Verbal			
Affective			
Thinking			

ACTION PATTERN RATINGS									
SUPERVISION		STRUCTURE		VERBAL			TEMPO		
supervise	unsuper.	organized	unorgan.	loud	med.	soft	fast	med.	slow

DURATION

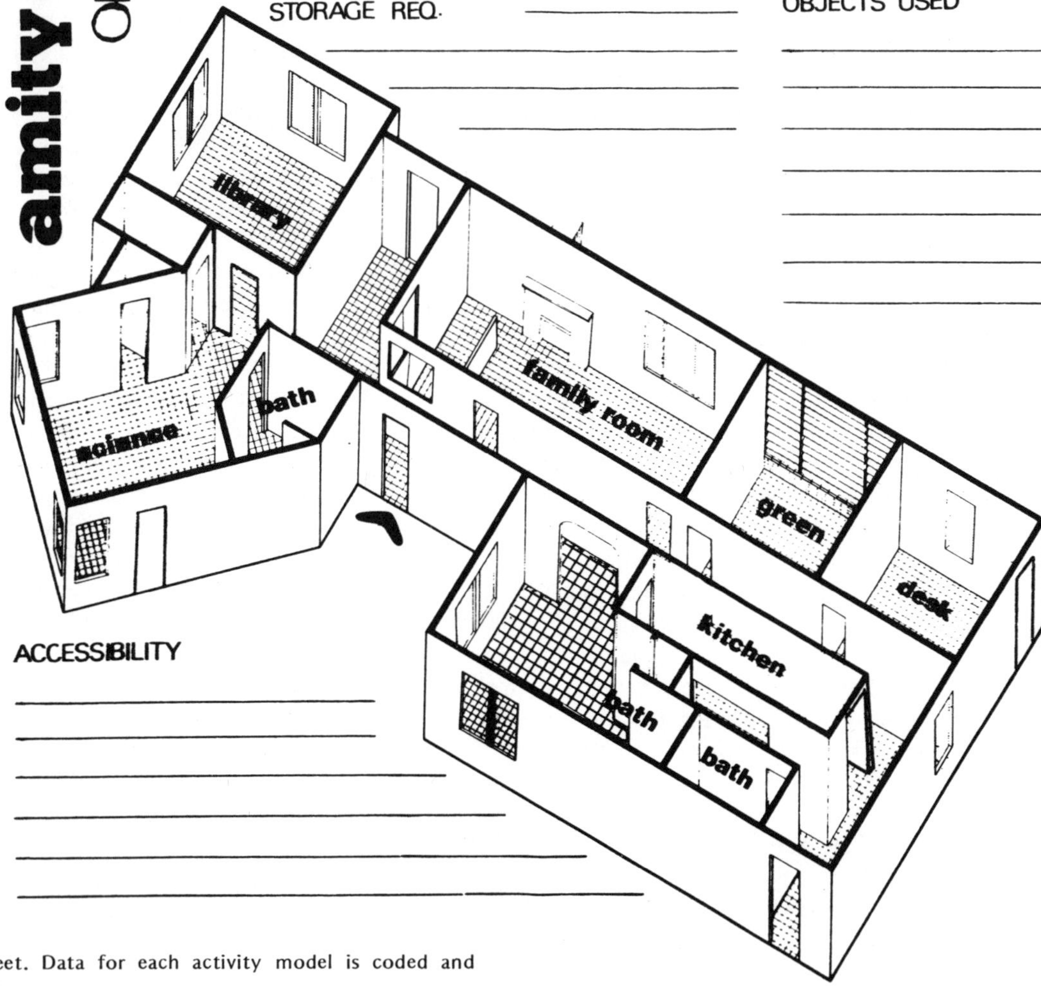

ACCESSIBILITY ______________________

NOTES:

Figure 2–1
Observation data sheet. Data for each activity model is coded and reported on this form.

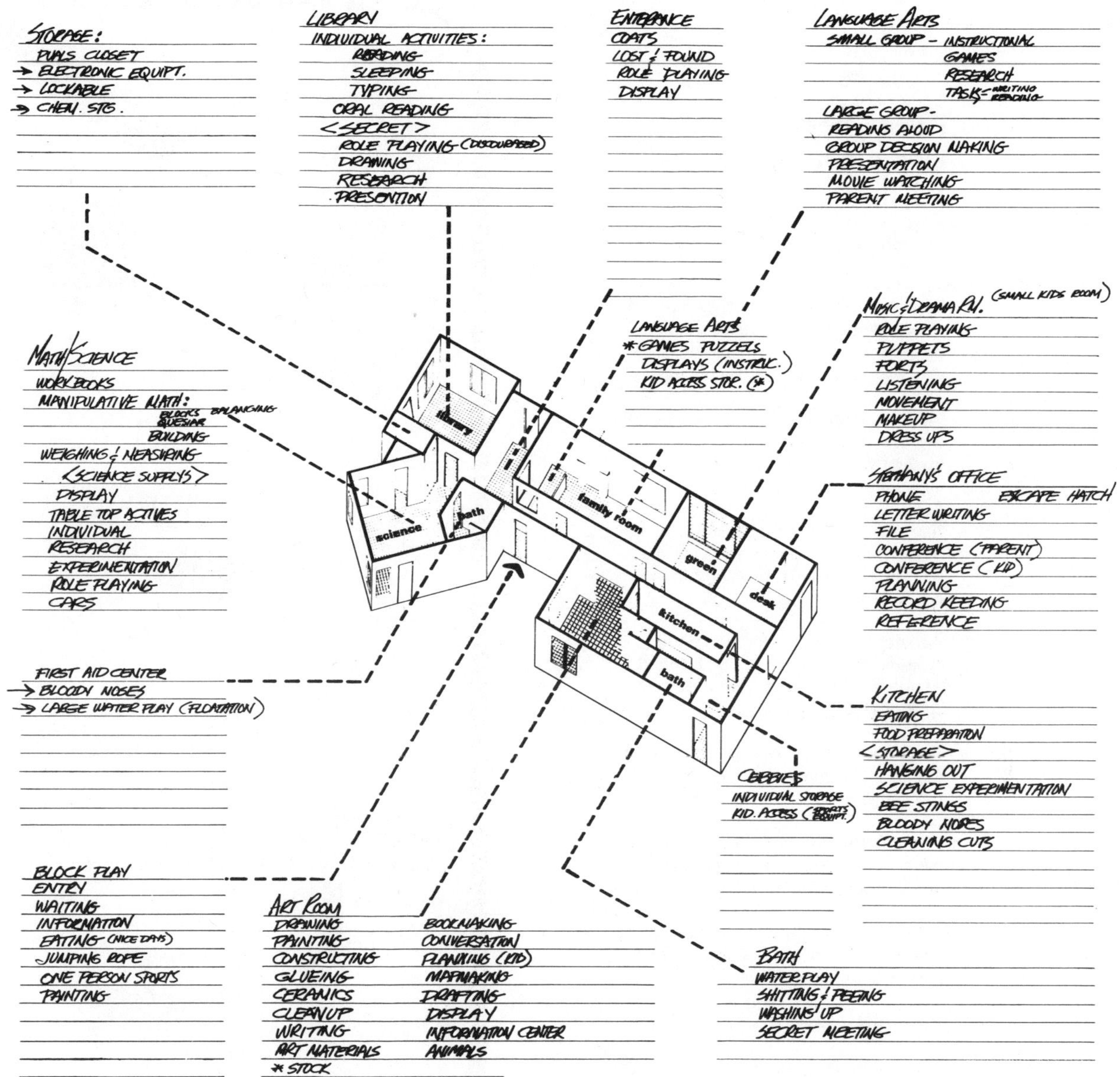

Figure 2–2
Activity locations. Each activity is located on the floor plan of the existing building.

Figure 2-3
Programming process. A definition of the processes and techniques utilized in this study.

Processes and Techniques Utilized in the Investigation

Systematic Observation of students within the present schools context, utilized to define the basic program activities and ellucidate essential setting characteristics.

Client Evaluation of the present facilities shortcomings to facilitate our further understanding of the schools requirements and desires.

Ranking and **Weighting Techniques** to establish the relative importance and relevance of criteria, relationships, and design requirements.

Activity Analysis to determine the requirements which must be fulfilled to provide for the successful performance of activity groupings.

Relationship Matrices utilized to examine dependencies and connections between programmaticly defined activities.

Client Evaluation of Program Development to insure conformity to the school's curricular objectives.

Anthropometric Analysis to develop dimensional and area standards between persons and objects likely to be used.

Behaviorally Based Design Requirements

Topological Diagrams to examine alternative activity setting arrangements consistent with programatically defined requirements and constraints.

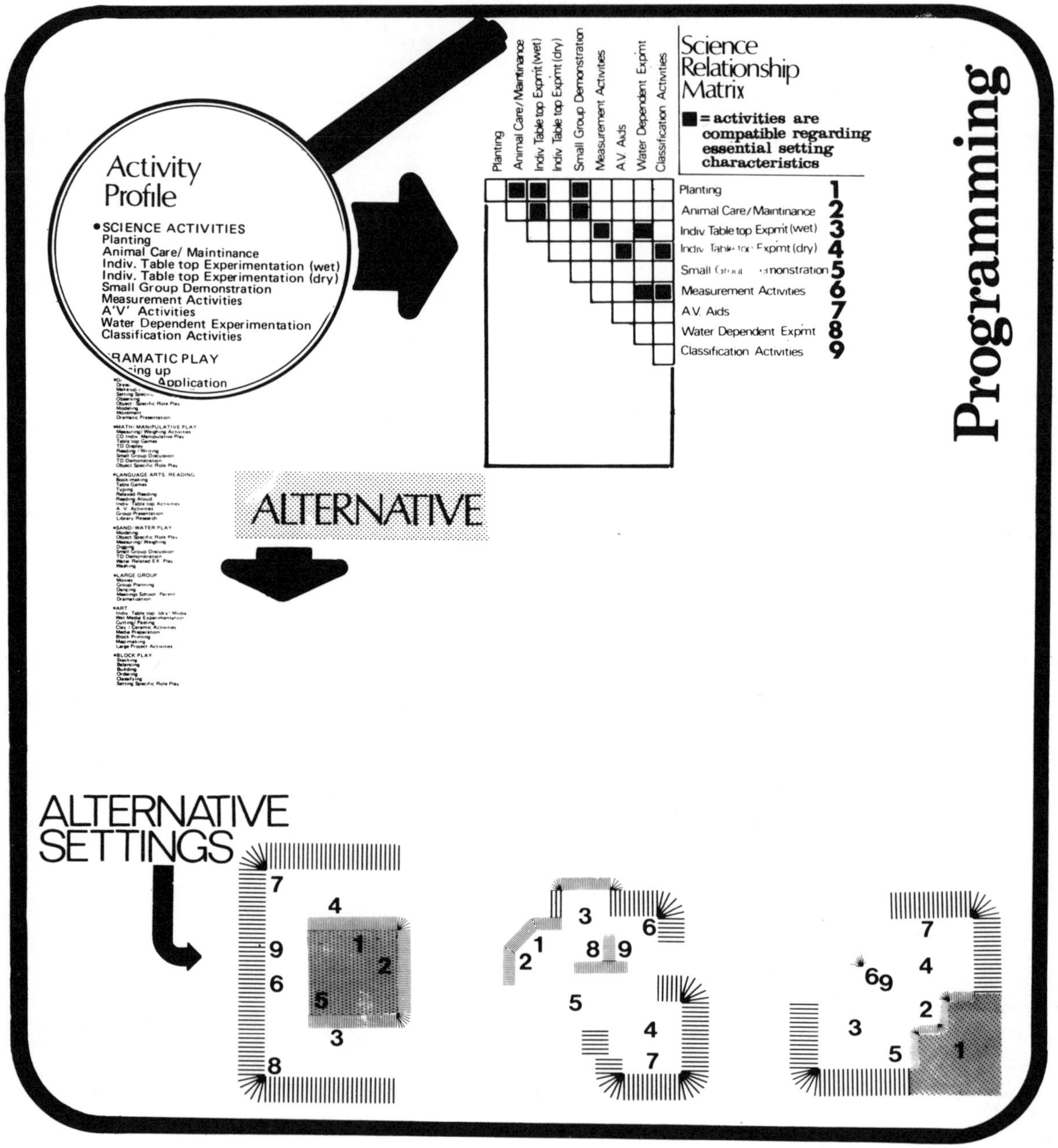

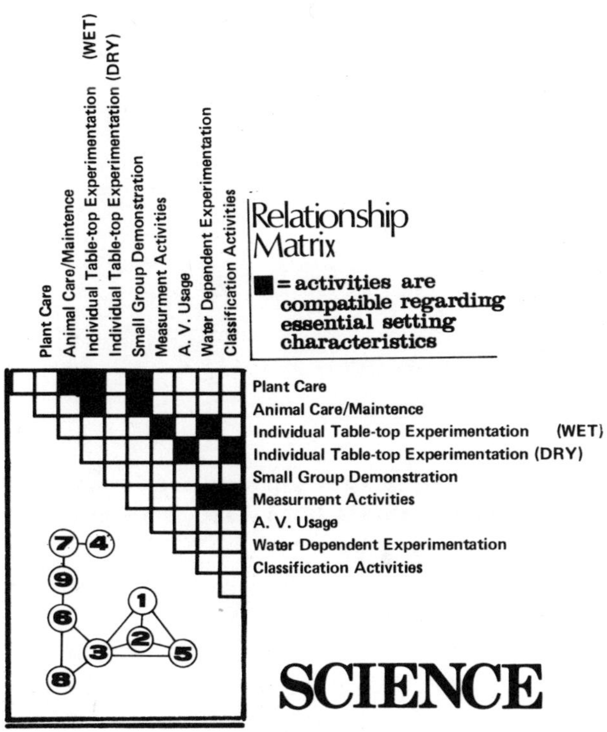

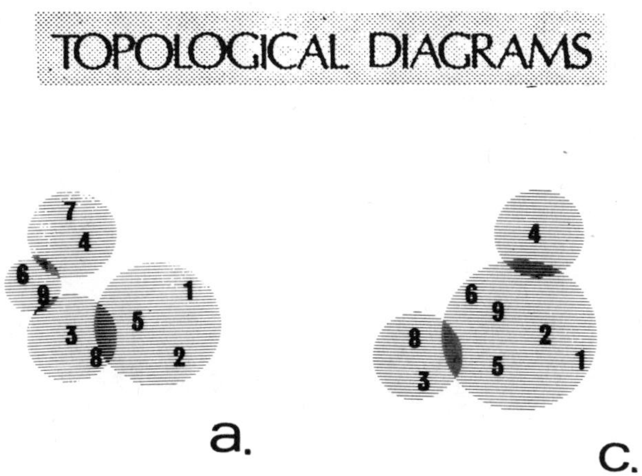

Figure 2-4
An outline of the transformation process for a typical activity center.

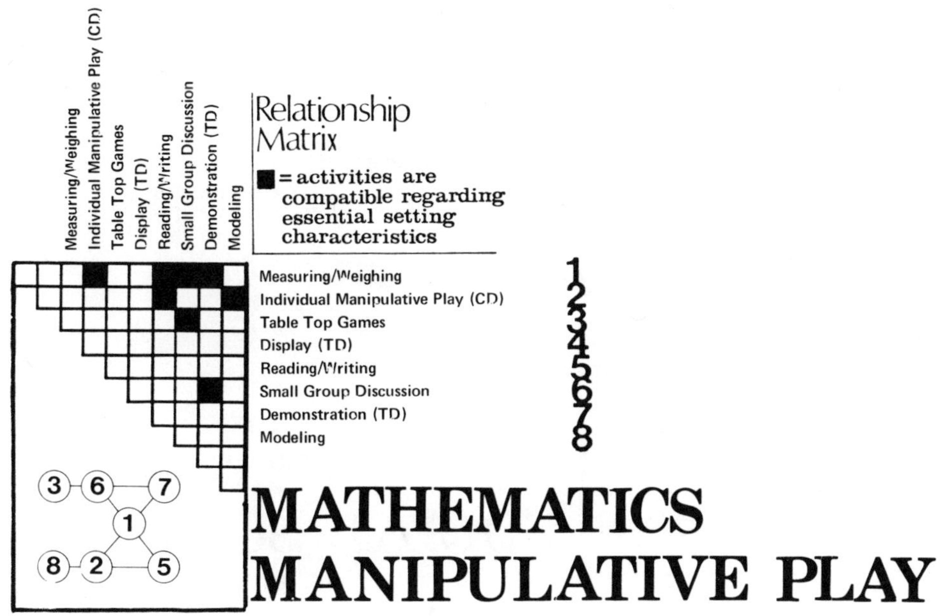

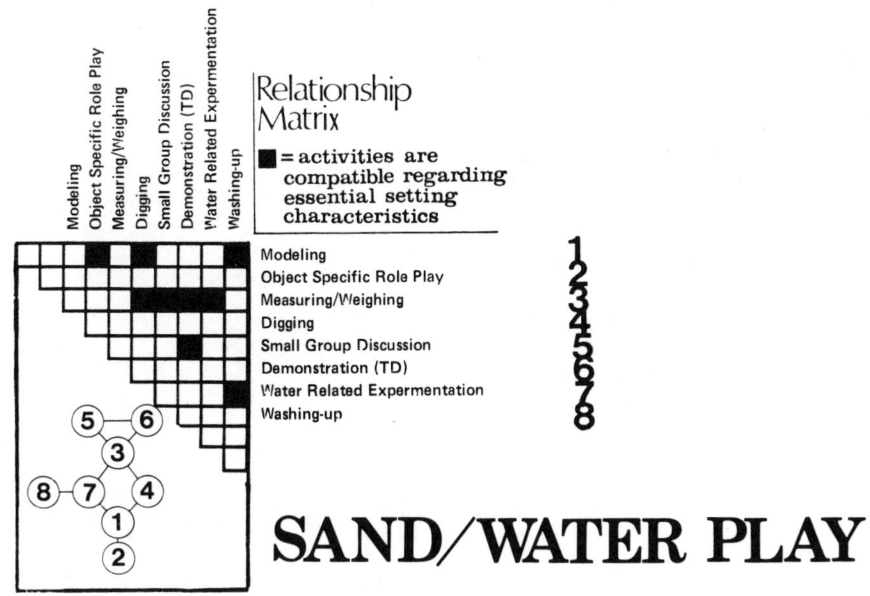

Figure 2-5
Activity centers. Six centers are described showing internal relationships.

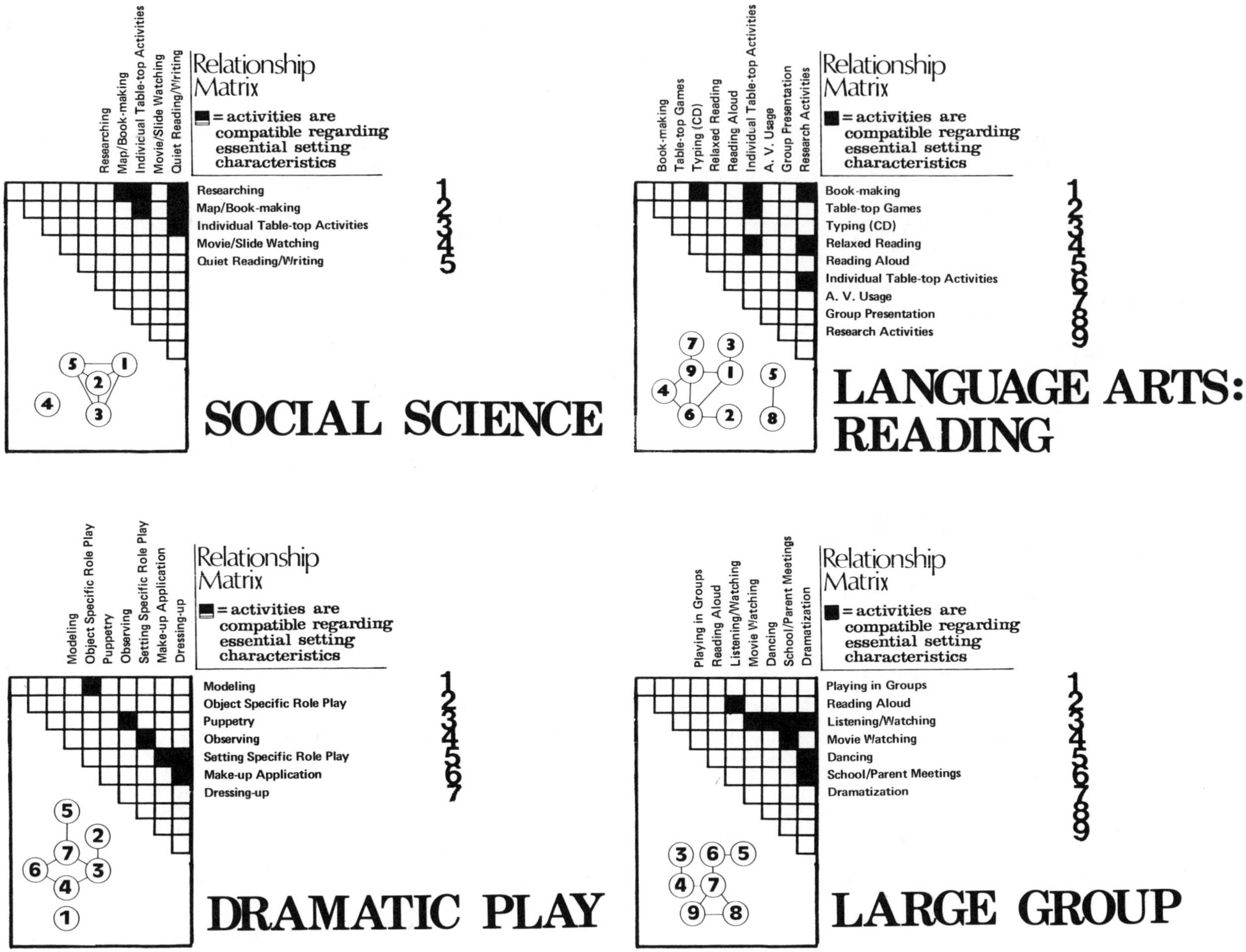

Figure 2–5 continued.

Primary Activity

MATHEMATICS

This activity involves thought in perhaps its most abstract form. In order to insure children's development of these cognitive skills a range of externalized, manipulative, and task oriented activities must be amply provide for.

Objectives

- visual discrimination
- concept formation
- problem solving
- skill development.
- language development

Participants

children
teacher
assistant

Molecular Activities

reading
manipulating
observing
ordering
relating
measuring
counting
classifying
writing
listening
displaying
constructing
taking apart
weighing
discussing
games

Design Requirements

1) must be easily accessed by teacher/assistants
2) Provide for quiet, individual play
3) Provide ample storage (child accessed) and work surfaces for manipulative toys
4) Provide flooring that encourages lounging
5) Provide a high level of lighting
6) Provide protection from circulation
7) Provide a range of work surfaces (vertical)
8) Provide for horizontal & vertical display
9) Provide for individually oriented and small group activities (teacher assisted)
10) Lesson distraction (vis. & acoustical) for individual activities
11) Provide comfortable & movable seating for teacher and children
12) Provide books in area to emphasize the particular concept being presented

Relationships

Figure 2–6
Activity center data sheet. The development of relationships and requirements for a typical activity center.

physical features and their arrangement within the learning center.

It is also necessary to arrange activities within the center to insure that each has its proper area and does not interfere with the operation of an adjacent activity. Thus the term adjacency becomes an important concept for spatial planning. Adjacencies are identified by selecting each activity in a center and placing it on a matrix as shown in figure 2-7. The matrix is a visual display of activity interactions and dependencies that form the basis for planning decisions.

From the notations on the matrix, a diagram can be constructed graphically to portray each centers' adjacency relationship. The final set of agreed upon adjacencies can become the basic rule for the physical plan of the learning environment.

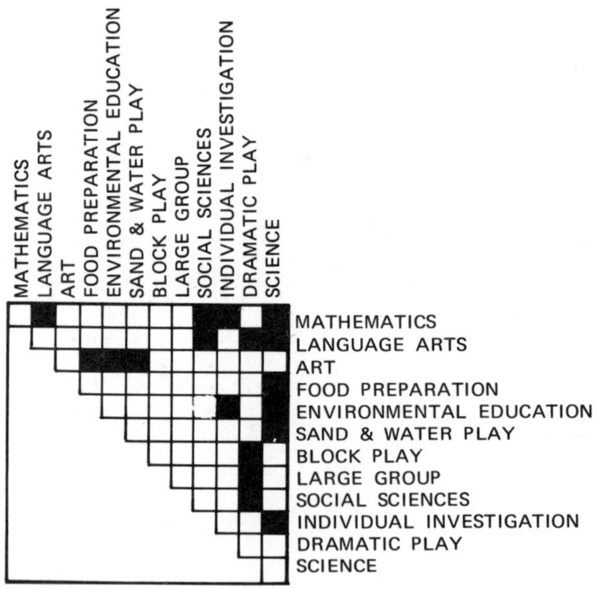

Figure 2-7
Activity center matrix. Relationship diagram describing the interconnection of all activity centers.

From the adjacency designations in the matrix (figure 2-7), a diagram displaying relationships can be developed by locating each activity around a circle as shown in figure 2-8. Then, each indication of a relationship shown in the matrix should be noted by drawing a line between the two activities. Each pair of matrix relationships should be included in the network diagram. The rule for unscrambling this network is to transcribe the same relationships between activities in a diagram without having lines intersect each other (figure 2-9). As the number of activities increases, the possibility of adhering to this rule becomes more difficult.

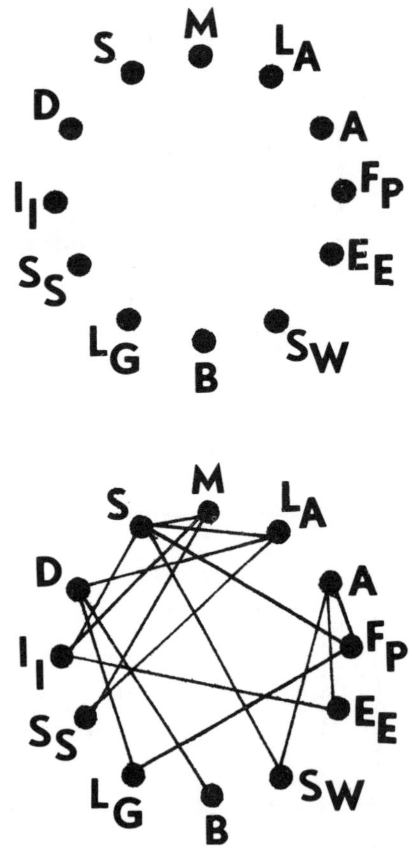

Figure 2-8
Activity net. The transposition of matrix relationships into a net diagram.

From the activity graph, it can be observed that there is a hierarchy of connections. Certain activity centers require a greater degree of spatial proximity than others. Activity centers such as science (S), math (M) and language arts (LA) have four or five connections to other centers. This dependency suggests that the location of these centers is particularly important when spatial planning occurs.

The following stage in the transformation process is the development of adjacency diagrams (figure 2-10) that reflect the desired relationships from the activity graph. During this phase of the process, spatial alternatives are generated that propose to fulfill the best fit requirement. The adjacency or venn-

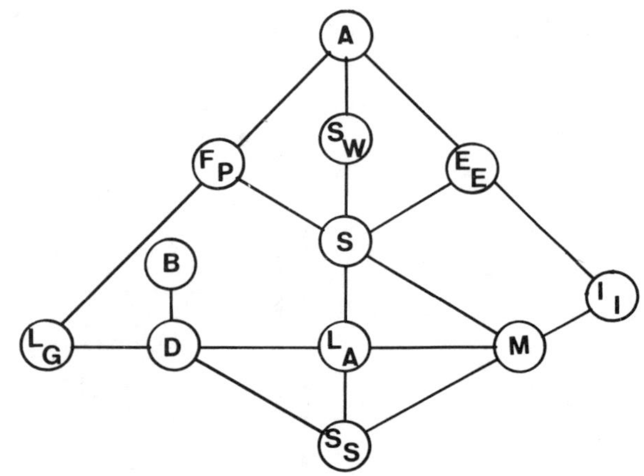

Figure 2-9
Activity graph. A diagrammatic description of the best possible relationships.

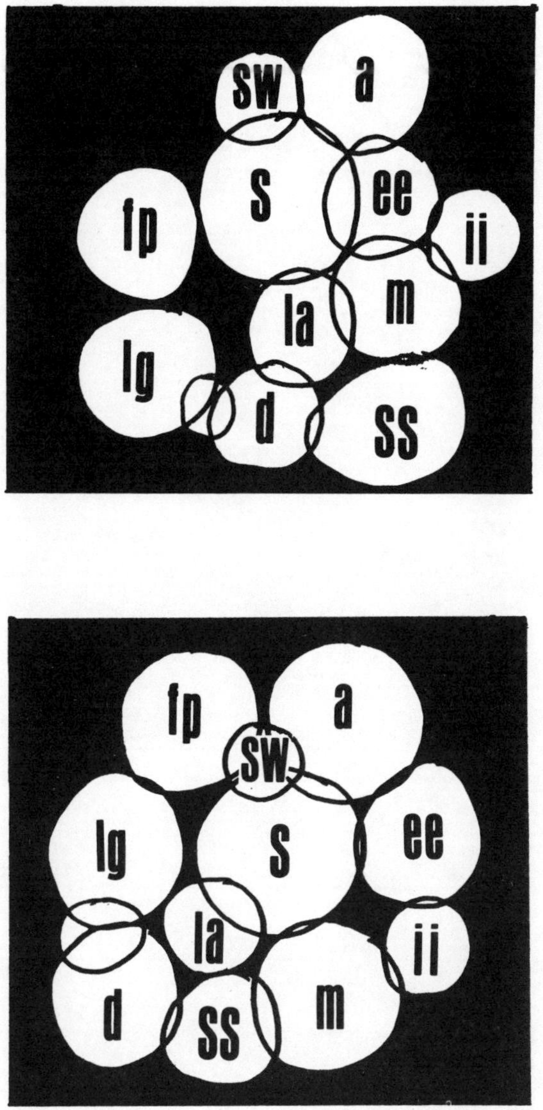

Figure 2-10
Adjacency diagrams. A description of alternative relationships between activity centers.

diagram is the first opportunity to examine the flow of people between activity centers. A more developed version of the adjacency diagram is referred to as a form diagram (figure 2-11). This is a less abstract but more spatial representation of the previous diagram. Form diagrams can approximate size-shape relationships and provide a suitable method for graphically portraying alternative spatial arrangements. While there has been a logical progression from verbal to graphic concepts, the form diagram only suggests a shape tendency since other types of information may considerably influence the modification of the shape (see figure 2-12).

Often, children's centers are the result of an adaptive use of facilities with certain limitations to the degree of renovation feasible. In these instances, it may not be possible to achieve the most desirable set of activity relationships; but the availability of an activity graph permits logical trade-offs to be made.

This activity-net diagram is the final step in the programming process. From this inventory of needs a place can be created that clearly fulfills the aspirations of all those affected by design decisions.

There is a growing belief that the objectives for learning are inextricably linked to the place where learning occurs. This recognition is fundamental in developing an awareness in children of the way the built environment can function to facilitate human desires. The everyday physical environment that children encounter is often ill-suited to their needs because they were never part of its creation, either directly or indirectly.

We have attempted to integrate a learning philosophy and style, observable behaviors, and programming concepts into a procedural approach that advocates people as the basic ingredient for planning change.

We have described the process for reflecting the purposes of a school into the place created to accommodate the schools activities. There has not been any attempt to elaborate upon the program because it

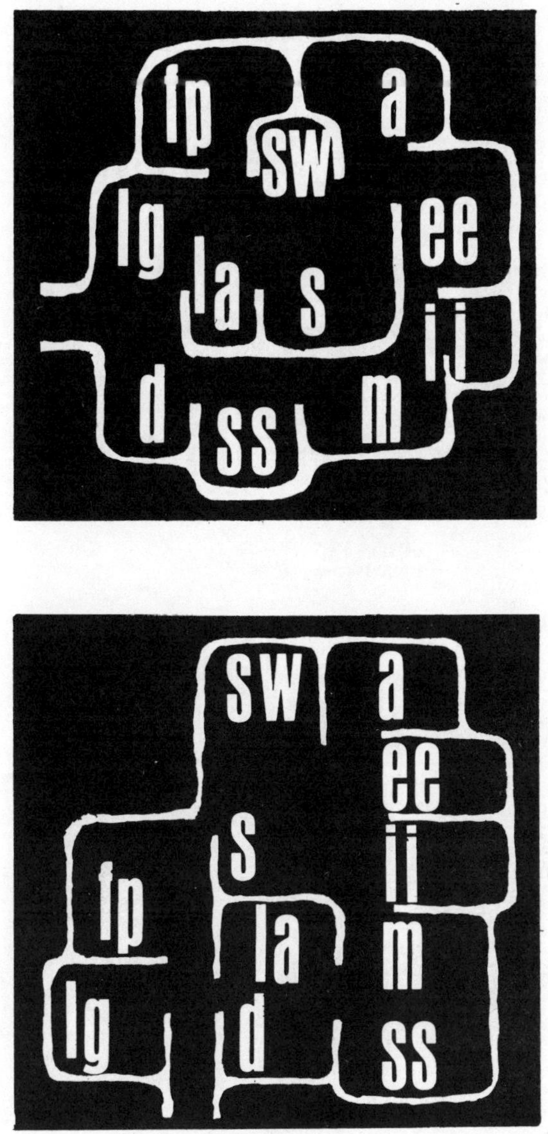

Figure 2-11
Form diagrams. A representation of the shape tendencies of the centers.

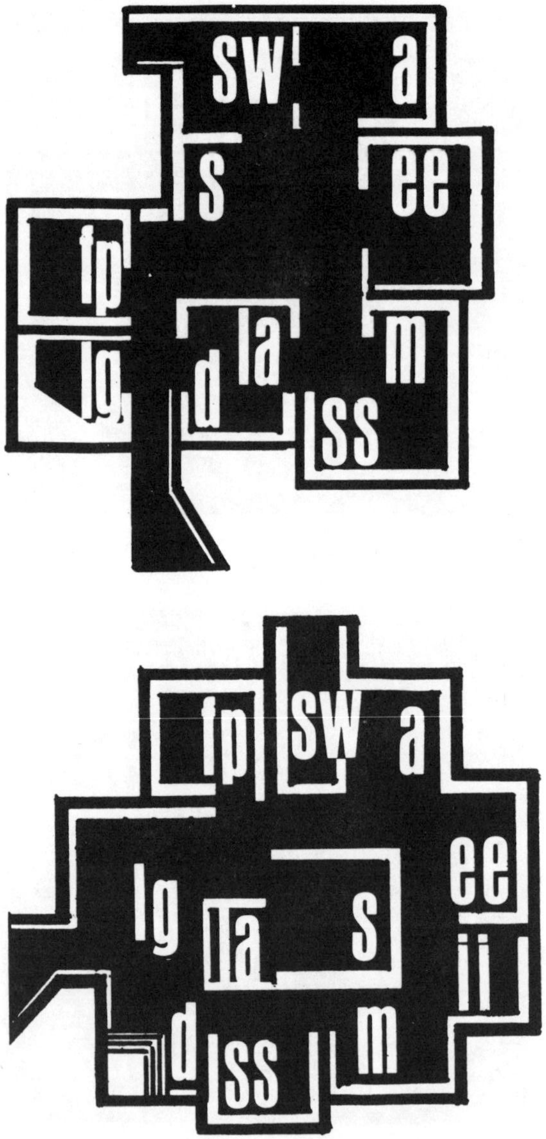

Figure 2–12
Floor plan. A measurable relationship between all physical elements in the center.

would only be pertinent to the users of the Amity Again school. Rather we have proposed concepts, methods, and procedures that are easily adaptable to other situations.

Our intentions were to illustrate further the need for insuring that the expertise of all those connected with the school could be incorporated into the design decision process. It is clear that the integration of a design decision process, with behavioral information modified by a group process for organizing this event, is a viable direction for involving people in the design of their environment.

REFERENCES

1. Neil Postman and Charles Weingartner, *The School Book,* Delacosti Press, N.Y., 1973.
2. Ivan Illich, *De-Schooling Society,* Harper and Row, Publishers, N.Y., 1970.
3. Postman and Weingartner, ibid.
4. Ibid.
5. Paul Gump and Brian Sutton-Smith, "Activity-Setting and Social Interaction: A Field Study," R. E. Herron and Brian Sutton-Smith, eds., in *Child's Play,* John Wiley and Sons, Inc., N.Y., 1971.
6. Dwayne E. Gardner, "An Ideal Environment for Learning," *Housing for Early Childhood Education,* Washington, D.C., 1968.

3

MURFREESBORO, N.C.
plan alternatives for the national register historic district

Henry Sanoff
Brad Smith
Larry Liberatore
David Polston

Community Development Group
School of Design
North Carolina State University
Raleigh, N. C.

June, 1976

This report was made possible through the support of the Murfreesboro Historical Association. Printing and publication was donated by Chowan College, Murfreesboro.

The participating members of the Community Development Group wish to acknowledge the assistance and cooperation provided by the Murfreesboro Historical Association, Historic Murfreesboro Commission, and the Town of Murfreesboro. We would also like to extend our appreciation to the State Division of Archives and History; Prof. Robert E. Stipe, University of North Carolina; Dan Discenza and Vickie Cable, graduate students; and Barbara Koenig, editor.

preface

In recent years, the small town of Murfreesboro, N. C. has, under the direction of the local historic association, rehabilitated and adaptively used several of its old, historically significant buildings. As a result, the community has not only benefited by the addition of space which has the potential to support a variety of activities, but also retains its important educational and cultural resources as reminders of the town's physical social and economic development. This concept of conservation through adaptive use is being applied in many places throughout the country and is widely recognized as a logical approach for gaining useable space while saving old buildings, particularly in small towns such as this one.

The first project to renovate and use an historic building in Murfreesboro began in the late Sixties. About the same time, a 12-block historic district was established near the center of town, and control over its development given to the Historic District Commission and the Murfreesboro Historical Association (MHA). Since then the MHA has sponsored two other public reuse projects, encouraged private individuals to restore six old homes for residential use, purchased several sites within the historic district for possible relocation of endangered buildings, and received Federal and State assistance for these and other projects. What was lacking, however, was a comprehensive plan to guide future growth for the historic district. Decisions to take action on an issue was more or less made as resources became available or as a crisis arose.

The Community Development Group (CDG) was asked to devise a plan that would enable the MHA to more effectively make decisions and direct the future of the historic district. CDG is a graduate architecture studio at the School of Design, North Carolina State University. Its purpose is to provide small communities and organizations who have limited funds with architecturally related services and expertise.

Since August, 1975, CDG has worked with the Murfreesboro Historical Association to develop the plan and design for adaptive uses in particular buildings. It was realized that a major portion of the town would be affected by any decisions made to develop the historic district, and that the historic district would in turn be affected by any changes occuring in the major portion of the town. Because of this, community input was also vital in organizing the goals of the MHA into a future plan.

Through several meetings with the MHA and with the aid of tools specifically designed to determine goal priorities and understand the intentions of the MHA members in terms of a general direction for the future, a goal-based, user-oriented plan was developed. In order to involve more of the community, a public hearing was set up to examine critical issues outlined in the plan, and to receive opinions and criticisms on the alternative planning directions possible for each issue.

The final plan is a result of the combination of data collected at these meetings as well as the data gathered through extensive reserach into existing environmental conditions, legal and technical constraints, etc. The nature of the plan allows flexibility as resources change or as the goals of the MHA change with time. It emphasizes the fact that the user is central to the planning process. By identifying the important issues, and outlining specific alternatives and implementation procedures, the users can change the plan as they feel it should change.

The importance of incorporating user participation into the planning process has become increasingly more evident in the past decade. Past experience with planning processes which have not included user participation have shown several disadvantages. One is that the planner must base his decisions on his past experiences and the conceptions he has of the client/user group. Unless the planner has a similar background to that of his client, there exists a high probability that his decisions will not reflect the client's values.

If the plan should fail in some way, the client/users will bear the consequences. This is a second disadvantage of a non-participatory planning process. The planner in the past has developed his plan, presented it to the community and then left for a new assignment. After completion of the plan, the consequences of the plan—its success or failure—depend on whether the client could provide the required financial resources at the required times. Often during the implementation stage, the client is not able to meet the required expenditures and the plan, with its promise of better things to come, is put on the shelf.

The planning process in Murfreesboro has produced a plan which reflects the goals of the client and is able to respond to future change. Its overall success will depend on the desires and motivation of the MHA and the community.

REASONS FOR SAVING OLD BUILDINGS

Saving old buildings is a growing concern of many small towns throughout the country. The increasing demand for residential and commercial development, coupled with the continual deterioration of older structures within a community, has raised questions concerning the importance of conserving old buildings. The dozens of reasons presented for preservation can be grouped into four main headings: cultural memory, successful proxemics, environmental diversity, and economic gain. Each of these four arguments are presented below in order to show the value of saving old buildings.

CULTURAL MEMORY

Buildings are tangible reminders of the accomplishments and growth that a community has made throughout its history. Most communities have grown during periods of prosperity. These times of development have usually produced different architectural styles in various sections of the community. The collection of these buildings throughout a community provides a physical record of the types of environments in which the community's forefathers worked, played, and lived out their lives. These different building types are also examples of architectural details and craftsmanship which were lost with the passing of the generations which produced them. Spider windows, exterior Victorian woodwork, and hand crafted six-panel doors with brass doorknobs are a few of the antique textures which past generations have left. These common details of the past are no longer a part of our modern buildings, and their value to the community increases each year as similar details in other communities are demolished. Not only is it important to save these examples of the past as signs of the community's cultural and historical resources, but it is especially important to save these reminders for the community's young and future generations. Saving old buildings is important to Murfreesboro's future generations, who will never have the opportunity to experience the history of Murfreesboro by walking through the Thompson house, or seeing the Vinson house, unless these and other old buildings in Murfreesboro are conserved.

SUCCESSFUL PROXEMICS

Successful proxemics is a complicated term for a simple idea. The word proxemics refers to the relationship that exists between people, the activities they engage in and the places where they perform these activities. Successful proxemics therefore occurs when there is a harmony or balance between people, activities, and environments.

Before the development of the car as the primary means of transportation and the development of land use zoning, most neighborhoods were examples of successful proxemics. Places of employment, schools, churches, and the neighborhood store were located within walking distance of one's home. Since the development of the automobile and zoning the concept of the small close knit neighborhood has been expanded to include a larger area than the traditional walking distance. People now have to drive to work, to schools, or to shopping areas which are located outside their residential areas. As a consequence of this expansion, the social interaction which once occurred as the result of walking to and from work or the neighborhood store has been lost. The feeling of a neighborhood identity has also begun to die due to the change in the neighborhood structure.

Murfreesboro still has neighborhoods which have successful proxemics. These neighborhoods, like others throughout the country, are having to adapt to the growing pressures for change. The two residential blocks just west of Second Street at the Georgia-Pacific plant are an example of a neighborhood in Murfreesboro which is having to cope with change. The residents of this area are employed by the Georgia-Pacific plant on the northeast corner of the neighborhood. There is a fish market on Second Street which serves not only as a place for buying and selling but also as a center for exchanging news about the neighborhood. Two blocks to the southwest is the central commercial district of the community, and the church on Broad Street and the minister's home are also located within the boundaries of the neighborhood. This neighborhood is feeling the pressure for change due to the growth of the Georgia-Pacific plant. New parking for employees who live outside of this neighborhood has eliminated the residences which once existed on the east side of Second Street and a historic home on Broad Street. If movements for growth and change are allowed to continue in the remaining two blocks of old homes then this successful neighborhood will slowly be destroyed. The future of this neighborhood and other successful neighborhoods such as the one on West Main Street depends, in part, on the conservation of the neighborhood's old buildings.

ENVIRONMENTAL DIVERSITY

The concept of environmental diversity is closely related to successful proxemics. Our present world is becoming increasingly more homogeneous in appearance and use. Residential subdivisions, shopping centers, and places of employment are now being mass produced in one pervasive "modern" architectural style. There are several reasons for this increasing standardization of building types. One is that fewer people today are directly involved in the planning and construction of the buildings they use. Building contractors and architects, with their expertise in construction, control the design of new buildings; thus this small group also controls the exterior appearance of all neighborhood and community growth. The construction industry is tending to produce standard building types as a result of the combined influence of specialization of construction labor, mass production of standardized building materials, and wide circulation of standard building plans. All of these measures are being employed in order to produce buildings which the public can afford.

Two more factors which have helped to produce homogeneous environments are the automobile and land use zoning, already mentioned in the successful proxemics argument. Before the development of the car it was difficult and expensive to get from one place to another, so local governments had to serve a wide variety of purposes. People could not jump into their cars and dash across town to buy a loaf of bread, or go to work. Zoning regulations, with restrictions on land use, building appearance and location of a building on its site are intended to produce an orderly environment. But the result is often homogeneity rather than any useful order of uses.

Standardization in the building industry, the automobile and land use zoning are helping to produce increasingly more homogeneous neighborhoods and communities. These influences make the preservation of older buildings and neighborhoods, with their diversity of building types, very desirable—not only for visual variety but also as an alternative to the homogeneity of new environments for living, working, and playing.

ECONOMIC GAIN

The economic gain that preservation can produce for a community is the fourth argument for saving old buildings. An adaptive use approach recycles old buildings to new uses which makes it economical to save them while retaining much of the structure's original character. In many cases, a building that is structurally sound can be rehabilitated at less than half the cost of erecting a new one for the same purpose. Rehabilitation attracts new residents or users to previously under-used structures. These investments in private rehabilitation will not only add to the cultural resources of the community but will also produce a higher market value for the improved property. This increased property value will in return produce a higher tax yield for the community. Preservation can also attract a small amount of tourism which will produce additional revenues for the community.

Both historical and non-historical buildings should be conserved in order to maintain the successful proxemics of the neighborhood.

2 CONSERVATION/ PRESERVATION TERMS

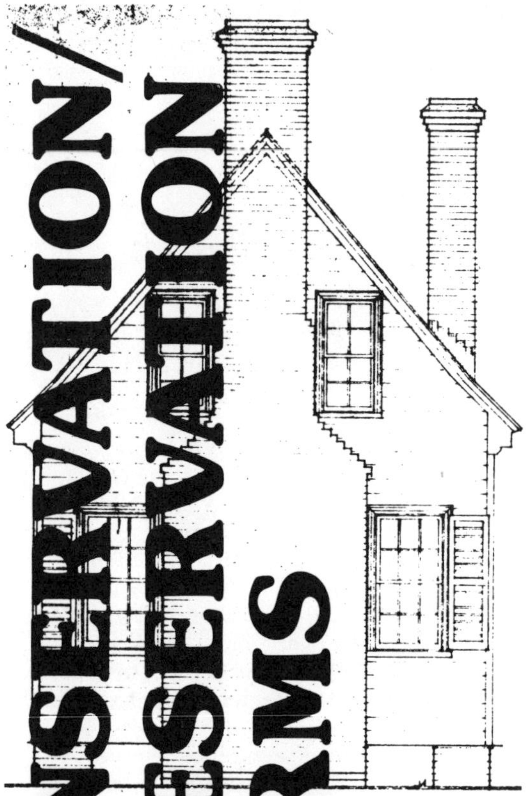

MURFREESBORO'S CONSERVATION/ PRESERVATION APPROACHES

The objective of architectural preservation is to conserve those buildings and sites which contribute to our environment and culture by their historic or architectural significance. Preservation projects may involve individual buildings or groups of buildings requiring various architectural approaches. There are a number of terms used in perservation work which are discussed below. Each of these terms have a special meaning and is used to describe an approach for saving old buildings. The approaches which have been used in Murfreesboro are presented as examples of the terms.

CONSERVATION: Conservation involves the rehabilitation of any building, historically significant or not, as a means for saving the structure. This may involve the modernization of structural and mechanical elements necessary to extend the useful life of the building. The architectural character of the exterior should be retained but there is no requirement to extend much effort to achieve strict historical accuracy or to precisely recreate the original structure.

PRESERVATION: Preservation involves either the restoration or rehabilitation of only historically significant buildings. Effort is required to retain the architectural character and integrity of the exterior and in many cases the interior. Preservation is the one general approach which encompasses all of the following terms.

REHABILITATION: Rehabilitation involves the preservation of a structure, site, or group of structures from futher decay or change through cleaning, strengthening, addition of modern elements and other similar methods. Rehabilitation includes the equipping of a building for an extended useful life with a minimum alteration of the original construction.

The Myrick house is an example of a private rehabilitation project in Murfreesboro

The Dr. Gary house, located on Wynn Street, is another example of a rehabilitation project

RESTORATION: Restoration involves the preservation of a structure, site, or group of structures in the condition or appearance of a specific period in time. Restoration requires a maximum amount of research and authenticity.

Several restoration projects have been implemented in Murfreesboro by both private and public concerns. The Roberts-Vaughan house was one of the first restoration projects.

The Rea Store, once the center of trade between Murfreesboro and the New England States, has been restored under the leadership of the Murfreesboro Historical Association.

Melrose (the left picture) is an example of a restoration project by a private individual. The Wheeler house (the right picture) is currently being restored by the Murfreesboro Historical Association.

ADAPTIVE USE: Adaptive use involves the preservation of a structure, site, or group of structures by giving them a contemporary function that will fill community needs. With adaptive use buildings are recycled to new uses, making it economical to save the building and retain much of its original character. Adaptive use can be coupled with either restoration or rehabilitation.

The Roberts-Vaughan house is now being used as the public library and headquarters for several community organizations.

The Rea Store is currently being used as a community museum.

RELOCATION: Relocation involves the movement of a building or part of a building to a new site as a means for saving the structure.

The Deanes house was relocated to Duncan Park and rehabilitated.

The Winborne Law Office will be relocated within the historic district.

RECONSTRUCTION: Reconstruction involves the use of research documents or archeological evidence to construct a replica of an earlier building or facility which no longer exists.

Archeological digs have located the foundations of the original outbuildings around the Wheeler house. These buildings may be reconstructed in the future.

REPRODUCTION: Reproduction involves the construction of a copy or imitation of an original building or building appearance without the use of research documents or archeological evidence.

HISTORIC DISTRICT ZONING: Historic district zoning involves the passing of an ordinance that designates an area in a community as an historic district. The purpose of the zoning is to (1) protect the historically significant buildings in an historic area; (2) to stabilize and improve property values in the area; and (3) to promote the use and preservation of historic buildings. Historic districts are governed by an Historic District Commission. This commission has the authority to determine the appropriate style, construction, location and use of structures on any site within an historic district. The Murfreesboro historic district zoning ordinance was passed in 1967. The district encloses a 12-block area in the northern part of the community. The Murfreesboro Historic District Commission was established with five members. The commission has the authority to control changes in the use and exterior appearance of any site within the historic district. The Roberts-Vaughan property on Main Street was spot-zoned as an historic district in 1970

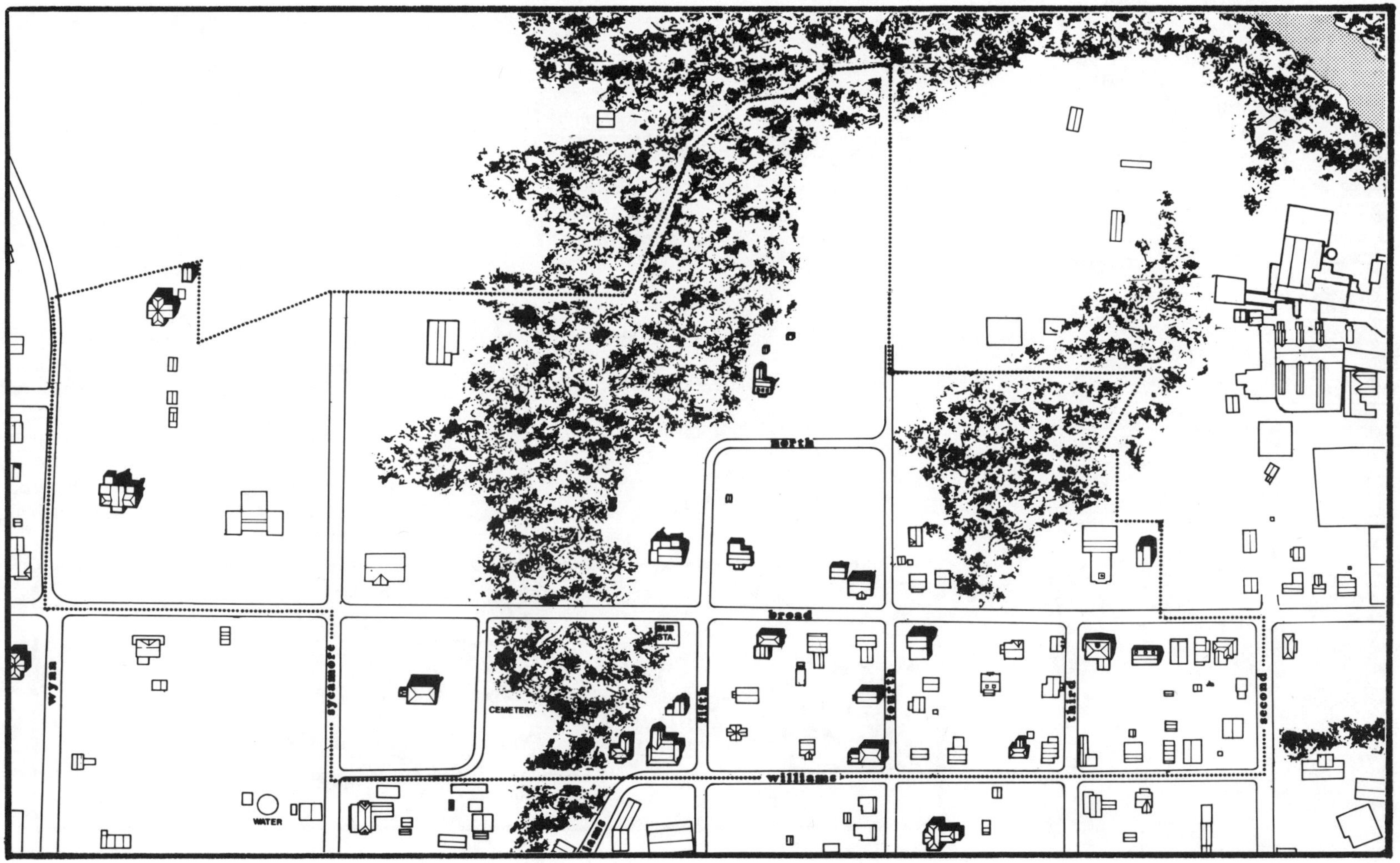

history of murfreesboro

During the 18th and 19th centuries, Murfreesboro experienced a richness and vitality unique to the northeastern part of North Carolina. Situated at a strategic point of navigation on the Meherrin River, the small community first served as a King's Landing, inspecting both incoming and outgoing goods for the British Crown. It later gained prominence as a major center of exchange for goods from New England, Europe, and West India, and the rich naval stores and agricultural products from the northeastern Carolina region. The residents themselves enjoyed continued prosperity through related businesses and services which, in turn, led to the physical and economic growth of the town.

The prosperity of the town was ultimately reflected in the richness and cultivated appearance of its structures--residential, commercial, and educational--many of which still stand today. Murfreesboro is, in fact, fortunate to have so many of its historic structures still existing. Although not totally free from neglect and random destruction, these elements potentially serve as tangible reminders of the patterns of physical, social, and economic growth the town has experienced. The location and arrangement of such historic resources as the rich Victorian homes along West Main St., the grouping of historic residential and commercial buildings in the center of the historic district, the proximity of the river and old town landings to the center of town, and so on, are representative of the type and extent of development Murfreesboro has experienced.

It is widely recognized that these elements should be preserved and actively used by the community. Aside from the obvious fact that many of the remaining 'antique' structures are of historic value, they have the potential of satisfying the community's needs for space and visual variety as well. In order to direct future decisions that will protect and use these resources, it is necessary to recognize their present physical condition, as well as the context in which they now exist.

One of the first tasks in organizing to plan for the future is the recognition of existing environmental conditions as they relate to the target area of the plan. Although the locally recognized historic district is of primary concern, activities or changes that occur in other parts of town will affect, or will be affected by, the planning decisions made. Therefore, research into these areas as well as the historic district is essential.

April, 1863–C. S. Corps of Engineers

Historic Boundaries

As can be seen below, the Historic District currently on the National Register is considerably larger than the area locally zoned as an historic district. What is important to recognize is the fact that buildings included in the National Register Historic District are given less protection from demolition than buildings within the local historic district.

Of particular concern are those buildings on West Main St., whose present state of disrepair, and possible demolition for highway commercial development, calls for immediate action.

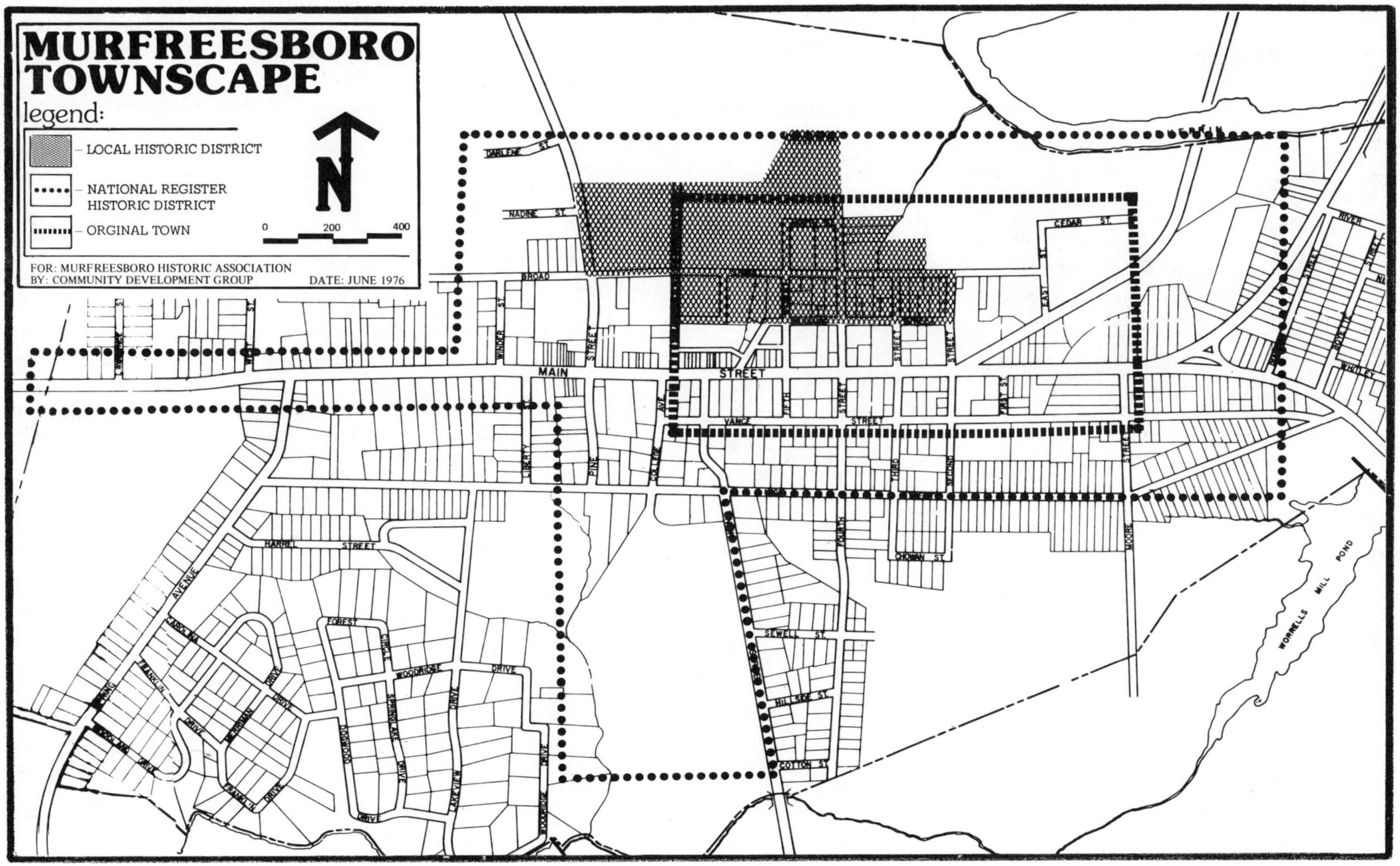

Historic District & Buildings Survey

In order to collect and record an accurate profile of old, significant buildings existing in and around Murfreesboro, a two-page survey was developed. Its format organized the information into three general categories:

Identification–including the location, present owner, and a recent photograph.

Classification--according to past and present uses, and the future intentions of the owner.

Description--including the present structural condition, and its significance to the town.

Cataloguing the information made it easily accessible and provided the basis for furthur research, some of which is described on the following pages.

HISTORIC DISTRICT & BUILDINGS SURVEY HOUSE # 35

community development group DATE 12/1/75

IDENTIFICATION

NAME:
Historic SEWELL HOTEL
Common SEWELL HOTEL

LOCATION:
Street Address 110 LIBERTY MURFREESBORO

In HD___ Out HD ✓

Original Location?
Yes___ No ✓ --if no, where? W. MAIN - COLLEGE TEXACO LOT

DATE ORIG. BLDG.: 1880's

PRESENT OWNER:
Name H.L. EVANS JR Age ___
Address 110 LIBERTY MURFREESBORO

ARCH. STYLE: GR. REVIVAL

TAKEN 10-1-75

CLASSIFICATION

OWNERSHIP: Association___ Public___ Private ✓

STATUS:
Occupied? No___ Yes ✓ --if yes, by whom? (owner)/tenant

Previous Repair (since 1967)--
Nothing___
Normal Maintenance___
Something (restor/(rehab)/renov) ✓
In progress___
Completed ✓ date________

Zoning RESIDENTIAL

Total Assessment________

Present Use--
commer___museum___entertain ✓
agric___educa___priv res ✓
relig___gov't___other________
none___(prev use HOTEL)

Sale being considered? No ✓ Yes___
If yes, is sale in progress?
Yes___ No___
Interested party________

PROPOSALS FOR BUILDINGS:
Proposed repair (owner)/inter. party)--
Nothing___
Normal Maintenance ✓
Something (restor/rehab/renov)

Proposed Use ((owner)/inter. party)--
commer___museum___entertain___
agric___educa___priv res ✓
relig___gov't___other________
none___(prev use________)

To be moved? No ✓ Yes___--if yes, where?________

Proposed occupant________

HISTORIC DISTRICT & BUILDINGS SURVEY HOUSE # 35

community development group DATE 12/1/75

DESCRIPTION

Structural condition--
Standard ✓
Substandard___
Dilapidated___

In good repair ✓
In need of minor repair___
In need of major repair___
Dilapidated___

Number of Stories 2
Basement? Yes___ No ✓

Facade Materials--
Clapboard ✓
Brick___
Shingle___
Aluminum Siding___
Board & Batten___
Asbestos Siding___
Pressed Tin___
Other________

Roof Form--
Gambrel___
Hip___
Shed___
Other________
Turetted___
Mansard___
Conical___
Shallow Hip___
Flat___
Simple Gable ✓

Predominant Color RED

Alteration of Original Exterior--
Unaltered___
Altered ✓ Explain:

Number of Pre-1920 Dependencies 1

Previous Uses:

Present Uses:

Acreage .36A.

Site Characteristics-- WELL LANDSCAPED; ADJACENT TO TEXACO STATION

Signficant Features--
Architecturally:

Historically: ONCE A PROMINENT HOTEL, THE HOUSE IS THE ONLY REMAINING SECTION

Other:

Historically & Environmentally Significant Buildings

Historically/architecturally significant buildings are buildings that were built before 1900, are associated with some important historical event or person, or considered to be extraordinary examples of a particular architectural style. A good number of them, ranging in style and period, exist within a relatively small area of town, and suggest a pattern of physical and economic development by their location.

Environmentally/socially significant buildings are buildings that fit in well with the buildings around them, whether they are 100 or 10 years old. They exist all over the town, but are most important in those areas with historically significant buildings. New residential developments can absorb most any kind of building design, but areas with historically significant buildings can only absorb buildings consistent with or sympathetic to the design of the historically significant buildings. Incongruous design will result in the destruction of the character given to the area by the historically significant buildings.

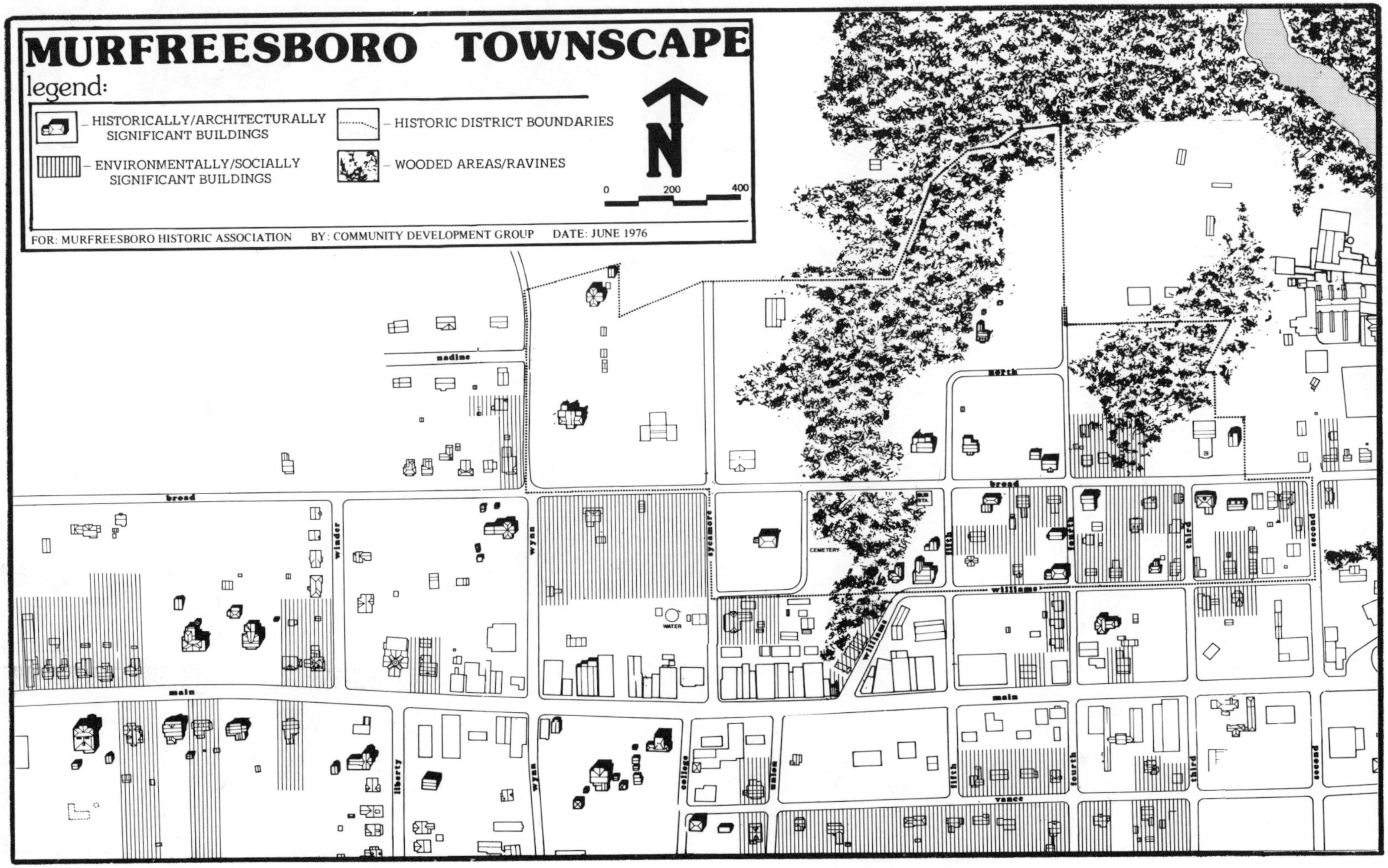

Building Conditions

Of the buildings identified as being significant to the history of Murfreesboro, half are now in need of major repair. Although most are not presently endangered by destruction, and several are now in the process of repair and restoration, some do need immediate protection from further decay and eventual destruction. Several of those along W. Main St. are of particular concern due not only to their present physical state, but also the pressures for commercial strip development. All structures presently in need of major repair, if left unaltered, will eventually decay to the point of being structurally unsound and beyond repair.

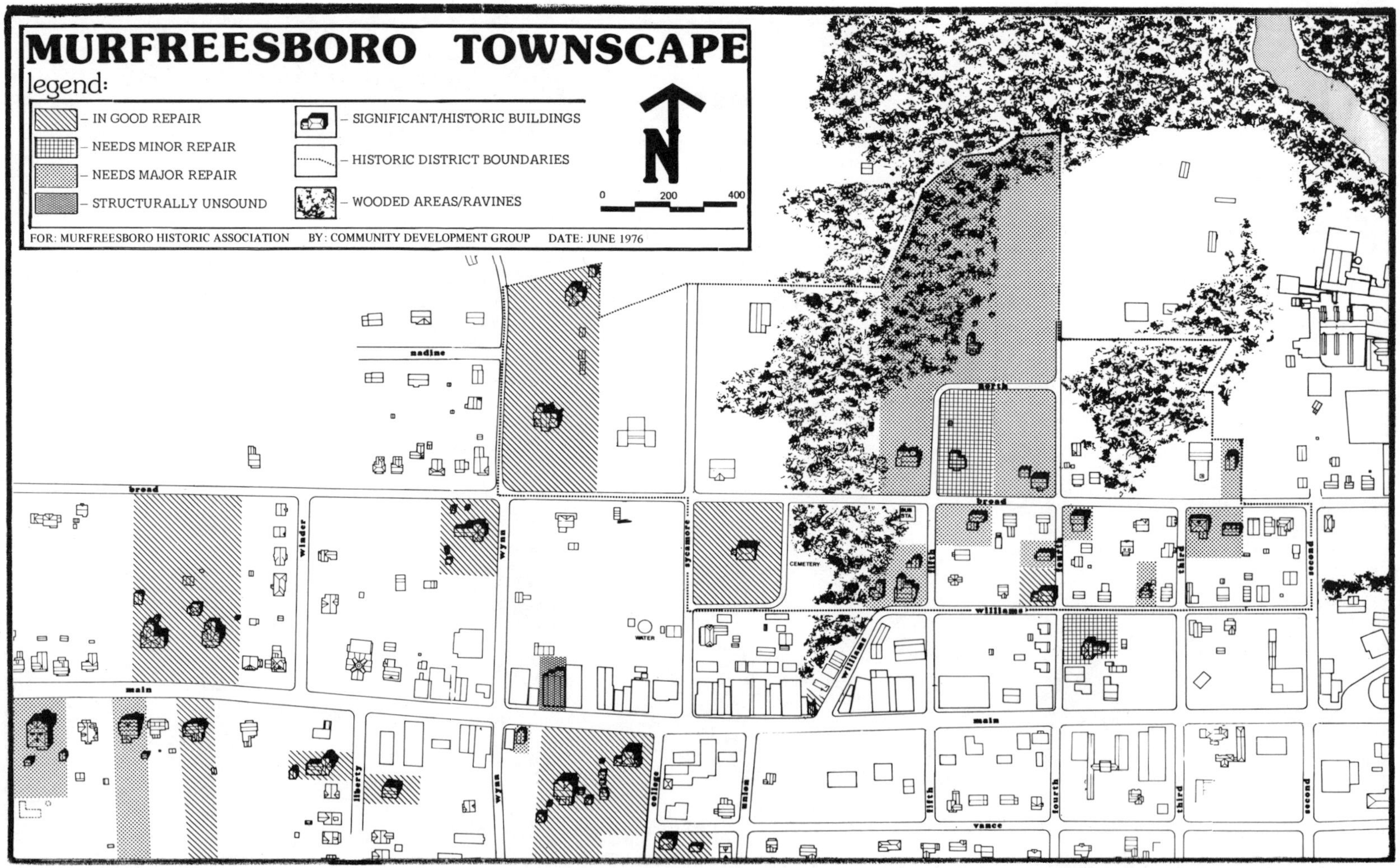

Building/Land Use

Several types of land and building uses exist within the historic district and the areas immediately surrounding it. The southwestern part of the historic district is adjacent to the CBD while the northeastern part is adjacent to an industrial area. In fact, the area serves as a place of employment for many of the residents living in the east end of the historic district. The district itself is physically divided by the ravine, and contains buildings for public, commercial, and residential use.

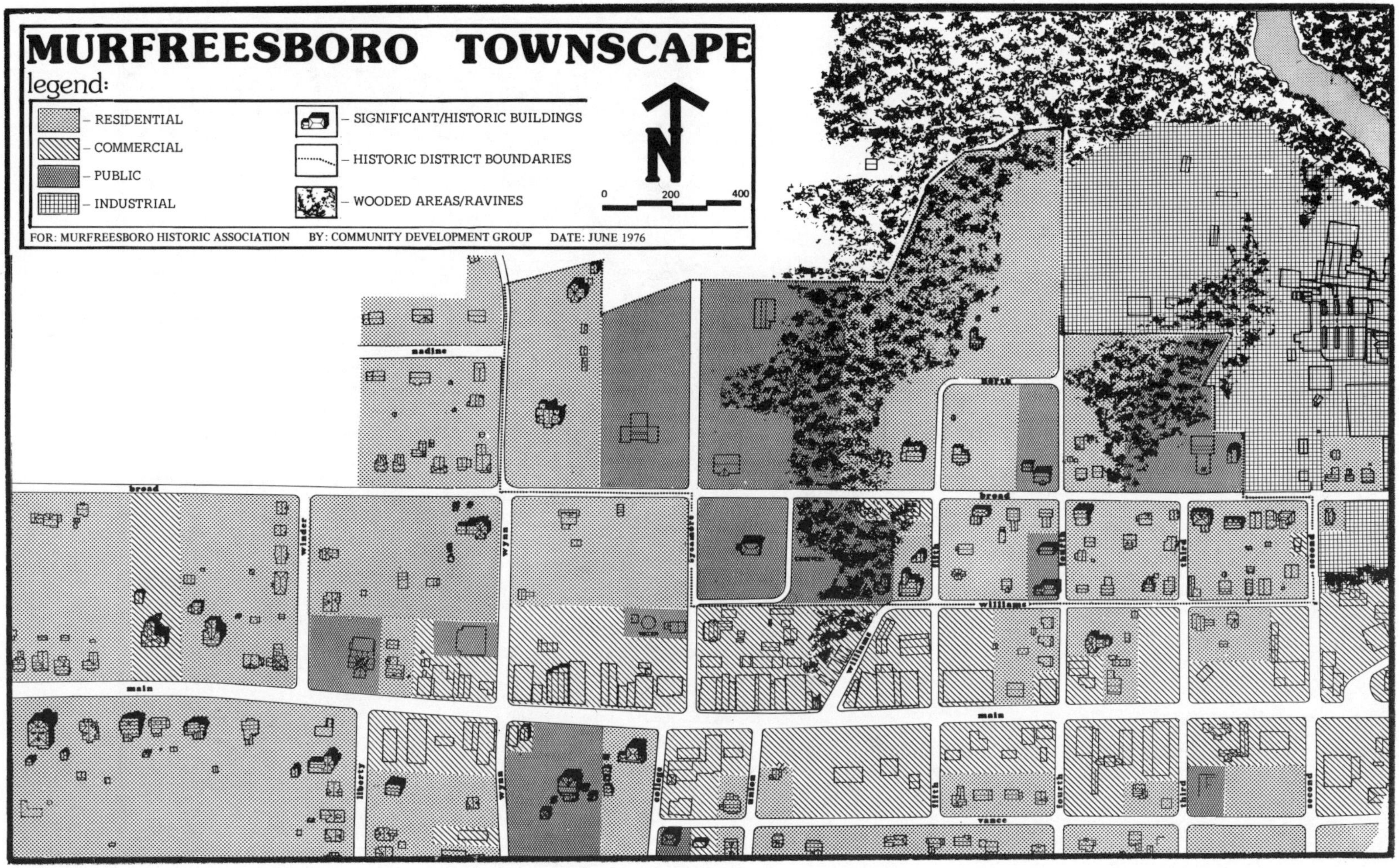

Character Areas

It has been noted that certain features of a particular area, or micro-neighborhood, act in defining that area, and distinguish it from others adjacent to it. Eight such subdivisions, or character areas, are evident in and around the historic district.

1) low density; three large, historic homes situated on large, landscaped lawns.

2) very low density of buildings; public use; physical/visual barrier of ravine.

3) high density historic structures; few 'full-time' residents.

4) high density; very active neighborhood; mostly non-historic homes.

5) predominantly commercial; low density makes it necessary for customers to drive.

6) high density commercial; proximity allows customers to walk between businesses.

7) largest group of historic homes on Main St.; physically isolated on both sides by commercial.

8) medium density residential; significant and non-significant homes; commercial infiltration.

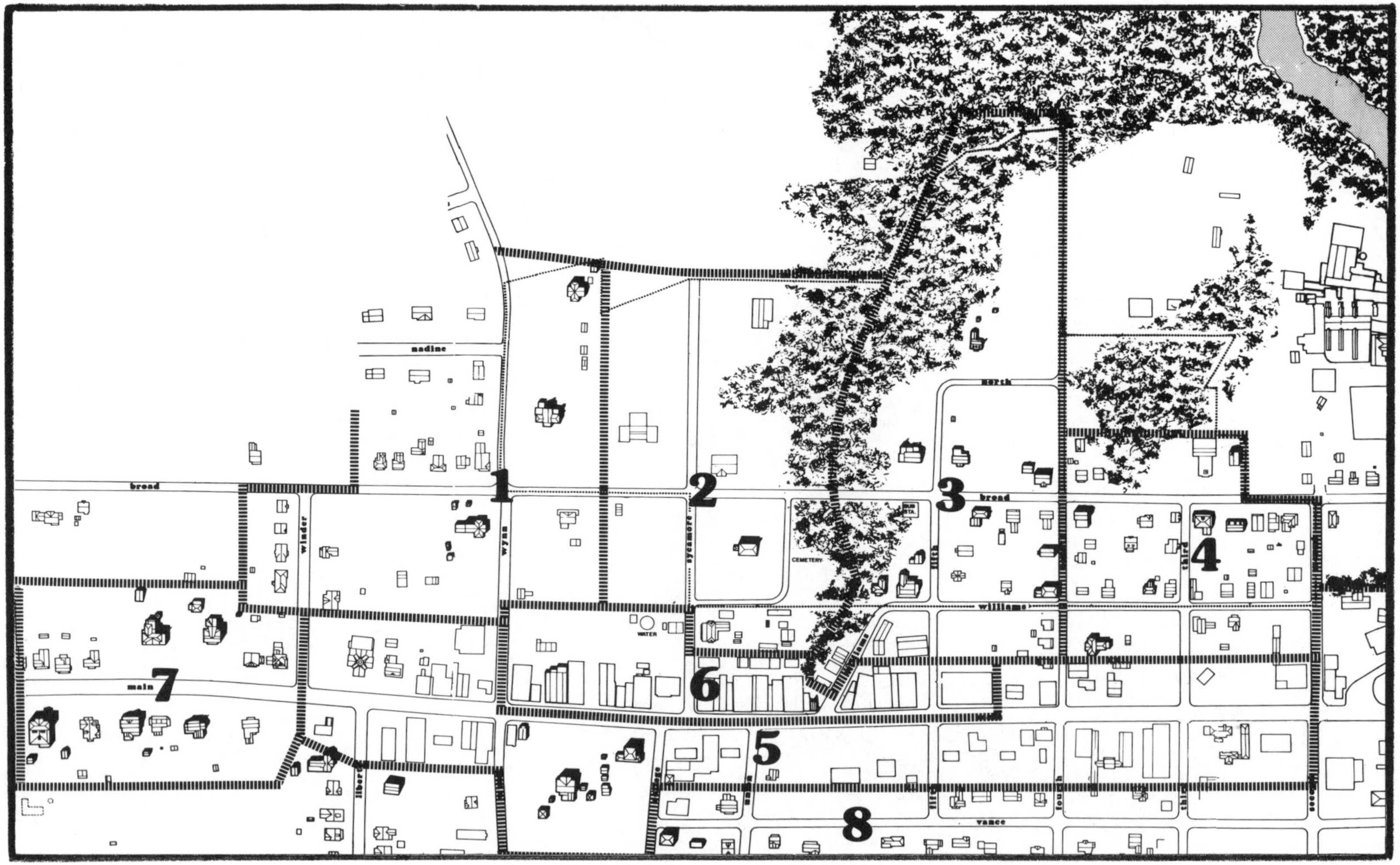

4 IDEAL FUTURE

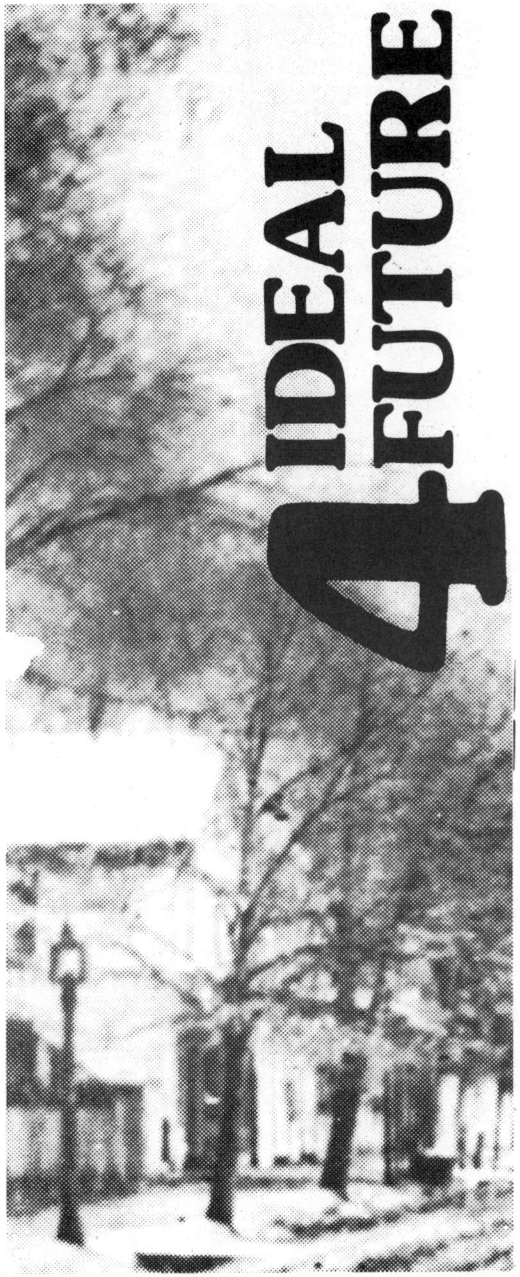

The goal list described in the following chapter was molded into a picture of the town the way the MHA would like it to be. This is what is called the "Ideal Future". The MHA can use this scenario to discuss its plans, to advertise about the town, for public debate, or just to keep track of the MHA's feelings about the direction of its work. The "Ideal Future" is broken down into sections according to the different parts of the town, plus a few words about the entire town.

Downtown business will improve as more shops and stores locate in the CBD

The **Central Section** of the historic district will be the location for most of the museums and authentically reconstructed homes. Here buildings will be open to the public and visitors will enjoy walking among reminders of the town's past. Besides the public uses, there will be private residences, and also some commercial and professional offices.

Fourth St., north of Williams Street

The **East End** of the district will continue to be primarily a residential community. There will be some new buildings, some historic homes relocated here, and some new faces in the neighborhood. But basically this will remain a community of smaller residences interspersed with some historic homes. The most noticeable change will be the dramatic improvement in the appearance and quality of the buildings and grounds through private rehabilitation aided by the MHA.

Residence on Third Street

The **West End** of the historic district will continue to be the public zone of the city. The use of this area will increase as the ravines, the Hertford Academy and the Community Center all see more activity.

Both ravines will be more heavily used as they are opened up for the enjoyment of their unique topography and wildlife. The low portion of the west ravine will become the crossroads between the historic district and the new commercial development in the water tower block. This new commercial area will tie in with and support the business of the stores and shops on Main Street. The "strip development" east and west of downtown will be limited and more businesses will locate downtown, due in part to the efforts of the MHA.

The MHA sees that it has an obligation that extends beyond the borders of the historic district. They understand that a building doesn't have to be historic to have value to the town, that the quality of an area is also worth preserving. Thus the MHA will work to promote the conservation and rehabilitation of all valuable commercial, social, or residential areas of town. They know that a small town has the unique opportunity to control its growth and direct its future. The MHA will involve itself in the decisions about this future.

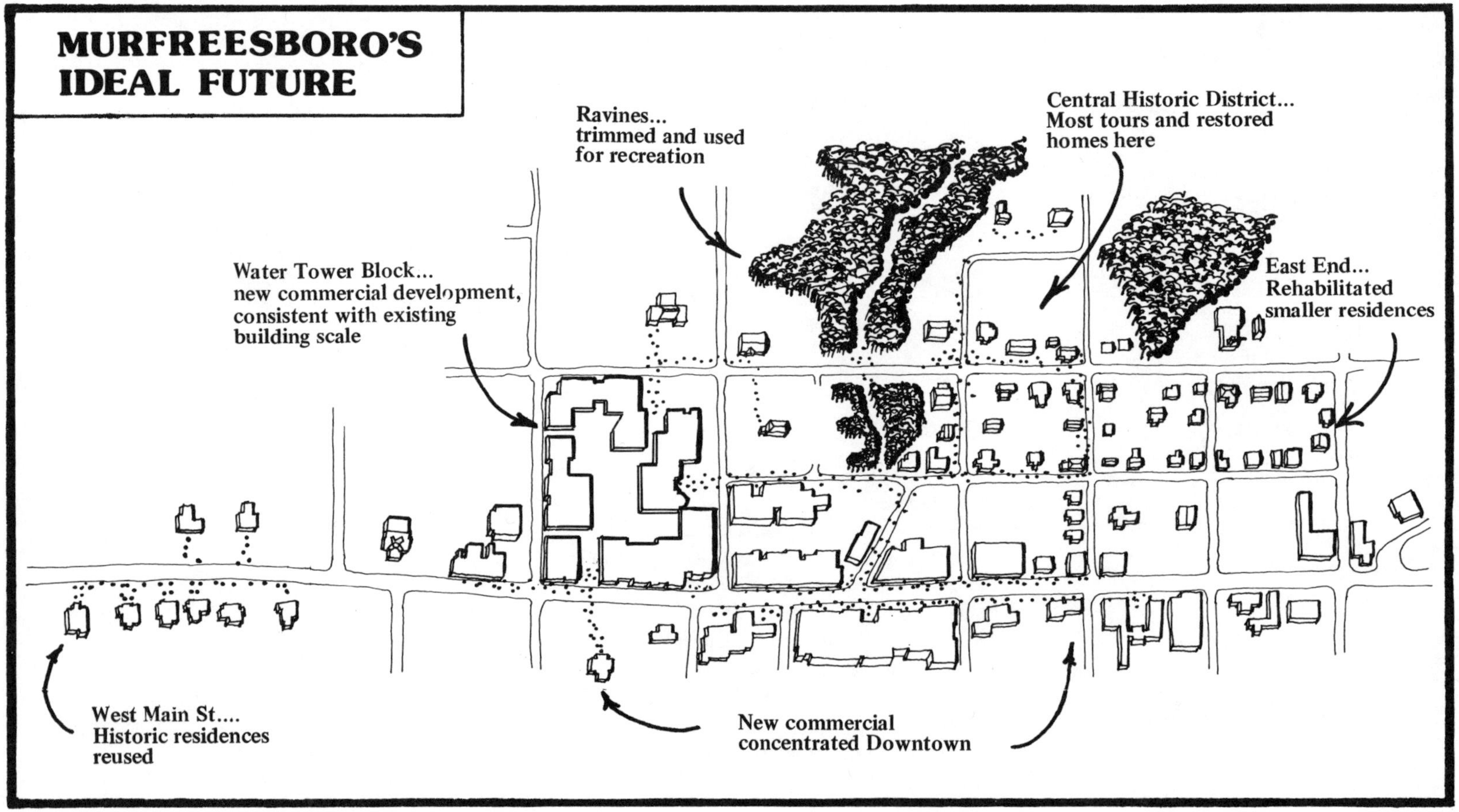

5 HOW TO USE THE PLAN

introduction

This chapter introduces the preservation/conservation plan for the historic district and town of Murfreesboro. It describes how the plan was developed, and functions as a planning workbook for the MHA to use as necessary.

WHY PLAN?
Why does the MHA need a preservation/conservation plan? The answer is simple—the MHA needs to know how to spend its limited resources most effectively.

Why does the town need a preservation/conservation plan? This answer is also simple—the town needs to know how to save its dwindling historically significant resources, how to conserve its economically valuable resources, and how to promote those qualities that make people want to live in Murfreesboro.

PLANNING MODELS
A few years ago, most plans were of the physical (master) planning model, where the policies and suggestions for action were based on physical information only (such as land use, building conditions, etc.). Recently, this model has given way to the goal-based rational planning model, where the policies and suggestions for action are based on social information as well as physical (such as the goals of the client/user group, census information, etc.).

The planning model utilized in Murfreesboro, though, adds a new element. We call it the goal-based **developmental** planning model, where the policies and suggestions for action are based on a constantly changing physical and social information base. That is, as the ideas, opinions and goals of the MHA, land use, census information, etc. change; so will the policies and suggestions for action change.

identification of goals

Utilization of a goal-based developmental plan requires that the goals of the client/user group be specified. Goals are a series of statements that describe the purpose of the client/user group (in this case, the MHA) and guide their actions. A change in goals of the client/user group normally means a change in what the client/user group sees as important, what policies they pursue, and what suggestions for policy implementation are made.

A gaming strategy was utilized to determine the goals of the MHA. After a series of general discussions with the townspeople and the MHA, and with CDG's own input, a list of 14 goals were drawn up that reflected a wide range of possible goals. These goals were discussed by the MHA, one at a time, and compared against each other until a ranking was developed. This ranking, from most important to least important, is as follows:

Goals of the MHA

1) Preserve the history of Murfreesboro.
2) Preserve architecturally and historically important buildings.
3) Boost the community interest and civic pride.
4) Control change in the historic district.
5) Use river and ravines to full potential.
6) Restore many homes to period authenticity.
7) Influence restoration in other parts of town.
8) Involve more people in the MHA's plans.
9) Improve business along Main St.
10) Eliminate the non-historic homes from the historic district.

how to determine if your goals have changed

During plan implementation, it is important that the MHA determine if the goals of the MHA have changed in any way. If they have, this will not only change the exact **general and specific suggestions for implementation** that should be followed, but possibly the **alternative plan.**

In order to determine if a change in goals has occured, follow these instructions–

Instructions

1) Assemble the MHA.
2) Pass out a sheet of paper on which is printed the current list of goals
3) Pass out a blank sheet of paper, and instruct each person to write from 5 to 7 goals, not included in the current list, they feel should be added to the goal list.
4) Break up into groups of 5 people each, and have each group come to a consensus on 5 goals that should be added.
5) Come together into one large group and have each group discuss and argue for their list of 5 goals. The MHA should come to a consensus on 5 goals that should be added.
6) On the back of the blank sheet of paper, instruct each person to rank the list of current goals and the 5 goals that should be added from most important to least important (assigning the number 1 to the most important, 2 to the second most important, etc.).
7) For each goal, add up the rank numbers each person assigned to that goal. The goal with the lowest total score is now the most important goal, the goal with the second lowest total score is now the second most important goal, etc. The 5 least important goals should be dropped from the list. This new list is now the most current list of goals of the MHA.

qualities/ importance of area

For each geographic area of town there exist certain qualities that are unique to that particular area. For example, qualities unique to the West Main St. area include–

–They (the residential buildings) are the only group of significant buildings left on Main St.
–They stop the commercial stripping.
–They are the only hint of Murfreesboro's significant buildings to the traveller passing through town.

Identification of a list of qualities depends upon the current list of goals. As the goals change, so will the qualities. Qualities also change over time. Some may be lost (due to plan implementation or unexpected physical changes within the area), some new ones may appear that were never obvious before. As the goals change, the MHA should go through each target area in the following chapter and add, delete, or change **qualities/importance of area** to fit the new goal list.

how to change the area qualities

If a change in goals has occured (see **How to Determine if Your Goals Have Changed**), it is very probable that the important qualities of each target area as seen by the MHA have also changed. If they have, this will not only change the exact **general and specific suggestions** that should be followed, but possibly the **alternative plan.**

In order to change the **qualities**, one person on the Board of Directors of the MHA should follow these instructions–

Instructions

1) Open up this report to **Appendix e**, and read the current goal list.
2) On a blank sheet of paper, write down the new or changed goals from the current goal list.
3) Open this report to the first target area– **West Main St.**
4) Write from 5 to 10 qualities, not included on the current **quality** list for this area, you feel should be added to the **quality** list and that somehow relates to the new or changed goals under 2) above. Repeat for each area of the plan.
5) Delete those **qualities** in each area that somehow are no longer related to the current goal list.

alternative plans: policy & general suggestions for implementation

Each **alternative plan** consists of two parts—the general **policy** and **suggestions for implementation.** The general **policy** is a statement describing an ideal future for each particular area of town. For example, for the **Williams St–South Side** area, the general **policy** for the first **alternative plan** is 'The historic district should be expanded to include the half-block to the south of Williams St. between Sycamore St. and Second.'

The **suggestions for implementation** are a group of statements describing our suggestions on how the general **policy** can be realized. For example, for the **Williams St.–South Side** area, a **suggestion** for implementing the **policy** mentioned above is 'encourage rehabilitation of the commercial businesses on the Williams St. extension.'

The report is so set up to allow the MHA to move their support from one **alternative plan** to another as the goals of the MHA change. This requires the MHA to make decisions concerning what **policy**, and what **general and specific suggestions** within that **policy**, to pursue.

determining the policies that enhance the area qualities & satisfy the related goals

Normally, there are two or three different **policies** that can be pursued within each target area in the following chapter. The first **policy** is a statement of the policy the MHA should pursue as long as the original list of goals (the list printed in **Appendix e**) remains current. As soon as this list of goals changes (see **How to Determine if Your Goals Have Changed**), the **policy** the MHA should pursue might also change (see **Procedure for Choosing the 'Right' Plan**).

Theoretically, the **policies** describe a general ideal future that is wholly based on enhancing the target area **qualities** and satisfying the goals. The first **policy** enhances the current **area qualities** and satisfies the current goals better than the other **policies.** The other two or three **policies** describe alternative, yet equally possible, ideal futures that are wholly based on enhancing different **area qualities** and satisfying different goals. These alternative **policies**, after a period of time during which the goals of the MHA and the **area qualities** have both changed, might better satisfy the new goals and new **area qualities** than did the first **policy**. It's also very probable that certain parts of each **policy**, making up a new **policy**, might be the best **policy** to pursue (see **Procedure for Choosing the 'Right' Plan**).

The **effects** of each **policy** are given to indicate what ideal future will be acquired if the **policy** is followed.

determining the suggestions for implementation that enhance the area qualities & satisfy the goals

Theoretically, the **general and specific suggestions for policy implementation** result from combining verbs with objects. For example, the verb 'restore' applied to an object(s) 'the W. Main St. group of significant buildings' makes up the **general suggestion** 'restore the W. Main St. group of significant buildings'. Such **general and specific suggestions**, when the object is a particular object in a particular area of town, will either result in the enhancement of the **qualities** of the area or not. For example, the **general suggestion** 'restore the W. Main St. group of significant buildings' results in the enhancement of the **quality** 'They (the residential buildings) are the only group of significant buildings left on Main St.'.

If the **general suggestions** do not result in the enhancement of the area, they are dropped from consideration. If they do result in enhancement, they must also satisfy the goals. Using the same example, 'restore the W. Main St. group of significant buildings' satisfies the goal 'Preserving architecturally and historically important buildings.'

priority

Each target area of town in the following chapter is arranged in order of priority. That is, an **alternative plan** in the first target area, the **W. Main St.** area, should be chosen, and implementation begun before implementation is begun on any of the other areas. This assumes that the MHA has an unlimited source of money to spend. Of course, this is not the case. Oftentimes, the MHA receives money from private individuals or groups to be spent on a particular object or in a specific location. In such a case, the MHA should go ahead and utilize the money for the stipulated purpose only if its impossible to convince the donor that the money should be spent according to the plan priorities.

Within each **alternative plan**, the **suggestions for implementation** are arranged according to their own priority system. There are 3 categories of priority—[1] or high priority, [2] or middle priority, and [3] or low priority. At the end of each **suggestion** is one of the above numbers in brackets. This indicates the relative importance of each **suggestion**. Those **suggestions** with a [1] should be implemented first, those with a [2] should be implemented only after all the [1]s are implemented, and those with a [3] should be implemented only after all the [2]s are implemented.

procedure for choosing the 'right' plan

For each geographic (target) area, the following questions must be asked by the MHA to determine its **policy** (which **alternative plan** to pursue) for that area. The questions should be asked every time implementation of any **general or specific suggestions** is considered (Note: the 'right' plan is that **plan** that best fits the current goals and current **area qualities**)—

Procedure

1) Has the existing goal list changed in any way (to determine, see **How to Determine if Your Goals Have Changed**)?
   Yes—
   —write down the new goal list
   —change the goals under **Goals Related to Area** for each area in the following chapter accordingly
   —change, add or delete **qualities** under **qualities/importance of area** for each area in the following chapter as the new goals have an effect on what the MHA sees as important
   —continue with question 2)
   No—
   —go ahead with implementation of the **general or specific suggestions**
2) Do any of the existing **alternative plans** best satisfy the new goal list?
   Yes—
   —choose that **alternative** to pursue, and implement **suggestions for implementation** that will realize that particular **alternative**
   No—
   —ask question 3)
3) Does any combination of existing **alternatives** best satisfy the new goal list?
   Yes—
   —decide what parts of each **alternative** to pursue
   —write down the new **policy** that satisfies these parts
   —implement only those existing **suggestions for implementation** that will realize that particular **policy**
   No—
   —ask question 4)
4) Does some entirely new **alternative** best satisfy the new goal list?
   Yes—
   —write down the new **policy**
   —make sure the new **policy** enhances the **area qualities** (see **Determining the Policies that Enhance the Area Qualities and Satisfy the Goals Related to Area**)
   —compose new **suggestion for implementation** that will carry out that particular **policy**. New **suggestions for implementation** should be developed by the method described earlier under **Determining the General and Specific Suggestions for Policy Implementation that Enhance the Area Qualities and Satisfy the Goals.**
   No—
   —discontinue planning for this area

Example

1) Yes. The only difference is that a new goal is added—'To increase new construction on Main St.'.
2) Yes. The new goal list is best satisfied by **alternative 3**—The W. Main St. significant buildings should be moved to the historic district. Implement the **general and specific suggestions** under this **alternative**.

6 THE PLAN

introduction

AREAS/ISSUES
This chapter contains the planning suggestions for each of the identified target areas/issues within Murfreesboro. By area we mean a particular geographic entity defined by physical and/or social boundaries or certain characteristics that make this area unique and separate from others directly adjacent. By issue we mean those topics of concern and importance to more than one area.

There are a total of 15 areas/issues, all chosen because of their importance to the development of a comprehensive preservation/conservation plan for Murfreesboro. Eight of the areas/issues center on the historic district in particular, the other seven on areas/issues outside of the historic district. The fifteen areas/issues, in the order in which they appear in the following pages, are:

- **A) West Main Street**
- **B) Regulation of Development–Murfreesboro**
- **C) Visual Quality–Murfreesboro**
- **D) Individually Significant Buildings**
- **E) Image–Historic District**
- **F) East End–Historic District**
- **G) Corner of Williams Street & Fifth Street**
- **H) Corner of Broad Street and Fourth Street**
- **I) Williams Street–South Side**
- **J) Central Section–Historic District**
- **K) Pedestrian Traffic–Historic District**
- **L) West End–Historic District**
- **M) Main Street Commercial–CBD & East Main Street**
- **N) Vance Street**
- **O) Ravines/River**

AREAS/ISSUES–PRIORITY
As mentioned in the previous chapter under **priority**, the order of the above areas/issues indicates the order of implementation. A should be implemented first, then **B** second, and so on. But, depending on the particular **alternative plan** followed under each area/issue, implementation of one of the areas/issues further down the priority list could be considered. For example, if under **B** the town goes ahead and expands the historic district to the boundaries of the National Register Historic District, certain areas/issues, whose major problems have been solved in the expansion, might be shifted further down the priority list. It's up to the MHA to determine how priorities have changed, if indeed they have changed at all.

The priority list above was chosen based on the **alternative plan** the MHA is expected to follow for each area/issue. This expected **alternative plan** though, is not, in all cases, the plan that the MHA should follow. Again, for issue B, it is expected that the MHA will follow **Alternative Plan 2** (encouraging the city to implement an Historic Properties Commission) but suggested that they follow **Alternative Plan 1** (encouraging the city to expand the existing historic district to the boundaries of the National Register Historic District). If the MHA did follow **Alternative Plan 1**, implementation of all the rest of the areas/issues plans would be considerably easier.

For all areas/issues, the first **alternative plan** is the alternative that should be followed if the present goals of the MHA do not change.

If, at any time, a significant building is going to be torn down (or any other equally critical thing occurs), the MHA should shift the above priorities in order to save the building.

A

WEST MAIN STREET

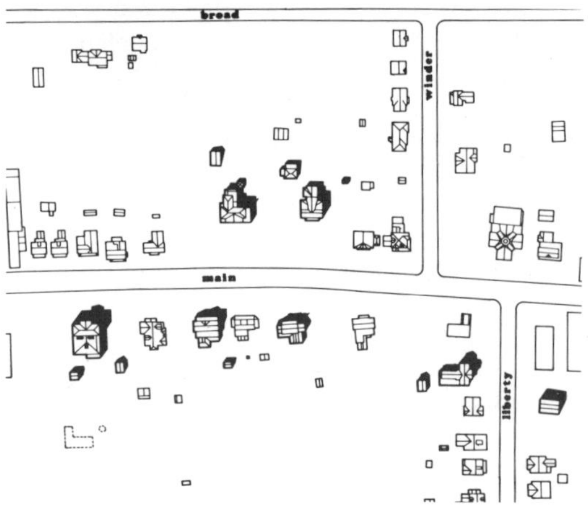

goals related to area

ADD & DELETE AS THE GENERAL GOALS OF THE **MHA** CHANGE (see **PROCEDURE FOR CHOOSING THE 'RIGHT' PLAN** in the previous chapter)

* Preserve architecturally and historically important buildings
* Boost the community interest and civic pride
* Restore many homes to period authenticity
* Influence restoration in other parts of town
* Involve more people in the MHA's plans
* Improve business along Main St.
* Bring tourist traffic to Murfreesboro

qualities/importance of area

ADD & DELETE AS THE QUALITIES OF THE AREA CHANGE

* Distinct residential character

West Main Street

* Contains one of only two Victorian buildings in Murfreesboro.

Vinson house

* Contains a large number of trees and greenery that (1) visually separates the CBD from the west end of town, and (2) reduces noise and dust from vehicular traffic.
* Potential for adaptively reusing buildings commercially is great.
* Large rear yards.

* Contains one of the most environmentally significant buildings in Murfreesboro.

Warren house

* One block from the CBD.
* Large front yard setbacks.
* Residential neighbors across street, behind and to the east.
* Room exists for relocating significant buildings from elsewhere and for building new residential.
* Tourist possibilities.
* Wide range of architectural styles present.
* Contains the largest, most finely detailed, Greek Revival building in Murfreesboro

Uriah Vaughan house

* Only group of significant buildings left on Main St.

Old Episcopal Church—site now occupied by a car wash

Lassiter Hotel—site now occupied by a supermarket

* Stops the commercial stripping that is occurring at all other places along Main St.
* Represents a different era in Murfreesboro's development—the post Civil War era.

* Acts as a gateway to the town from the west.

W. Main St.—heading east into the historic area

W. Main St.—leaving the historic area, entering the CBD

* Contains one of the most historically significant buildings in Murfreesboro.

Walter Reed house

* Only hint of Murfreesboro's significant buildings to the traveller passing through town.

choosing the 'right' plan

EVERYTIME IMPLEMENTATION OF ANY **GENERAL** OR **SPECIFIC SUGGESTION** IS CONSIDERED UNDER ANY OF THE FOLLOWING **ALTERNATIVE PLANS**, GO THROUGH **PROCEDURE FOR CHOOSING THE 'RIGHT' PLAN** IN THE PREVIOUS CHAPTER

implementing the 'right' plan

IF THE **MHA** HAS DIFFERENT, BETTER, OR MORE EXACT **SUGGESTIONS** FOR IMPLEMENTING ANY OF THE FOLLOWING **ALTERNATIVE PLANS**, GO THROUGH THE **DESIGN GUIDELINES** IN THE FOLLOWING CHAPTER TO DETERMINE IF THE **SUGGESTIONS** ARE ACCEPTABLE

alternative plan 1

SCRATCH OUT ALL **SUGGESTIONS** THAT ARE NO LONGER PERTINENT OR HAVE ALREADY BEEN IMPLEMENTED

policy

West Main St. should be zoned historic district

effect

Will protect all significant buildings within the area for a minimum of 90 days.

general suggestions for implementation

* Encourage the planning board to zone the following area historic district [1] :

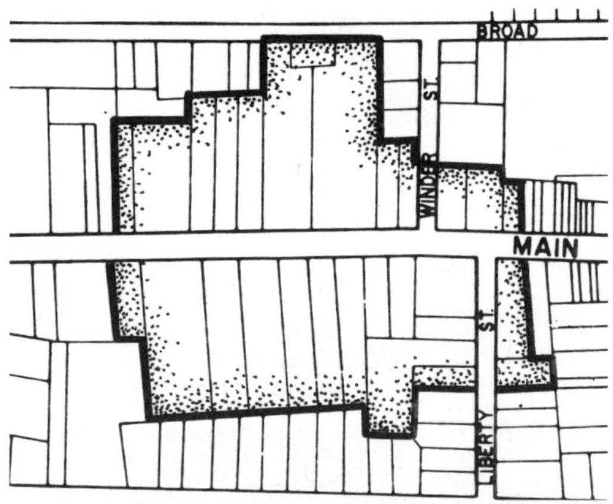

* Nominate the Vinson and Uriah Vaughan houses, Garrett funeral home, and Lawrence (Barnes) house to the National Register of Historic Places. [1]
* Ask the planning board to prepare, or have the MHA prepare, a report on the quality of W. Main St. [1]
* Encourage owners of property within area to attend the public hearing(s) concerning the proposal to zone the area historic district. [1]

After historic district zoning has been secured:

* Encourage the planning board to adopt the practice of negotiating with developers, and other property owners, who request zoning changes on property outside the areas covered in this report. If the developer owns property within the W. Main St. area and will agree to grant to the MHA or to the city a preservation easement on this property--or will agree to relocate an obnoxious use to a site outside of the area discussed in this report--the developer's zoning change application should be accepted.[1]
* Encourage elderly owners of significant buildings and land to grant a preservation easement to the MHA in order to avoid paying the Federal Estate Tax. [1]
* Develop and distribute guidelines suggesting do's and don'ts for remodeling and repairs of exteriors (see **Remodeling Old Houses Without Destroying Their Character** by George Stephen for a good example). [1]

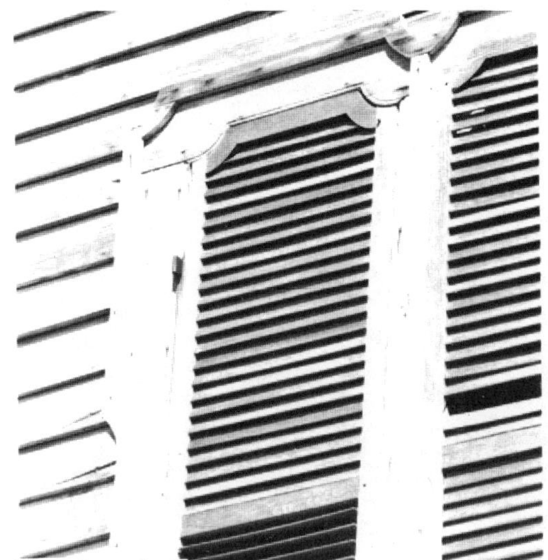

Vinson house detail

* Encourage the owners of all empty lots in the W. Main St. area to develop with moderate or high income residences, or the MHA should consider buying empty lots for relocating significant buildings for residential use. If unsuccessful, encourage new commercial development or the relocation of significant buildings for commercial use. [1] New buildings and relocated significant buildings should conform to the **Design Guidelines** in Chapter 7.

* Encourage all owner/tenant occupants of all residences within area to remain. If unsuccessful, encourage others to move in. If unsuccessful again, encourage commercial use for the significant buildings and non-significant buildings. [1]

* Encourage the owner/tenant occupants of all significant commercial buildings within area to remain. [1]

Joseph G. Rea house

* Encourage single-family residential use of the empty significant buildings. If unsuccessful, encourage some compatible commercial use. [1] The total square footage of the three empty significant buildings are:
  –Vinson house: 2475 sq. ft.
  –Uriah Vaughan house: 6600 sq. ft. (rough estimate, doesn't include basement of rear section)
  –Walter Reed house: 1950 sq. ft.

If unsuccessful again:
(1)buy and restore the Vinson house for moderate or high income residential use, then sell/rent with the added requirement that it be accessible to tour groups.
(2)buy and rehabilitate/restore the Uriah Vaughan house for residential use (and the Yeates Law Office for residential or commercial use), then sell/rent with the added requirement that it be accessible to tour groups.
(3)buy and rehabilitate the Walter Reed house for moderate income residential use, then sell/rent with the added requirement that it be accessible to tour groups.

Follow this priority when buying, rehabilitating/restoring, and selling/renting the above buildings–
(1)Buy the Vinson house. [1]
(2)Restore the Vinson house exterior. [1]
(3)Buy the Uriah Vaughan house.[1]
(4)Restore the Uriah Vaughan house ext.[1]
(5)Buy the Walter Reed house. [2]
(6)Restore the Vinson house interior, then sell or rent. [2]

Vinson house–interior detail

(7)Rehabilitate/restore the Uriah Vaughan house interior. [2]
(8)Rehabilitate/restore the Yeates Law Office exterior. [2]

Yeates Law Office

(9)Rehabilitate the interior and exterior of the Walter Reed house, then sell or rent. [3]
(10)Rehabilitate/restore the Yeates Law Office interior, then sell or rent both the Uriah Vaughan house and the Yeates Law Office, either together or separately. [3]

* Restore/rehabilitate all additions to the significant buildings unless dilapidated or not needed for existing or intended use.

* Encourage single-family residential use of the empty non-significant buildings. [1]

* Specific uses of all new development, and all adaptively used buildings, should be compatible with residential activities–for instance highway commercial would not be a compatible land use. The MHA should compose a listing of all possible land uses compatible with historic district uses and activities. Encourage the zoning board to review this list and add it to the town's zoning ordinance as a guide to permitted and prohibited land uses in the historic district. [1]

* During building rehabilitation/restoration, exterior trim which has architectural interest or value, such as that on the Gingerbread house, should be restored even when not original to the building. [1]

* Discourage owners of any commercial development or commercial adaptive reuse in this area from putting parking lots in front of the buildings. They should go behind. [1]

* Encourage preservation of the greenery within area. [1]

* Encourage the city to lay a concrete sidewalk along the south side of W. Main St. connecting Rich's supermarket to the Texaco service station on the southwest corner of W. Main St. and Liberty St. The sidewalk should be the same distance from the curb as the sidewalk on the north side of W. Main St. [1]

* If acceptable to the occupants of the area, and if it will not adversely affect the existing greenery to any great extent, encourage the utility companies to put all telephone and electrical poles and wires, along the north and south sides of W. Main St., underground (except street light poles). [2]

specific suggestions for implementation

SUGGESTIONS PERTAINING TO SPECIFIC BUILDINGS–

* Encourage rehabilitation, where possible, of the non-significant residences on the north side of W. Main St. west of the funeral parlor. [1]

* Encourage the owner of the Warren house (between the Vinson and Walter Reed houses) to rehabilitate and sell/rent as a residence. If unsuccessful, encourage the owner to sell it to someone interested in adapting it for a compatible commercial use. [1]

* Encourage the owner of the Vinson house to immediately board up windows and repair roof so that the interior doesn't deteriorate further and break-ins are minimized. [1]

* Encourage owner of the long thin commercial building across the street from Rich's supermarket to buy lot directly west of building, rehabilitate exterior so that front of building faces west, rehabilitate interior into 3 or 4 retail commercial spaces that can be rented (each with its major entrance on the west side), and install a parking lot west of building in front of the stores. New commercial development, in a compatible architectural style, could be developed west of this building. [1]

SUGGESTIONS PERTAINING TO LANDSCAPING–

* Encourage owner of the Uriah Vaughan to clear greenery northwest of the building so that it is easily visible to travellers coming into town from the west. [1]

* Encourage owner of the brick residence west of the Vinson house to trim trees on the boundary between the two so that the Vinson house is clearly visible to travellers coming into town from the west. [1]

* Encourage the owners of all significant buildings to clean up and trim the front yards. [1]

* Encourage the owners of lots on the south side of W. Main St. to continue farming the area behind the buildings until such time the buildings can be rehabilitated/restored and sold. [1]

SUGGESTIONS THE MHA SHOULD MAKE TO OTHERS–

* Encourage the city to install unobtrusive, well-designed markers on or near all significant buildings. If located on the building itself, the markers should be of an easily visible color (but not an overly contrasting color) and should only include the street number as verbal information. All markers should be the same color, should be no more than 12 inches in any dimension, and the shape of the marker must be simple (round, square, etc.). If located near the building, the marker should stand no no farther than 12 inches above ground level, should be no more than 12 inches in any dimension, and should be located within 5 feet of a street. Again, the markers should be the same color, of a simple shape, and in addition, of a durable material (metal, plastic, stone, etc.). [2]

* Encourage the city to ask the State of N.C. to speed up the by-pass around Murfreesboro. [2]
  EFFECT–this will eliminate truck traffic through the W. Main St. area, making it a more desirable place to live.

* Encourage the city to install unobtrusive, well-designed markers on W. Main St., at the west and east boundaries of the W. Main St. historic district. The markers should read: **West Main Street Historic District**. [2]

* Encourage the Patio and the Texaco service station to replace their signs with smaller, more unobtrusive, signs (follow the regulations under Article VIII [Conditional Uses Permitted]–Section 5 of the General Ordinances for the Town of Murfreesboro). [1]

* Encourage the city to replace the existing street lights in the W. Main St. area with street lights contemporary to the 1880's and 1890's. [2]

OTHERS–

* Consider the effects of a street running parallel to W. Main St., half way between High St. and W. Main St., connecting to Liberty St. on the east and Spring St. on the west. [1]
  EFFECTS–such a street would allow the owners of the extremely long lots on the south side of W. Main St. to divide them in half and develop the southern half with new

residences. This might convince the owners of these lots to save the significant buildings if it can be shown that a substantial profit can be made without destroying these buildings. It would also give the occupants of the significant buildings on W. Main St. a vehicular entrance from the south.

* Inform all owners and tenants within the W. Main St. area that a certificate of appropriateness is required for most large changes to the exterior of a building or on the property itself. [1]

alternative plan 2

SCRATCH OUT ALL **SUGGESTIONS** THAT ARE NO LONGER PERTINENT OR HAVE ALREADY BEEN IMPLEMENTED

policy

The W. Main St. area should be permanently zoned C-3 (Commercial-Office District)

effect

Will halt highway commercial development of W. Main St.

general suggestions for implementation

* Buy, at any cost, and restore the Vinson house for museum use. Check into all funding sources, including State grants-in-aid, to purchase and restore building. [1]

Vinson house as it might look restored

* Encourage interested parties to buy and restore/rehabilitate the Uriah Vaughan, Walter Reed, and Warren houses for any permitted uses within the C-3 zoning district. [1]

After the Vinson house has been bought:

* Ask the planning board to zone the area in **Alternative 1** C-3 (Commercial-Office District). [1] Once passed, be ready to fight against any zoning changes to this C-3 area.

* Encourage owners of area to attend the public hearing(s) concerning the proposal to zone the area C-3. [1]

After C-3 zoning has been secured:

* Develop and distribute guidelines suggesting do's and don'ts for remodeling and repairs of exteriors (see **Remodeling Old Houses Without Destroying Their Character** by George Stephen for a good example). [1]

* Encourage the owners of all empty lots on W. Main St. to develop with any use permitted in the C-3 zoning district. [1] New construction should conform to the **Design Guidelines** in Chapter 7.

* Encourage the owner/tenant occupants of all significant buildings being used commercially to remain. [1]

* Encourage all owner/tenant occupants of all residences, significant and non-significant, within area to remain. If unsuccessful, encourage others to move in. If unsuccessful again, encourage any use permitted in the C-3 zoning district for both significant and non-significant buildings. [1]

* Encourage any use permitted in the C-3 zoning district for the empty significant buildings. The total square footages of the three empty buildings are:
  –Vinson house: 2475 sq. ft.
  –Uriah Vaughan house: 6600 sq. ft. (rough estimate, doesn't include basement of the rear section)
  –Walter Reed house: 1950 sq. ft.

  [1] If unsuccessful:
  (1)buy and rehabilitate/restore the Uriah Vaughan house and Yeates Law Office for any uses permitted in the C-3 zoning district, then sell or rent.

Yeates Law Office in 1890

  (2)buy and rehabilitate/restore the Walter Reed house for any uses permitted in the C-3 zoning district, then sell or rent.
  (3)buy and rehabilitate/restore the Warren house for any uses permitted in the C-3 zoning district, then sell or rent

  Follow this priority when buying, rehabilitating/restoring, and selling/renting the above buildings (if such is necessary):
  (1)Buy the Vinson house. [1]

(2)Restore the Vinson house exterior. [1]
(3)Buy the Uriah Vaughan house. [1]
(4)Restore the Uriah Vaughan exterior. [1]
(5)Restore the Yeates Law Office exterior. [1]
(6)Restore the Vinson house interior, then sell or rent. [1]
(7)Buy the Walter Reed house. [2]
(8)Rehabilitate/restore the Uriah Vaughan house interior, then sell or rent. [2]
(9)Rehabilitate/restore the Yeates Law Office interior, then sell or rent (either with or without the Uriah Vaughan house. [2]
(10)Buy the Warren house. [3]
(11)Rehabilitate/restore the interior and exterior of the Walter Reed house, then sell or rent. [3]
(12)Rehabilitate/restore the interior and exterior of the Warren house, then sell or rent. [3]

* Rehabilitate/restore all additions to the significant buildings unless dilapidated or not needed for existing or intended use. [1]

* Encourage any use permitted in the C-3 zoning district for the empty non-significant buildings. [1]

* Discourage owner of any commercial development or commercial adaptive reuse in this area from putting parking lots in front of the buildings. They should go behind the buildings. [1]

* During building rehabilitation/restoration, exterior trim which has architectural interest or value, such as that on the Gingerbread house, should be restored even when not original to the building. [1]

* Encourage the city to adopt Article VIII (Conditional Uses Permitted)—Section 5 of the General Ordinances of the Town of Murfreesboro for all signs for permitted and conditional uses within the area. [1]

* Encourage preservation of the greenery within the area. [1]

specific suggestions for implementation

SUGGESTIONS PERTAINING TO SPECIFIC BUILDINGS—

* Encourage owner of the long thin commercial building across W. Main St. from Rich's supermarket to buy lot directly west of building, rehabilitate exterior so that front of building faces west, rehabilitate interior into 3 or 4 office (or any other permitted use in the C-3 district) spaces that can be rented (each with its major entrance on the west side), and install a parking lot west of building in front of these spaces. New commercial/office development, in a compatible architectural style, could be developed west of this building. [1]

SUGGESTIONS PERTAINING TO LANDSCAPING—

* Encourage the owner of the Uriah Vaughan house to clear greenery so that the building will be clearly visible to travellers coming into town from the west. [1]

* Encourage owner of the brick house west of the Vinson house to trim the trees on the boundary between the two so that the Vinson is clearly visible to travellers coming into town from the west. [1]

SUGGESTIONS THE MHA SHOULD MAKE TO OTHERS—

* Consider the effects of a street running parallel to W. Main St., half way between High St. and W. Main St., connecting to Liberty St. on the east and Spring St. on the west. [1] EFFECT—such a street would allow the owners of the extremely long lots on the south side of W. Main St. to divide them in half and develop the southern half with new residential and the half on W. Main St. with commercial. This might convince the owners of these lots to save the significant buildings if it can be shown that a substantial profit can be made without destroying these buildings. It would also give the occupants of the significant buildings on W. Main St. a service/parking vehicular entrance from the south.

OTHERS—

* Encourage the city to lay a concrete sidewalk along the south side of W. Main St. connecting Rich's supermarket to the Texaco service station on the southwest corner of W. Main St. and Liberty St. [2]

alternative plan 3

SCRATCH OUT ALL SUGGESTIONS THAT ARE NO LONGER PERTINENT OR HAVE ALREADY BEEN IMPLEMENTED

policy

The W. Main St. area should be zoned C-2 (Highway Commercial) only after all vacant buildings on W. Main St. have been relocated to the existing historic district

effect

Will preserve the most significant buildings on W. Main St. Will open up the W. Main St. area for for highway commercial development.

general suggestions for implementation

* Encourage the planning board to inform the owners of the Vinson, Uriah Vaughan and Walter Reed houses that each of their lots can be zoned C-2 (at their request) if they agree to donate the significant building on the lot to the MHA and pay the expense of moving the building to the historic district.

B
REGULATION OF DEVELOPMENT– MURFREESBORO

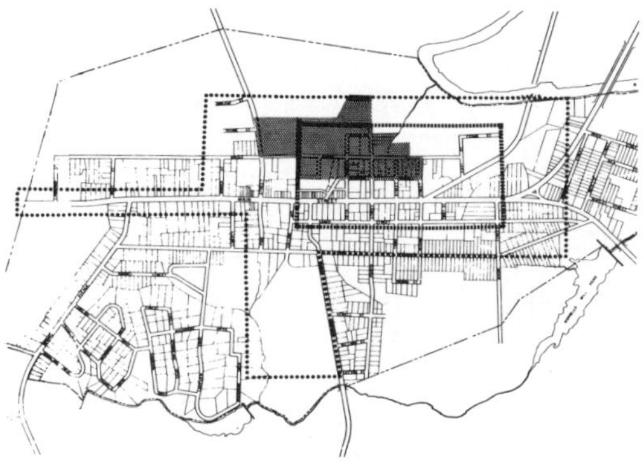

goals related to area

ADD & DELETE AS THE GENERAL GOALS OF THE MHA CHANGE (see **PROCEDURE FOR CHOOSING THE 'RIGHT' PLAN** in the previous chapter)

* Preserve architecturally and historically important buildings
* Boost the community interest and civic pride
* Control change in the historic district
* Influence restoration in other parts of town
* Involve more people in the MHA's plans
* Improve business along Main St.

____

qualities/importance of area

ADD & DELETE AS THE QUALITIES OF THE AREA CHANGE

* The history of Murfreesboro's development is reflected within different geographic areas.
* The west ravine is still a major physical and visual barrier, dividing the northern part of the town in half.
* The Meherrin River is still a major contributor to the town's character.
* Chowan College still contributes to the town's character, as it and other educational institutions have in the past.
* The Central Business District (CBD) remains an active commercial center.
* Much of the original open space/rural character of the town still exists.

Area between the Gingerbread and Wheeler houses

* The West Main St. group of buildings, reflecting Murfreesboro's development in the mid and late 1800's, is still a definable residential neighborhood.

W. Main St.

* The original town of Murfreesboro, as laid out by William Hardy Murfree in 1787, is still pretty much intact.

____

choosing the 'right' plan

EVERYTIME IMPLEMENTATION OF ANY **GENERAL** OR **SPECIFIC SUGGESTION** IS CONSIDERED UNDER ANY OF THE FOLLOWING **ALTERNATIVE PLANS**, GO THROUGH **PROCEDURE FOR CHOOSING THE 'RIGHT' PLAN** IN THE PREVIOUS CHAPTER

implementing the 'right' plan

IF THE **MHA** HAS DIFFERENT, BETTER, OR MORE EXACT **SUGGESTIONS** FOR IMPLEMENTING ANY OF THE FOLLOWING **ALTERNATIVE PLANS**, GO THROUGH THE **DESIGN GUIDELINES** IN THE FOLLOWING CHAPTER TO DETERMINE IF THE **SUGGESTIONS** ARE ACCEPTABLE

alternative plan 1

SCRATCH OUT ALL **SUGGESTIONS** THAT ARE NO LONGER PERTINENT OR HAVE ALREADY BEEN IMPLEMENTED

policy

The existing historic district should be expanded to the boundaries of the National Register Historic District.

effect

Will protect most historic resources in Murfreesboro.

general suggestions for implementation

* The National Register Historic District includes (see page 11):

CORNER	LATITUDE	LONGITUDE
NW	36°26'43"	77°06'40"
NE	36°26'43"	77°05'24"
SE	36°26'21"	77°05'24"
SW	36°26'21"	77°06'40"

i.e.—the National Register Historic District includes 630 acres; and delineates a rectangle, with the west boundary an E-W line at the point where the town line meets West Main St., the north boundary a E-W line at the point where Darlene St. meets Wynn St., the east boundary a N-S line where Diamond St. meets U.S. 158, and the south boundary line an E-W line where Cotton St. meets Union St.

* Acquire all materials relating to the National Register Historic District application (quality information). [1]

* Submit an application with above materials to the City Council requesting that the National Register Historic District boundaries also be used for the local historic district. [1]

* Encourage owners of property within the area to attend the public hearing(s) concerning the proposal to zone the entire National Register Historic District area a historic district. [1]

* Ask the planning board to expand the legal description of the existing historic district ordinance to include specific performance requirements for deciding upon certificates of appropriateness (see Chapter 7—**Design Guidelines**). Regulations for:
  —walls of continuity
  —relationship of landscaping
  —height-width proportion of building's front facade
  —height-width proportion of openings within the facade
  —rhythm of spacing of buildings on streets
  —relationship of materials
  —relationship of textures
  —relationship of color
  —relationship of architectural details
  —relationship of roof shapes
  —scale

UNITS OF SCALE

VERTICAL HORIZONTAL

  —relationship of ground cover
should all be included in the ordinance (see the Hillsborough or Salisbury, N.C. historic district ordinance for example). [2]

* After the historic district becomes adopted through a zoning change:
  —Encourage the Historic District Commission to consider the specific effects of new construction on Main St. on:
    —W. Main St., between Liberty St. and Spring St.
    —the existing historic district north of Williams St.
    —the Vance St. properties
    —the CBD buildings on the north side of Main St. between Wynn St. and Fifth St.
  before granting a certificate of appropriateness. [2]

specific suggestions for implementation

SUGGESTIONS THE MHA SHOULD MAKE TO OTHERS–

* Talk to owners within the National Register Historic District and ask them to support expansion of the local historic district. [1]

* Encourage the City Council to include on the Historic District Commission members representing a wider cross section of the Murfreesboro community. [1]

* Encourage the Hertford County Historic District Commission to designate historic district zoning for any historically significant areas within the county. [2]

* Encourage the public to attend all meetings or hearings of the Historic District/Properties Commission. [1]

OTHERS–

* Start a campaign to educate the townspeople to the advantages of expanding the local historic district boundaries to include all of the National Register Historic District. [1]

* Start a campaign to educate the townspeople about the town's qualities as mentioned under **qualities/importance of area**. Utilize:
  –movies
  –slide shows
  –tours
  –talks by developers who have reused old buildings economically
  –others [1]

alternative plan 2

SCRATCH OUT ALL **SUGGESTIONS** THAT ARE NO LONGER PERTINENT OR HAVE ALREADY BEEN IMPLEMENTED

policy

An Historic Properties Commission should be designated for the town.

effect

Will protect the most historically important buildings within Murfreesboro for a minimum of 90 days.

general suggestions for implementation

* Ask the City Council to designate the Historic District Commission as the Historic Properties Commission, in accordance with N. C. enabling legislation (see **Appendix c**). [1] EFFECT–will authorize the Historic District Commission to–
  –recommend to the City Council structures, sites, areas, or objects within the city limits to be designated as 'historic properties'.

David Barnes house in 1880

  –acquire fee-simple ownership (or any lesser included interest including preservation easements or restrictive covenants) to any such 'historic properties'; to hold, manage, restore, and improve same; and to exchange and dispose of same by sale or lease in a manner that will conserve the property.
  –restore, preserve, and operate such 'historic properties'.
  –conduct an educational program on 'historic properties' within Murfreesboro.
  –contract with the State, the U.S. government, any agency of either, or with any other organization.

* After the Historic District/Properties Commission is established:
  –Encourage the Commission to adopt an ordinance designating all buildings on the National Register of Historic Places as 'historic properties'. [1] Buildings within Murfreesboro currently on the National Register of Historic Places include:
  –Rea Store
  –Wheeler house
  –Myrick house
  –Melrose house
  –Hertford Academy
  –Roberts-Vaughan house
  –Dr. Issac Pipkin house
  –Rose-Bower house
  –Walter Reed house
  –the Columns
  –Winborne house
  –Encourage the Commission to apply National Register criteria to those historically significant buildings considered eligible for the National Register of Historic Places. Encourage the Commission to adopt an ordinance designating those buildings found to satisfy the requirements for the National Register of Historic Places as 'historic properties'. [1]

C

VISUAL QUALITY~ MURFREESBORO

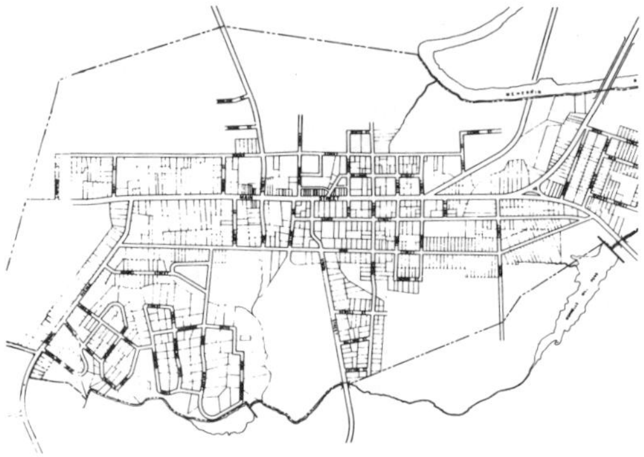

goals related to area

ADD & DELETE AS THE GENERAL GOALS OF THE MHA CHANGE (see **PROCEDURE FOR CHOOSING THE 'RIGHT' PLAN** in the previous chapter)

* Preserve the history of Murfreesboro

* Boost the community interest and civic pride

* Improve business along Main St.

____

____

____

____

____

qualities/importance of area

ADD & DELETE AS THE QUALITIES OF THE AREA CHANGE

The town of Murfreesboro has:

—buildings which exhibit a wide range of architectural examples—including the Federal, Greek Revival, Victorian and Contemporary styles, major works of real architectural distinction (David Barnes house, Myrick house, Uriah Vaughan house) and unusual decoration (Gingerbread house).

—buildings which have a wide range of surface qualities—including color (red, white, yellow), texture (brick, clapboard, board and batten), and shape (Victorian).

Vinson house detail

—a number of interesting single forms—including utilitarian objects (lamps), decorative objects (spider windows), and curiosities (gates).

Gate on Union St.

—buildings with a wide range of architectural relationships—including street facade sequences (south side of W. Main St.; Peter Williams—Murfree Law Office; CBD), and boundaries (fences).

—a wide variety of land and plant forms—topographic features (ravines; Meherrin River), styles of planting design (David Barnes yard; on Union St. south of Vance St. and north of High St.), groups of trees (the two ravines within the historic district; along the river; the ravine between First St. and Moore St.; at David Barnes house; Melrose), unusual plants (in ravines, especially huge pines in west ravine).

—a large number of viewpoints and landmarks—including forms which dominate or symbolize (the water tower, Uriah Vaughan house, Roberts-Vaughan house), pleasant scenes (Williams St. extension; area directly west of the Wheeler house), panoramic views (corner of Wynn St. and Darlene St.; U.S. 258 bridge

Meherrin River—looking west from U. S. 258 bridge

over Meherrin River), and long vistas (Broad St.; central Main St. and W. Main St.; Vance St.).

—different kinds of urban open spaces—including informal recreation areas (west ravine), parks and public spaces (Riverside Park, Roberts-Vaughan house, High St. Middle School), cemetaries (Thompson, Southall, Wise's), intersections (Williams St. and Main

Southall Cemetery (Hertford Academy in background)

St., Broad St. and Fourth St.), and alleys and backways (driveway between Hertford Academy and Southall cemetery).

—several sites of historic significance—including sites associated with major personalities (Indian Queen Tavern—Lafayette was entertained here in 1825; Walter Reed home—he lived here with his father), sites of single important events (campground and fairground—where a Confederate bivouac was located from 1861 to 1865), and sites of recurring events (Wesleyan Female Institute, Chowan College, Lassiter Hotel, Murfree Law Office).

—a number of social activity sites—including sites of recurring activities (Old Masonic Hall, Hertford Academy) and sites associated with major groups (Roberts-Vaughan—MHA, First Baptist Church).

—a range of visual communication types—including verbal messages (sign in front of Roberts-Vaughan house) and conventional

Roberts—Vaughan house

symbols (barber's sign on Williams St. extension).

choosing the 'right' plan

EVERYTIME IMPLEMENTATION OF ANY **GENERAL** OR **SPECIFIC SUGGESTION** IS CONSIDERED UNDER ANY OF THE FOLLOWING **ALTERNATIVE PLANS**, GO THROUGH **PROCEDURE FOR CHOOSING THE 'RIGHT' PLAN** IN THE PREVIOUS CHAPTER

implementing the 'right' plan

IF THE **MHA** HAS DIFFERENT, BETTER, OR MORE EXACT **SUGGESTIONS** FOR IMPLEMENTING ANY OF THE FOLLOWING **ALTERNATIVE PLANS**, GO THROUGH THE **DESIGN GUIDELINES** IN THE FOLLOWING CHAPTER TO DETERMINE IF THE **SUGGESTIONS** ARE ACCEPTABLE

alternative plan 1

SCRATCH OUT ALL **SUGGESTIONS** THAT ARE NO LONGER PERTINENT OR HAVE ALREADY BEEN IMPLEMENTED

policy

A Community Appearance Commission (CAC) should be established.

effect

Will enhance and improve the visual quality and aesthetic characteristics of Murfreesboro.

general suggestions for implementation

* Ask the City Council to appoint a Community Appearance Commission (CAC), in accordance with N.C. enabling legislation. [1] EFFECT—will authorize the CAC to:
  - —initiate, promote and assist in the implementation of programs of general community beautification.
  - —seek to coordinate the activities of individuals, agencies and organizations, public and private, whose plans, activities and programs bear upon Murfreesboro's appearance.
  - —provide leadership and guidance for individuals and public and private organizations in matters of area or community design and appearance.
  - —make studies of the visual characteristics and problems of Murfreesboro, including surveys and inventories, and to recommend standards and policies of design for the entire area, any portion or neighborhood, or any project.
  - —prepare both general and specific plans for the improved appearance of Murfreesboro or any part. The plans shall set forth desirable standards and goals for the aesthetic enhancement of Murfreesboro or any part.

* Encourage the City Council to include in the ordinance establishing the CAC the following powers [1]:
  - —to request from any public organization its plans for public buildings, facilities, or projects to be located within Murfreesboro.
  - —to review these plans and to make recommendations to the appropriate agency, the Murfreesboro planning board, or City Council regarding the aesthetic suitability of the plans.
  - —to formulate and recommend to the City Council or planning board the adoption or admendment of ordinances that will serve to enhance the appearance of Murfreesboro and its surrounding areas.
  - —to direct the attention of city officials to needed enforcement of any ordinance affecting the appearance of Murfreesboro.
  - —to seek voluntary adherence to the standards and policies of its plans.
  - —to enter in the performance of its official duties upon private lands.
  - —to promote public interest in, and understanding of, its recommendations, studies, and plans, and to that end to prepare, publish and distribute to the public such studies and reports as will advance the cause of improved Murfreesboro appearance.
  - —to conduct public hearings and meetings.

* Encourage the City Council to appropriate money for the CAC's use. [1]

* Encourage private agencies, foundations, organizations, individuals, the State or federal government, or any other source to contribute money to the CAC. [1]

* Encourage the City Council to form a joint city/county CAC. [1]

* Once the CAC is established:
  - —Encourage the CAC to make a survey of the design resources within Murfreesboro. [1]
  - —Encourage the CAC to recommend to the City Council that the existing historic district be expanded to the boundaries of the National Register Historic District. [2]
  - —Encourage the City Council to add to the zoning ordinance:
    - —an admendment stating that historic district zoning boundaries should be through the middle of a block, not along a street. [1]
    - —an amendment stating that certain visually obnoxious land uses should not bound an historic district. This amendment must not affect the location of historic district boundary lines during historic district zoning implementation, nor should this prevent alternative solutions besides removal of conflicting land uses. [1]
    - —an amendment establishing a sign ordinance, limiting size, location, and co of signs. [2]

specific suggestions for implementation

SUGGESTIONS PERTAINING TO SPECIFIC BUILDINGS—

* Restore the façades of the Peter Williams house, Murfree Law Office, Thompson house, Winborne Law Office, Dormer house, Lassiter house, and Howell house. [1–3]

* Offer to restore the exterior facade of the Vinson house if the owner agrees to keep it the house in its present location. [1]

SUGGESTIONS PERTAINING TO LANDSCAPING—

* All properties owned by the MHA should be kept trimmed and neat. [1]

SUGGESTIONS THE MHA SHOULD MAKE TO OTHERS—

* Encourage the CAC to seek voluntary adherence to its standards and policies from:
  —Georgia-Pacific
  —the building materials supply establishment south of Williams St. between Williams St. extension and Fifth St.
  —East Main St. commercial owners [2]

alternative plan 2

SCRATCH OUT ALL **SUGGESTIONS** THAT ARE NO LONGER PERTINENT OR HAVE ALREADY BEEN IMPLEMENTED

policy

A general educational program on the design resources of Murfreesboro should be implemented.

effect

Will increase citizen environmental awareness.

general suggestions for implementation

* Conduct a survey, such as the following utilized in Doylestown, Pa., of the design stimuli (resources) in Murfreesboro. [1]

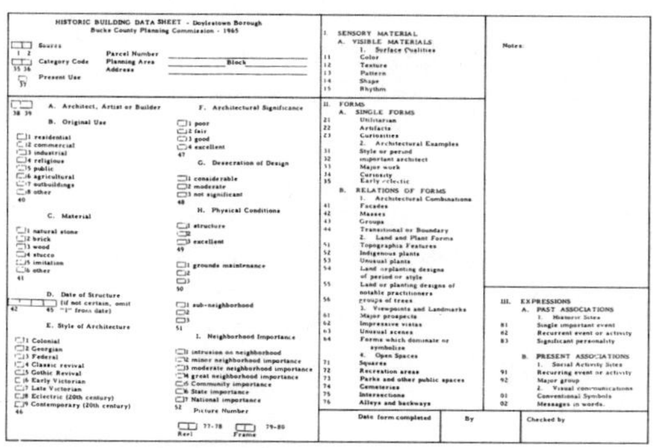

HISTORIC BUILDING DATA SHEET - Doylestown Borough
Bucks County Planning Commission - 1965

Source (1 2)
Category Code (35 36)
Present Use (37)
Parcel Number
Planning Area ______ Block ______
Address

A. Architect, Artist or Builder (38 39)

B. Original Use
1 residential
2 commercial
3 industrial
4 religious
5 public
6 agricultural
7 outbuildings
8 other
40

C. Material
1 natural stone
2 brick
3 wood
4 stucco
5 imitation
6 other
41

D. Date of Structure (if not certain, omit "1" from date) (42 45)

E. Style of Architecture
1 Colonial
2 Georgian
3 Federal
4 Classic revival
5 Gothic Revival
6 Early Victorian
7 Late Victorian
8 Eclectric (20th century)
9 Contemporary (20th century)
46

F. Architectural Significance
1 poor
2 fair
3 good
4 excellent
47

G. Desecration of Design
1 considerable
2 moderate
3 not significant
48

H. Physical Conditions
1 structure
2
3 excellent
49
1 grounds maintenance
2
3
50
1 sub-neighborhood
2
3
51

I. Neighborhood Importance
1 intrusion on neighborhood
2 minor neighborhood importance
3 moderate neighborhood importance
4 great neighborhood importance
5 Community importance
6 State importance
7 National importance
52 Picture Number
Reel 77-78 Frame 79-80

I. SENSORY MATERIAL
A. VISIBLE MATERIALS
1. Surface Qualities
11 Color
12 Texture
13 Pattern
14 Shape
15 Rhythm

II. FORMS
A. SINGLE FORMS
21 Utilitarian
22 Artifacts
23 Curiosities
2. Architectural Examples
31 Style or period
32 important architect
33 Major work
34 Curiosity
35 Early eclectic
B. RELATIONS OF FORMS
1. Architectural Combinations
41 Facades
42 Masses
43 Groups
44 Transitional or Boundary
2. Land and Plant Forms
51 Topographic Features
52 Indigenous plants
53 Unusual plants
54 Land or planting designs of period or style
55 Land or planting designs of notable practitioners
56 groups of trees
3. Viewpoints and Landmarks
61 Major prospects
62 Impressive vistas
63 Unusual scenes
64 Forms which dominate or symbolize
4. Open Spaces
71 Squares
72 Recreation areas
73 Parks and other public spaces
74 Cemeteries
75 Intersections
76 Alleys and backways

Notes:

III. EXPRESSIONS
A. PAST ASSOCIATIONS
1. Historic Sites
81 Single important event
82 Recurrent event or activity
83 Significant personality
B. PRESENT ASSOCIATIONS
1. Social Activity Sites
91 Recurring event or activity
92 Major group
2. Visual communications
01 Conventional Symbols
02 Messages in words.

Date form completed | By | Checked by

Doylestown, Pa., survey sheet

* Encourage an interested Chowan College class to do the design resources survey. [1]

* Encourage the city government and all Murfreesboro's residents to conserve these design resources. [1]

* Encourage the public schools in Murfreesboro to implement an environmental education program. The program should educate students about the built environment and its relationships to people and to the natural environment. [1]

* Conduct a series of programs, open to the public, on the design resources of Murfreesboro. Such programs might include [1]:
  —Slide programs about the design resources in Murfreesboro (and/or about specific parts of the design resources).
  —Participatory projects, to get the town's people involved in the actual conservation of the design resources. Projects might include: litter pick-up; identification of plants/trees by markers; development of a large map, located somewhere in the center of the town, locating the major design resources of Murfreesboro; etc.
  —Workshops on specific methods the general public can use to conserve design resources.

* Make a movie about the conservation of design resources in general to be circulated to communities around the state. [2]

* Develop a series of brochures concerning specific design resource elements to be circulated around the state. [2]

D
INDIVIDUALLY SIGNIFICANT BUILDINGS

Worrell's Grist Mill

Borland house

Captain Lawrence house

goals related to area

ADD & DELETE AS THE GENERAL GOALS OF THE **MHA** CHANGE (see **PROCEDURE FOR CHOOSING THE 'RIGHT' PLAN** in the previous chapter)

* Preserve the history of Murfreesboro
* Preserve architecturally and historically important buildings
* Boost community interest and civic pride
* Control change in the historic district
* Restore many homes to period authenticity
* Influence restoration in other parts of town
* Involve more people in the MHA's plans
* Eliminate the non-historic homes from the historic district

____

____

____

____

____

____

____

____

____

____

____

____

qualities/importance of area

ADD & DELETE AS THE QUALITIES OF THE AREA CHANGE

* The buildings are cultural, educational, and historic resources to the community
* There still exists an abundance and variety of buildings representative of the town's past growth
* Many are potentially adaptable for re-use to satisfy community needs
* These buildings, once they are allowed to be destroyed, are not replaceable; a piece of history would be lost.

Lassiter Hotel—site now occupied by a supermarket

* Several specific areas have a high concentration of historic structures, including the central section of the historic district and W. Main St.
* Most of these historic structures have been owned by prominent Murfreesboro families, through several generations, up to the present.

* Among these buildings are many examples of fine architectural details from the Federal, Victorian, and Greek Revival styles.

Vinson house–interior stairway

choosing the 'right' plan

EVERYTIME IMPLEMENTATION OF ANY **GENERAL** OR **SPECIFIC SUGGESTION** IS CONSIDERED UNDER ANY OF THE FOLLOWING **ALTERNATIVE PLANS**, GO THROUGH **PROCEDURE FOR CHOOSING THE 'RIGHT' PLAN** IN THE PREVIOUS CHAPTER

implementing the 'right' plan

IF THE **MHA** HAS DIFFERENT, BETTER, OR MORE EXACT **SUGGESTIONS** FOR IMPLEMENTING ANY OF THE FOLLOWING **ALTERNATIVE PLANS**, GO THROUGH THE **DESIGN GUIDELINES** IN THE FOLLOWING CHAPTER TO DETERMINE IF THE **SUGGESTIONS** ARE ACCEPTABLE

alternative plan 1

SCRATCH OUT ALL **SUGGESTIONS** THAT ARE NO LONGER PERTINENT OR HAVE ALREADY BEEN IMPLEMENTED

policy

Immediate recognition and protection of all historically significant buildings should be promoted. Owners should be convinced to rehabilitate and adaptively use the buildings in their present locations, or sell to someone who will do so.

effect

Immediate action, expecially on those buildings in the most danger of destruction, will prevent furthur decay, and will prevent, or at least postpone, any plans for demolition. Leaving them in their present location will have the effect of:

1) more accurately reflecting the historic growth patterns of the town;

2) enabling people to see this development through informal contact (passing by), and not necessarily through planned contact (taking a tour);

3) maintaining or enhancing the character of the area in which these buildings are located;

4) maintaining a place in the memories of those people who actually experienced using the buildings when they were younger.

general suggestions for implementation

* Survey all buildings in the local historic district. in the National Register Historic District, in Murfreesboro, and in Hertford County, in that order, to determine their present structural condition and significance. Use the form **Historic District and Buildings Survey** to catalog all information. [1]

* First work to restore buildings already owned by the MHA, but take measures to protect all endangered buildings immediately. [1]

Thompson house

* Apply for state and federal matching grants for the rehabilitation of individual buildings. [1]

* Initiate campaigns to raise money: flea markets, social gatherings, guest lecturers, etc. [1]

* Of key importance is issue visibility. Make clear to the community the advantages of restoration and adaptive re-use, as well as the intentions of the MHA. [1]

* Begin a community-wide campaign to save a specific building from destruction. [2]

* Increase public awareness through media, public meetings, compaigns to raise money, etc. [1]

* Encourage present owners of historic buildings to maintain or repair to standard conditions the structures and properties they own. [1]

specific suggestions for implementation

SUGGESTIONS PERTAINING TO SPECIFIC BUILDINGS–

* Take into consideration all historic buildings within the community that are in need of major repair. Evaluate each building separately and decide, by concensus of the MHA members, a rank ordering of the buildings considered most important to the community. Also determine a similar list of buildings, from those most endangered by destruction to those least endangered. Using these lists, take immediate action on buildings of top priority–that is, take measures to at least protect, and eventually rehabilitate, the most important and the most endangered buildings. [1]

* Encourage exterior clean-up operations on buildings in need of major repair so that the public will see some signs of improvement. [1]

Vinson house

* Develop and distribute guidelines suggesting do's and don'ts for remodeling and repairs of exteriors (see **Remodeling Old Houses Without Destroying Their Character** by George Stephen for a good example). [1]

Vinson house–exterior detail

* For specific suggestions for building use or re-use, see the area/issue in this chapter dealing with the area where the building is located.

* Encourage placement of a special plaque, or a unique street number, on all historic buildings within the community, so these buildings can be clearly identified by others.[2]

* Petition for the expansion of the historic district boundaries to encompass a larger number of historic structures. See the issue, **Regulation of Development–Murfreesboro,** in this chapter. [1]

* See other suggestions for increasing control over historic buildings in the issue section, **Regulation of Development–Murfreesboro.**

* Inform the owners of buildings that are identified on the National Register of Historic Places of the 50% property tax deferral.

* Refer to **The Code of the Town of Murfreesboro,** Chapter 6, Article 5, Section 6-29. Amend the section to include a special clause pertaining to historic structures that provides for negotiation with the building inspector to obtain variances. [2]

SUGGESTIONS THE MHA SHOULD MAKE TO OTHERS–

* Encourage building owners to take a more active role in the MHA's plans by inviting them to meetings, asking them to participate in the actual implementation of plan alternatives, etc. [1]

* Send to the owners of significant buildings a copy of the proceedings from each MHA meeting. [2]

* Solicit, by interview, the attitudes of significant buiding owners. [1]

* Find potential buyers for historic homes by advertising through widely circulated media. [2]

* Encourage owners or occupants of buildings adjacent to historic ones to maintain, or upgrade their properties. Offer advice, contribute monies, labor, or other resources for a neighborhood clean-up program; give information on various state and federal self-help programs. [1]

alternative plan 2

SCRATCH OUT ALL SUGGESTIONS THAT ARE NO LONGER PERTINENT OR HAVE ALREADY BEEN IMPLEMENTED

policy

The most valuable historic buildings that are in the greatest danger of destruction should be moved to locations within the historic district, either by individuals who will use the building in an appropriate way, or by the MHA, who would rehabilitate and then sell or rent.

effect

Although at an added expense, buildings considered most valuable to the community would be preserved. Moving a building should be one of the last considerations to save a building from demolition.

general suggestions for implementation

* Educate owners to the historic importance of their building to the community. [1]

* Many times an owner may not consider it worthwhile to move a building. Campaign throughout the community to get wide-spread support, raise monies, etc. [1]

* Ask an owner who is sensitive to the historic value of his building to donate the structure to the MHA on the condition that the MHA will move the building. [1]

* Ask on owner to donate money toward the cost of moving the building as well. [2]

* Enlist the help of civic organizations, college groups, or high school students to prepare the structure for the move. [1]

* Continue to utilize present means of moving buildings and offer a higher fee if the job can be done quickly. [1]

Winborne Law Office

* Refer to **Design Guidelines**, in the following chapter, in order to aid in choosing a suitable location for the building. Determine how well the structure will complement the quality and character of structures already existing in the area surrounding the intended site. If the building is compatible, determine where to locate it on the site itself using the guidelines outlined in the same chapter. [1]

E

IMAGE~ HISTORIC DISTRICT

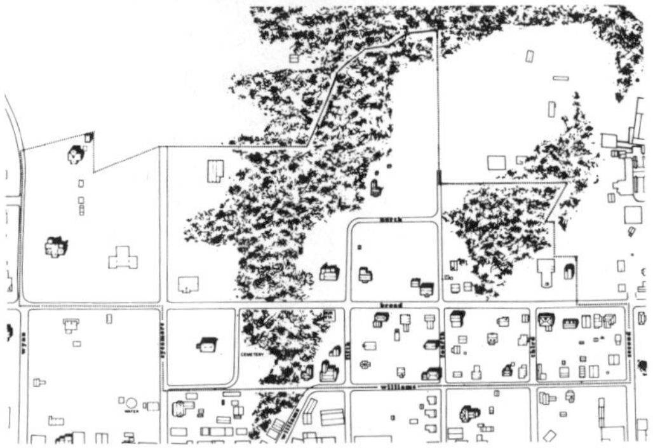

goals related to area

ADD & DELETE AS THE GENERAL GOALS OF THE MHA CHANGE (see **PROCEDURE FOR CHOOSING THE 'RIGHT' PLAN** in the previous chapter)

* Preserve the history of Murfreesboro
* Boost community interest and civic pride
* Control change in the historic district

qualities/importance of area

ADD & DELETE AS THE QUALITIES OF THE AREA CHANGE

* The range of densities in the central section suggest both the urban and the rural character of the early town.

Area between the Gingerbread and Wheeler houses

* There is a dense, active, existing neighborhood in the east end.

Williams Street—east end of historic district

* There are thickly wooded ravines cutting the historic district into sections.

West ravine—looking east on Broad Street

* There is a public area with sparse density excellent for public activities.

West end of historic district—city hall in background

* In the central section and the east end, the transition from the historic district to Main St. is, in most cases, visually unnoticeable.
* In the west end (the public area), the transition from the historic district to Main St. is visually, very noticeable.
* Trees and plants isolate the central section of the historic district from Georgia-Pacific.
* The industrial image of Georgia-Pacific overpowers the residential image of the buildings on and near the corner of Second St. and Broad St.
* The central section has the highest concentration of historically/architecturally significant buildings in Murfreesboro.
* The central section also has the highest concentration of 2-story and brick buildings.
* There are many trees and shrubs throughout the central section and the east end.
* The predominant colors on the buildings in the historic district are red and white.
* Ravines on both sides visually isolate the central section north of Broad St.
* Open area north of west end (north of Dr. Gary building) makes the west end seem more dense in comparison.

Area north of Dr. Gary house on Wynn Street

* Excellent transition between the historic district and the residences on the west side of Wynn St.
* Open space of the water tower block fits in well with the open space of the west end.

Water tower block–Dr. Issac Pipkin house in background

* The historic district has a number of landmarks: Melrose, the City Hall, Hertford Academy, Wheeler house, the Rea Store, the First Baptist Church Study.

choosing the 'right' plan

EVERYTIME IMPLEMENTATION OF ANY **GENERAL** OR **SPECIFIC SUGGESTION** IS CONSIDERED UNDER ANY OF THE FOLLOWING **ALTERNATIVE PLANS**, GO THROUGH **PROCEDURE FOR CHOOSING THE 'RIGHT' PLAN** IN THE PREVIOUS CHAPTER

implementing the 'right' plan

IF THE **MHA** HAS DIFFERENT, BETTER, OR MORE EXACT **SUGGESTIONS** FOR IMPLEMENTING ANY OF THE FOLLOWING **ALTERNATIVE PLANS**, GO THROUGH THE **DESIGN GUIDELINES** IN THE FOLLOWING CHAPTER TO DETERMINE IF THE **SUGGESTIONS** ARE ACCEPTABLE

alternative plan 1

SCRATCH OUT ALL **SUGGESTIONS** THAT ARE NO LONGER PERTINENT OR HAVE ALREADY BEEN IMPLEMENTED

policy

The four areas within the historic district--the east end, the central section, the ravines, and the west end–should be distinct, but each area should be visible from other areas.

effect

Will give the historic district visual and physical diversity.

general suggestions for implementation

* Central Section of the historic district: the MHA should maintain, and encourage others to maintain, an historic image. Historic setbacks, historic landscaping, historic signage, historic paint colors should be re-created. (see the **Design Guidelines** in Chapter 7) This central section should include a mixture of residential and commercial/public buildings. The owners of the non-significant buildings should be encouraged to remain, and rehabilitate their residences. Empty lots should be bought by the MHA and used for relocating significant buildings. The density of buildings south of Broad St. should be at the average town density, the buildings north of Broad St. should be at rural density. Williams St. between the Williams St. extension and Fourth St. should contain mostly commercial/public buildings. The remainder of the central section should contain mostly residential buildings. The corner of Broad St. and Fourth St. should be mostly single-family, low or moderate-income residential, and the corner of Williams St. and Fifth St. should be mostly commercial/public. Visual and physical connections across the west ravine to the west end of the historic district should be made.

* East End of the historic district: should be seen as merely a part of town where conservation (rehabilitation) occurs; setbacks, landscaping, etc. similar to the town norm. It is not part of the central section historic image. The east end should contain mostly residential buildings. Some of the east end residential buildings should be rehabilitated, and some (the historically significant buildings) should be restored. New single-family residences aimed at low-income families should be encouraged. These should fit in well with the buildings around them in terms of design. materials, etc. Historically/architecturally significant buildings should not be relocated here unless it will be for single-family, low-income residential use. Georgia-Pacific should be prevented from demolishing any more residences in or outside the historic district. No commercial development should be allowed on the south side of Williams St. that will have a detrimental effect on residential activities.

* West Ravine: encourage the city/owners to thin trees in the west ravine (see **Ravines/River** and **Central Section–Historic District**)

* West End of the historic district: either more active public use of the area should be encouraged. or more private use in the form of new or relocated significant residential buildings. In either case, the west ravine should be visually opened to the central section of the historic district. An alternative, more active, use of the Hertford Academy should be encouraged, and more active use of the community center should be encouraged (along with a different, more appropriate, exterior facade). The visual connection to a potential shopping center development in the water tower block should always be considered before anything is done in the west end. Residential use of the open space north of the west end should be encouraged.

* All of the above four areas–the east end, the central section, the west ravine, and the west end–have distinct and separate images. These images should be reinforced whenever action is taken in these areas. The **Design Guidelines** outlined in Chapter 7 should be carefully followed; and the MHA should develop and distribute guidelines suggesting do's and don'ts for remodeling and repairs of exteriors (See the following 3 publications for good examples–**Remodeling Old Houses Without Destroying Their Character** by George Stephen [for $3.95 from the Preservation Bookstore; National Trust for Historic Preservation; 740–748 Jackson Place, N. W.; Washington, D. C. 20006], **Revitalizing Older Houses in Charlestown** by George Stephen [for $.50 from the Superintendent of Documents, U. S. Government Printing Office, Washington, D. C. 20402], and **Water Street Study, Chillicothe, Ohio: Suggestions for Rehabilitation** by Visnapuu & Gaede, Inc. [for $1.50 from the Preservation Bookstore, National Trust for Historic Preservation]. Included in these publications are rehabilitation criteria for such elements as roofs; eaves; walls and surfaces; windows; storefronts, shop windows and entry doors; colors; rear and side walls; signs; materials; doorways and plantings. These criteria should be used by the Historic District Commission when granting certificates of appropriateness.). [1]

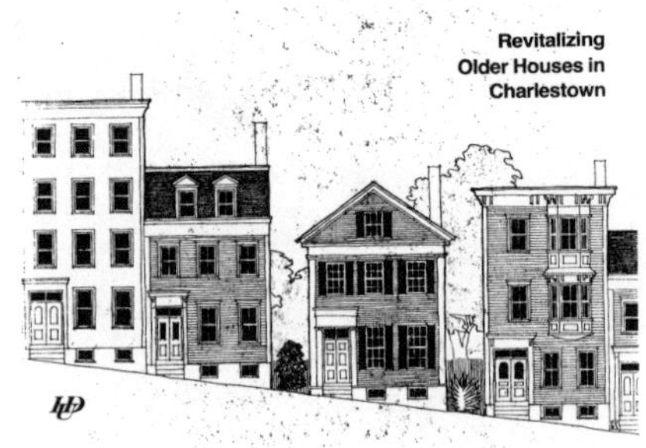

* The half block to the south of Williams St., between Sycamore St. and Second St., should be added to the historic district. The part between the west ravine and Sycamore St. should conform to the image of the west end. The part between the west ravine and Fourth St. should conform to the image of the central section. The part between Fourth St. and Second St. should conform to the image of the east end. [1]

specific suggestions for implementation

SUGGESTIONS PERTAINING TO SPECIFIC BUILDINGS–

* The significant buildings should be identified with signs on or near the buildings, according to the criteria outlined in **Central Section–Historic District**. [1]

* Signage criteria should be developed for commercial/public buildings in the historic district. Signs should not utilize imitation Colonial motifs. [1]

SUGGESTIONS PERTAINING TO LANDSCAPING–

* Significant buildings should be connected with pedestrian paths (see **Pedestrian Traffic–Historic District**). [2]

* Develop and adopt a planting scheme (including a list of appropriate plants) for the central section. [1]

OTHERS–

* Encourage the Historic District Commision to develop and adopt criteria specifying acceptable exterior colors for the central section. [1]

* Determine the best locations for visual connections between the west end and central section of the historic district and Main St. Proceed to carry out these visual connections. [2]

alternative plan 2

SCRATCH OUT ALL **SUGGESTIONS** THAT ARE NO LONGER PERTINENT OR HAVE ALREADY BEEN IMPLEMENTED

policy

The entire historic district should have a homogeneous image.

effect

Will be large enough to attract tourists.

general suggestions for implementation

* The entire historic district should maintain an historic image (the same one as applicable for the central section in **alternative 1**) using historic setback, historic landscaping, historic signage, historic paint colors (see the **Design Guidelines** in Chapter 7). [1]

* The entire historic district should include a mixture of residential and commercial/public buildings. Williams St. between the Williams St. extension and Second St. should contain mostly commercial/public buildings. The remainder of the district should contain mostly residential buildings (except on Broad St. between the City Hall and the west ravine). [1]

* The owners of the non-significant buildings should be encouraged to remain and rehabilitate their residences. Empty lots and empty buildings should be bought by the MHA or someone else and used for relocating significant buildings or building new single-family, moderate or high-income residences. The density of buildings south of Broad St. should be at the average town density; the buildings north of Broad St. should be of rural density. [1]

* The corner of Broad St. and Fourth St. should be mostly single-family, moderate or high-income residential; and the corner of Williams St. and Fifth St. should be mostly commercial/public. [1]

* Visual and physical connections across both ravines should be made. [2]

* The boundaries of the historic district should be marked. Consider alternative methods of marking the boundaries–such as with signs, trees, materials used on sidewalks, street lights, picket fences, etc. [1]

Deanes' house in 1890

* The historic image of the historic district should be reinforced whenever action is taken. The **Design Guidelines** outlined in Chapter 7 should be carefully followed; and the MHA should develop and distribute guidelines suggesting do's and don'ts for remodeling and repairs of exteriors (see the three publications referred to in **alternative 1**. Included in these publications are rehabilitation criteria for such elements as roofs; eaves; walls and surfaces; windows; storefronts, shop windows and entry doors; colors; rear and side walls; signs; materials; doorways and plantings. These criteria should be used by the Historic District Commission when granting certificates of appropriateness.). [1]

F

EAST END~ HISTORIC DISTRICT

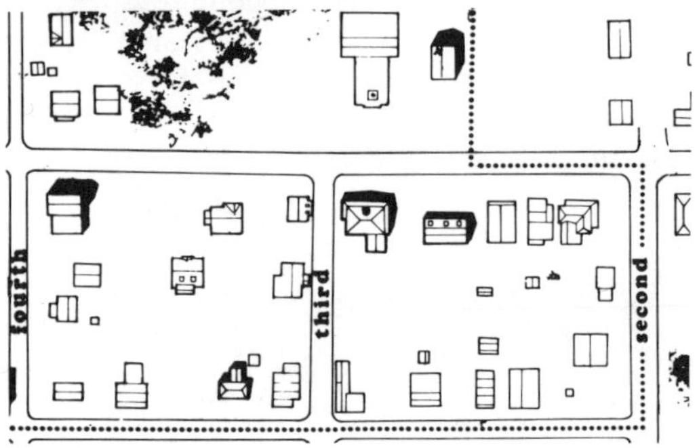

goals related to area

ADD & DELETE AS THE GENERAL GOALS OF THE **MHA** CHANGE (see **PROCEDURE FOR CHOOSING THE 'RIGHT' PLAN** in the previous chapter)

* Involve more people in the MHA's plans
* Boost the community interest and civic pride
* Control change in the historic district
* Influence restoration in other parts of town
* Eliminate the non-historic homes from the district

____

____

____

____

____

____

____

____

____

____

____

____

____

____

____

____

qualities/importance of area

ADD & DELETE AS THE QUALITIES OF THE AREA CHANGE

* The only active neighborhood in the historic district

Second Street – between Broad and Williams Street, looking south

* Many of the town's oldest citizens live here
* Good mix of residential and commercial uses
* Some historic homes, but mostly smaller homes; many with nice residential character
* Part of the original town as laid out by William Hardy Murfree in 1787

Williams Street—east end of historic district

* The home of active social groups
* East ravine located here
* Buildings are all similar in scale and in setback

Broad Street—east end of historic district

* Many residents walk to work
* Area is bordered by the Georgia-Pacific industrial plant, the town's principle employer

Broad Street—east end of historic district

* This lively, active community is potentially a great asset to the conservation effort

____

choosing the 'right' plan

EVERYTIME IMPLEMENTATION OF ANY **GENERAL** OR **SPECIFIC SUGGESTION** IS CONSIDERED UNDER ANY OF THE FOLLOWING **ALTERNATIVE PLANS**, GO THROUGH **PROCEDURE FOR CHOOSING THE 'RIGHT' PLAN** IN THE PREVIOUS CHAPTER

implementing the 'right' plan

IF THE **MHA** HAS DIFFERENT, BETTER, OR MORE EXACT **SUGGESTIONS** FOR IMPLEMENTING ANY OF THE FOLLOWING **ALTERNATIVE PLANS**, GO THROUGH THE **DESIGN GUIDELINES** IN THE FOLLOWING CHAPTER TO DETERMINE IF THE **SUGGESTIONS** ARE ACCEPTABLE

alternative plan 1

SCRATCH OUT ALL **SUGGESTIONS** THAT ARE NO LONGER PERTINENT OR HAVE ALREADY BEEN IMPLEMENTED

policy

Support conservation of both the buildings and the community, with less concern for recreating the original townscape.

effect

The neighborhood will become part of the historic district without being destroyed or greatly altered.

general suggestions for implementation

* Restore the historic Edward Murphy house on Broad Street, owned by the MHA, for residential use. [1]
  EFFECT— Will serve as an example to the community, and will show the Association's good intentions. Will strengthen the neighborhood.
* Set up a "revolving fund" for rehabilitation loans to the owners of certain buildings, to be made according to the value of the structure to the community, its historic importance, and the needs of the owner (preference should be given to owner-occupied houses).
  —Old Masonic Lodge [1]
  —Bill Pruden House [1]
  —William Trader House [1]
  —other buildings [2]
* Encourage the city to apply for federal funds for this same work. [1]

* With a neighborhood group, co-sponsor a building maintenance, repair and clean-up campaign. [1]

  EFFECT– Besides involving residents in a neighborhood concern, minor repairs will help prevent deterioration.

* Promote a policy of constructing new buildings on empty lots and those with dilapidated structures. People who have been displaced by changes in the East End should have the first opportunity to rent or buy these newly constructed or relocated buildings. [2]

specific suggestions for implementation

SUGGESTIONS PERTAINING TO SPECIFIC BUILDINGS–

* Edward Murphy house–Rehabilitate for rental, or sale with guarantees about preserving the exterior appearance. The sale procedure could be a lease with an option to buy at a very low price if the lessee agrees to carry out the rehabilitation himself. [1]

* William Trader house– The owner of this building should be encouraged to clean up and care for the grounds, as well as make repairs, to make the building rentable. [1]

* Old Masonic Lodge–This early nineteenth-century building deserves rehabilitating. The work could be backed by the advice, organization and funds of the MHA. The MHA knows how to organize volunteer help, and this might be the most effective method. [1]

* Bill Pruden house–This house would be a good place to initiate the revolving fund loans, where the MHA would promote and help finance rehabilitation by the owner-occupant.

Bill Pruden house

SUGGESTIONS PERTAINING TO LANDSCAPING–

* The MHA should use its talents for organizing public interest, volunteers and money to set up a Home Maintenance, Repair and Clean-Up Campaign. Sponsored along with a neighborhood group, the campaign could comprise contests, volunteer help for the needy, tours.[1]

* The grounds are quite important to the overall appearance of the East End. The MHA should promote an appearance program that focuses on several topics. [1] For instance:

  Soil and ground cover–Under-nourished and trampled soil should be rebuilt. Green grass would make a world of difference.

  Planting–Shrubs and trees shouldn't obscure the buildings, and in the front yard, shrubbery should generally be no more than waist high.

  Off-street parking–Cars should be parked as far back as possible from the street.

  Assorted junk–Should be cleaned up or screened.

  Fences–Natural material (wood, stone) and as low as possible in the fronts of buildings.

SUGGESTIONS THE MHA SHOULD MAKE TO OTHERS–

* Georgia-Pacific should be made aware of its responsibilities as a good neighbor to the historic district, and should limit its traffic to Second Street, try to keep speeds down, and screen its parking lots, especially the one at the northwest corner of Broad St. and Second St. [1]

alternative plan 2

SCRATCH OUT ALL **SUGGESTIONS** THAT ARE NO LONGER PERTINENT OR HAVE ALREADY BEEN IMPLEMENTED

policy

More property should be acquired and buildings demolished to make way for relocated historic buildings.

effect

The East End will begin to look and function more like the central section of the historic district, with fewer, larger buildings and lots.

general suggestions for implementation

* Restore the historic Edward Murphy house on Broad Street, owned by the MHA, for residential use. [1]

  EFFECT– Will serve as an example to the community, and will show the Association's good intentions.

* Buy up less important homes at visible locations and demolish to make room for future relocation of historic homes. [1]
  EFFECT– Will make the area look more like the central section, especially since residential use will be reduced.

* Buildings in the town and in the county, which the MHA sees as valuable and endangered, should be listed in priority order and moved to the newly created lots in the historic district. [2]

* Sell or rent these relocated buildings as residences.
  EFFECT– Will maintain the residential character which is the best (and probably the only) use for the East End. [2]

specific suggestions for implementation

SUGGESTIONS PERTAINING TO SPECIFIC BUILDINGS–

* Old Masonic Lodge– This early nineteenth-century building deserves rehabilitating. The work could be backed by the advice, organization and funds of the MHA. The MHA tion knows how to organize volunteer help, and this might be the most effective method.

Old Masonic Lodge

* Edward Murphy house–Rehabilitate for rental, or sale with guarantees about preserving the exterior appearance. The sale procedure could be a lease with an option to buy at a very low price if the lessee agrees to carry out the rehabilitation himself. [1]

* William Trader house– The owner of the building should be encouraged to clean up and care for the grounds, as well as make repairs, to make the building rentable. [1]

* Bill Pruden house–This building could be rehabilitated either by the MHA or by the private owner with the aid of the MHA. [1]

* Acquire land for open space and for relocation of significant buildings by demolishing existing buildings at corner sites, such as the NE corner of Williams and Fourth Sts., and the NE corner of Williams and Third Sts. [1]

* Acquire and demolish buildings that are not historically or environmentally significant (see Chapter Three). [1]

SUGGESTIONS PERTAINING TO LANDSCAPING–

* The MHA should use its talents for organization of public interest, volunteers and money to help set up a Home Maintenance, Repair and Clean-Up Campaign. Sponsored along with a neighborhood group, the campaign could comprise contests, volunteer help for the needy, tours, etc. [1]

SUGGESTIONS THE MHA SHOULD MAKE TO OTHERS–

* Georgia-Pacific should be made aware of its responsibilities as a good neighbor to the Historic District, and should limit its traffic to Second Street, try to keep vehicle speeds down, and screen its parking lots, especially the one at the NW corner of Broad and Second Streets. [1]

alternative plan 3

SCRATCH OUT ALL **SUGGESTIONS** THAT ARE NO LONGER PERTINENT OR HAVE ALREADY BEEN IMPLEMENTED

policy

Most property and buildings should be acquired and demolished, and reconstruction of the East End to match the central section of the historic district should be attempted.

effect

This will produce a four-block central section for the historic district instead of a two-block one. Like the present central section, the East End will have limited use, mostly public, and will have much vacant land which could contain most of the historic buildings in the county. This, however, will be extremely expensive. The East End will be mostly empty and underused for many years to come, while the residential community presently located there will be destroyed.

general suggestions for implementation

* Acquire and demolish all buildings except those deemed historically important. [1]

* Since the number of significant buildings to be moved must be very great to have forced choosing this alternative, the MHA should consider forming its own house-moving company and/or moving a great number of buildings at one time. [1]

G

CORNER OF WILLIAMS & FIFTH

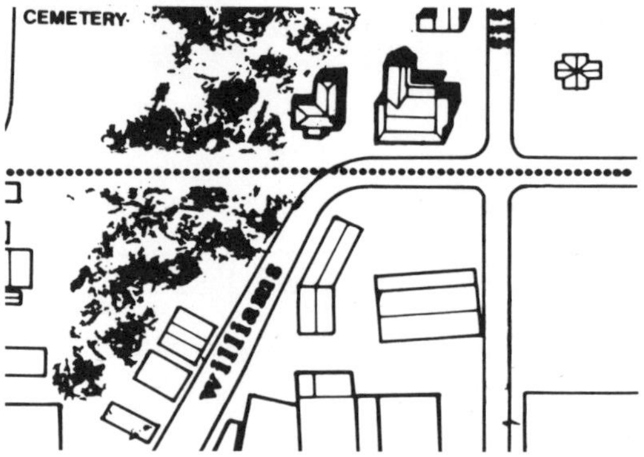

goals related to area

ADD & DELETE AS THE GENERAL GOALS OF THE **MHA** CHANGE (see **PROCEDURE FOR CHOOSING THE 'RIGHT' PLAN** in the previous chapter)

* Preserve architecturally and historically important buildings
* Boost community interest and civic pride
* Control change in the historic district
* Use river and ravines to full potential
* Restore homes to period authenticity
* Involve more people in the MHA's plans
* Improve business along Main St.

qualities/importance of area

ADD & DELETE AS THE QUALITIES OF THE AREA CHANGE

* Public/commercial activities were originally located here.
* The area is one block from Main St., offering easy access to goods and services.
* The area contains some of the oldest buildings in Murfreesboro.

Peter Williams house and Murfree Law Office

* The area contains important neighborhood commercial businesses.

Williams St. extension—buildings on the west side

* The area is on the southern boundary of the local historic district; thus, it is important visually.
* The area acts as a transition area from commercial to residential and vice versa, giving diversity to the area.
* There is little vehicular traffic through the area.
* The narrow setback of the buildings along Williams St. is conducive to public/commercial activity.
* There is much pedestrian activity within the area.

Williams Street extension

* There is easy access to the west ravine from the area.
* The area south of Williams St. across from the Peter Williams house has potential as open space or for possible relocation of significant buildings.
* The area is only one block from the City Hall and community center.
* The tourist attraction potential of the area is great.
* The area has a distinct residential character.
* The area is half within, and half outside, the local historic district.
* When the ravine is thinned, the Hertford Academy and other parts of the west end of the historic district will be easily visible.

Southall Cemetery and Hertford Academy as seen from across the ravine

* The area contains some of the most environmentally significant buildings in Murfreesboro (the buildings on the west side of the Williams St. extension).
* Williams St. extension is a good informal vehicular and pedestrian entrance to the historic district from Main St.

Williams Street extension

* The area is part of the original town as laid out by William H. Murfree in 1787.
* There is great potential within the area for adapting buildings for commercial reuse.
* The lot on the northeast corner of Williams St. and Fifth St. offers space for relocating a significant building, thus strengthening the focal character of this corner.
* The Williams St. extension offers the only view from Main St. directly toward a significant building within the historic district.
* The Belk-Tyler parking lot on Williams St. offers space for relocating a significant building, or building a new commercial structure, increasing this corner's importance.
* The narrowest stretch of the west ravine is next to the Murfree Law Office.

choosing the 'right' plan

EVERYTIME IMPLEMENTATION OF ANY **GENERAL** OR **SPECIFIC SUGGESTION** IS CONSIDERED UNDER ANY OF THE FOLLOWING **ALTERNATIVE PLANS**, GO THROUGH **PROCEDURE FOR CHOOSING THE 'RIGHT' PLAN** IN THE PREVIOUS CHAPTER

implementing the 'right' plan

IF THE **MHA** HAS DIFFERENT, BETTER, OR MORE EXACT **SUGGESTIONS** FOR IMPLEMENTING ANY OF THE FOLLOWING **ALTERNATIVE PLANS**, GO THROUGH THE **DESIGN GUIDELINES** IN THE FOLLOWING CHAPTER TO DETERMINE IF THE **SUGGESTIONS** ARE ACCEPTABLE

alternative plan 1

SCRATCH OUT ALL **SUGGESTIONS** THAT ARE NO LONGER PERTINENT OR HAVE ALREADY BEEN IMPLEMENTED

policy

The area should be used for public/commercial activities related to the original uses of the buildings, and to the activities within the central section of the historic district.

effect

Will offer the best transition from the commercial activities of the CBD to the public/residential activities of the central section of the historic district.

general suggestions for implementation

* Rehabilitate the Peter Williams house as a rental efficiency apartment. Remove all additions except the earliest, to be used for kitchen/bathroom/closets. Remove most of porch except the section between the two front windows. [1]
  EFFECT–only minimal expenditures would be required to convert the building from a residence to an information center for the historic district, if necessary.

Peter Williams house

* Encourage the use of the Murfree Law Office as a law office and law library owned and operated by a private individual. [1] If unsuccessful, encourage the town to purchase and use the building as a public library combining law and present library in the Roberts-Vaughan Center. [2]

* Encourage the city to construct an east-west pedestrian bridge (of wood construction) to connect Williams St. with the dirt road between Hertford Academy and Daniel's Nursery. The area can serve as the first step in recreational use of the ravine. [1]

Pedestrian bridge across the west ravine

* Encourage protection of the large, unusual trees from being cut down and from possible damage by pedestrians due to the location of the bridge. [1]

* Determine if the wood-frame building on the west side of the Williams St. extension is historically or architecturally significant. If it is, make a record of the building using the **Historic District and Buildings Survey**. [1]

Building on Williams Street extension

* Encourage the town to place a sign at the corner of Main St. and Williams St. extension that indicates the entrance to the historic district. [2]

* Encourage the owners of the brick buildings on the east side of the Williams St. extension to clean up the facades facing the Williams St. extension. [1]

* The cemetery should be protected from excessive pedestrian travel.

* For the old commercial buildings on the west side of the Williams St. extension, consider public/commercial uses, compatible with residential/public/commercial activities in the central section of the historic district. [1] For example, these buildings could be rehabilitated for antique shops, craft shops, etc. The MHA should consider buying the wood-frame building, if found to be architecturally or historically significant, rehabilitate and sell/rent for commercial use. [2]

Williams Street extension–as it could look

* Encourage owner of lot between Fifth St. and Williams St. extension to store building materials elsewhere. [1] If unsuccessful, buy and use for relocating one (or two if small) historically or architecturally significant building(s) for commercial/public use; or encourage the town to buy and use as public open space. [2]
  EFFECT–will allow more of the historic district to be visible from Main St. looking down the Williams St. extension; relocated building(s) would make corner busier, more conducive to commercial/public activities.

* Use the building on the northeast corner of Fifth St. and Williams St. for storage until such time that the lot is needed for relocation of a large significant building. [1] Such a significant building should be for commercial or public use, and should be an example of town design, rather than rural Hertford County design. If possible, do not remove any of the trees on this lot.

* Encourage Belk-Tyler to develop the north end of their parking lot for commercial/public use. Parking for both businesses could be located between the two. [2]

* Consider the possibility of the Murfree Law Office, Peter Williams and Ferguson houses being used together for some public/commercial function. [1]

alternative plan 2

SCRATCH OUT ALL **SUGGESTIONS** THAT ARE NO LONGER PERTINENT OR HAVE ALREADY BEEN IMPLEMENTED

policy

The area should be used for residential activities

effect

Will make area part of the lively, residential neighborhood in the central section and east end of the historic district.

general suggestions for implementation

* Peter Williams house–Rehabilitate as a rental efficiency apartment. Remove all additions except the earliest, to be used for kitchen/bathroom/closets. Remove most of the porch except the section between the two front windows. [1]

* Murfree Law Office–Restore as a single-family residence and sell. [1]

* Encourage the city to construct an east-west pedestrian bridge (wood construction) across the ravine to connect Williams St. with the dirt road between the Hertford Academy and Daniel's Nursery. [1]

* Encourage the owner/tenant occupants of the commercial businesses on the west side of the Williams St. extension to remain. Encourage the occupants to rehabilitate the exteriors, or encourage the city to look into federal funds for rehabilitating neighborhood commercial businesses (see **A Guide to Federal Programs** (1974) and **1976 Supplement: A Guide to Federal Programs** available from the National Trust for Historic Preservation bookstore). [1]

* Determine if the wood-frame building on the west side of the Williams St. extension is historically or architecturally significant. If it is, record the building using the **Historic District and Buildings Survey**. [1]

* Encourage owner of lot between Fifth St. and Williams St. extension to store building materials elsewhere. [1] If successful, the MHA should buy and use for relocating one (or two if small) historically or architecturally significant building(s) for residential use; or encourage the town to buy and use as public open space. [2]
  EFFECT–will allow more of the historic district to be visible from Main St. looking down the Williams St. extension; relocated building(s) would make the corner busier, more conducive to residential activities.

* Use the building on the northeast corner of Fifth St. and Williams St. for storage until such time that the lot is needed for relocation of a large, significant building. [1] Such a significant building should be for residential use, and should be an example of town design rather than rural Hertford County design. If possible, do not remove any of the trees on this lot.

* Encourage Belk-Tyler to plant a large barrier hedge between their parking lot and Williams St. and Fifth St. [2]

H

CORNER OF BROAD & FOURTH

goals related to area

ADD & DELETE AS THE GENERAL GOALS OF THE MHA CHANGE (see **PROCEDURE FOR CHOOSING THE 'RIGHT' PLAN** in the previous chapter)

* Preserve architecturally and historically important buildings
* Boost the community interest and civic pride
* Restore many homes to period authenticity
* Control change in the historic district
* Eliminate the non-historic homes from the historic district

____

____

____

____

____

____

____

qualities/importance of area

ADD & DELETE AS THE QUALITIES OF THE AREA CHANGE

* The corner is on the boundary between the central section of the historic district and the east end, thus it is important visually.
* The corner has the potential for tying together the central section and east end.
* The significant buildings express the life style of the 1800's.
* The corner contains one of the most historically significant buildings in Murfreesboro.

Wheeler house

* The corner contains one of the most environmentally significant buildings in Murfreesboro.

P. S. White house

* Broad St. is a minor east-west vehicular route with little traffic. At this corner, the area north of Broad St. has rural density characteristics, and the area south of Broad St. has town density qualities.
* The corner acts as a transition area from the historic activities of the central section to the residential activities of the east end.

* MHA owns land on three of the four corners.
* The road going north from this corner is unpaved.
* There are signs of previous archeological excavations at this corner.

Archeological excavation next to Wheeler house

* The corner has three 2-story buildings on it.
* The area has a building on every corner.
* The two buildings south of Broad St. have residential buildings next to them on Broad St.
* Low or moderate income residential use of the two buildings south of Broad St. could socially tie the east end to the central section.
* The east ravine, in the historic district, borders the east boundary of this corner.
* Whatever happens to this corner will have a significant impact on both the east end and central section.
* The corner has the potential for three residential buildings.
* Fourth St., the north-south road through this corner, could become a major entrance to the historic district.
* The two buildings south of Broad St. are very similar in style and design.

Lassiter house

P. S. White house

* The corner is part of the original town as laid out by William H. Murfree in 1787.
* The corner has a distinct residential character.
* The corner has tourist attraction potential.
* The narrow setback, along Broad St., of the buildings on this corner is conducive to residential activities.
* The corner contains some of the oldest buildings in Murfreesboro.
* The corner is two blocks away from Main St., offering easy access to goods and services.
* The corner is only two blocks from the community center and City Hall.

____
____
____
____
____
____
____

choosing the 'right' plan

EVERYTIME IMPLEMENTATION OF ANY **GENERAL** OR **SPECIFIC SUGGESTION** IS CONSIDERED UNDER ANY OF THE FOLLOWING **ALTERNATIVE PLANS**, GO THROUGH **PROCEDURE FOR CHOOSING THE 'RIGHT' PLAN** IN THE PREVIOUS CHAPTER

implementing the 'right' plan

IF THE **MHA** HAS DIFFERENT, BETTER, OR MORE EXACT **SUGGESTIONS** FOR IMPLEMENTING ANY OF THE FOLLOWING **ALTERNATIVE PLANS**, GO THROUGH THE **DESIGN GUIDELINES** IN THE FOLLOWING CHAPTER TO DETERMINE IF THE **SUGGESTIONS** ARE ACCEPTABLE

alternative plan 1

SCRATCH OUT ALL **SUGGESTIONS** THAT ARE NO LONGER PERTINENT OR HAVE ALREADY BEEN IMPLEMENTED

policy

The area should be predominately used for low and moderate income residences.

effect

Will connect the east end residential area with the central section.

general suggestions for implementation

* Wheeler house—currently being restored as a house-museum. If off-street parking must be

Wheeler house

provided, locate parking directly north of the main building, extending no farther west than the west facade of the main building. Any landscaping on the Broad St. side of the building must be carefully done to visually tie it with the buildings across Broad St. The landscaping on the east side of the building must be considered separately due to the different relationships between the Wheeler house and the residences across Fourth St. The open space to the rear and west of the Wheeler house should be preserved. No sidewalk should be put in front of the Wheeler on Broad St., and paving of Fourth St. should not occur. The MHA should put a sign on Broad St. to advertise the museum, once it is completed. [1]

* P. S. White house—(on the SW corner of Broad St. and Fourth St.) Rehabilitate as a low or moderate income residence, and either sell or rent when completed. Remove additions, including porch, if not needed by the prospective owner/tenant. Preserve the trees and shrubs on the lot. [1] A walkway, as mentioned in **Central Section—Historic District**. should be placed in front of the building and on its east side. [2] Landscaping behind the building should block out any view into the building from the second story of the Winborne Law Office. Landscaping in the front should tie the P. S. White house and the Wheeler house together, and landscaping on the east side should visually tie it with the Lassiter house.

* Lassiter house— Rehabilitate or restore as a low or moderate income residence, then sell or rent. Remove additions if not needed by prospective owner/tenant. Landscaping on the west side should visually tie it with the P. S. White house.

* Encourage the owner/tenant occupants of the residential buildings on the northeast corner of Broad St. and Fourth St. to remain. If unsuccessful, encourage others to move in. [1] If unsuccessful again, buy the residences and rehabilitate for residential use, if possible. [2] If this is not possible, and the lot is needed for relocating a significant building, demolish the residence and relocate a 2-story significant building on the lot. [2] The relocated significant building should be for low or moderate income residential use, should be of a design found in the town (not of a design found only in rural Hertford County), and either sold or rented after rehabilitation/restoration. If the lot is not needed for relocating a significant building, encourage a developer to buy the property and build new low or moderate income residences. [2] Such new development will be subject to the design criteria utilized by the Historic District Commision, since it is within the historic district. (see Chapter 7—**Design Guidelines**).

* Follow this priority when rehabilitating and selling/renting the above buildings—
  (1) Rehabilitate the P. S. White exterior and interior, then sell or rent. [1]
  (2) Rehabilitate the Lassiter exterior and interior, then sell or rent. [2]

specific suggestions for implementation

SUGGESTIONS PERTAINING TO SPECIFIC BUILDINGS—

* On significant buildings rehabilitate all exterior trim, even when not original to the building. [1]

SUGGESTIONS PERTAINING TO LANDSCAPING—

* Discourage Georgia-Pacific and other owners from land-fill or tree-cutting in the east ravine. [1]

SUGGESTIONS THE MHA SHOULD MAKE TO OTHERS—

* Encourage the city to install unobtrusive well-designed, markers on or near the Wheeler and Lassiter houses. The signs should conform to the criteria outlined in **Central Section—Historic District**. [2]

OTHERS—
* Encourage continued archeological research in this area. [1]

alternative plan 2

SCRATCH OUT ALL SUGGESTIONS THAT ARE NO LONGER PERTINENT OR HAVE ALREADY BEEN IMPLEMENTED

policy

The area should be predominately used for moderate income residences, and should retain an historic image.

effect

Will better relate to the central section of the historic district.

general suggestions for implementation

* Wheeler house—see **alternative 1**. [1]

* P. S. White house—Demolish this building and use lot for relocating a 2-story significant building. [2] Such a building should be used for moderate or high income residential use, should be of a design found in the town (not of a design found only in rural Hertford County), and should be either sold or rented after restoration. Perserve the trees and shrubs on the lot. Landscaping behind the relocated building should block out any view into the building from the second story porch of the Winborne Law Office. The **Design Guidelines** in Chapter 7 should be followed for relocation of a building on this lot.

* Lassiter house—Restore as a moderate or high income residence and sell or rent. Remove additions if not needed by the prospective owner/tenant. [1]

* Encourage the owner/tenant occupants of the residential buildings on the northeast corner of Broad St. and Fourth St. to remain. If unsuccessful, buy the residences and use lot for relocating a 2-story significant building. [2] Such a significant building should be for moderate or high income residential use, should be of a design found in the town (not of a design found only in rural Hertford County), and should be sold or rented after restoration. The **Design Guidelines** in Chapter 7 should be followed during relocation of a building to this lot.

* Follow this priority when buying/relocating, restoring, and selling/renting the above buildings:
  (1) Restore the Lassiter house exterior and interior, then sell or rent. [1]
  (2) Demolish the P. S. White house and relocate a significant building on the lot. Restore the exterior and interior of the relocated building, then sell or rent. [2]
  (3) Buy and demolish the residences on the northeast corner of Broad St. and Fourth St. (if empty), relocate a significant building on the lot, restore the exterior and interior of the significant building, and then sell or rent. [3]

Corner of Broad St. and Fourth St.—looking east on Broad Street

* Landscaping in front of and on the Fourth St. sides of the buildings on this corner should be visually consistent.

specific suggestions for implementation

SUGGESTIONS PERTAINING TO SPECIFIC BUILDINGS—
* Restore all exterior trim on the significant buildings, even when the trim is not original to the building. [1]

SUGGESTIONS PERTAINING TO LANDSCAPING—
* Discourage Georgia-Pacific and other owners from land-fill or tree-cutting in the east ravine. [1]

* Encourage the city to put down a walkway, as as mentioned under **Central Section—Historic District**, in front of the relocated building on the southwest corner of Broad St. and Fourth St. and on its east side. [1]

SUGGESTIONS THE MHA SHOULD MAKE TO OTHERS—
* Encourage the city to install unobtrusive, well-designed, markers on or near the significant buildings at this corner. The signs should conform to the criteria outlined in area **J—Central Section—Historic District**. [2]

I

WILLIAMS STREET~ SOUTH SIDE

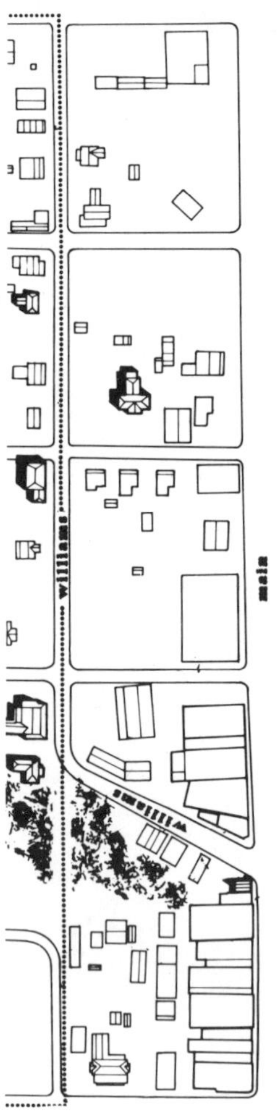

goals related to area

ADD & DELETE AS THE GENERAL GOALS OF THE MHA CHANGE (see **PROCEDURE FOR CHOOSING THE 'RIGHT' PLAN** in the previous chapter)

* Boost the community interest and civic pride
* Control change in the historic district
* Use river and ravines to full potential
* Influence restoration in other parts of town
* Involve more people in the MHA's plans
* Improve business along Main St.
* Eliminate the non-historic homes from the historic district

qualities/importance of area

ADD & DELETE AS THE QUALITIES OF THE AREA CHANGE

* Commercial activities were originally located here.
* Location one block from Main St. offers access to goods and services.
* Southern boundary of local historic district—important visually.

Williams Street

* Residential use on south side of Williams St. important to ease transition from commercial area to residential.
* The street contains some of the oldest buildings in Murfreesboro.

Peter Williams house and Murfree Law Office

* Williams St. extension is a possible formal entrance to the area.

* Two dead ends means little vehicular traffic.

* The narrow setback of the buildings along Williams St. is conducive to public/commercial activity.

Rea Store—Williams St.

* Much pedestrian activity along street.

* Easy assess to ravine from street.

* Daniel's Nursery increases the amount of plant life in the CBD.

* Commercial–residential transition gives diversity to street.

____

____

____

____

____

____

____

____

____

____

____

____

____

____

choosing the 'right' plan

EVERYTIME IMPLEMENTATION OF ANY **GENERAL** OR **SPECIFIC SUGGESTION** IS CONSIDERED UNDER ANY OF THE FOLLOWING **ALTERNATIVE PLANS**, GO THROUGH **PROCEDURE FOR CHOOSING THE 'RIGHT' PLAN** IN THE PREVIOUS CHAPTER

implementing the 'right' plan

IF THE **MHA** HAS DIFFERENT, BETTER, OR MORE EXACT **SUGGESTIONS** FOR IMPLEMENTING ANY OF THE FOLLOWING **ALTERNATIVE PLANS**, GO THROUGH THE **DESIGN GUIDELINES** IN THE FOLLOWING CHAPTER TO DETERMINE IF THE **SUGGESTIONS** ARE ACCEPTABLE

alternative plan 1

SCRATCH OUT ALL **SUGGESTIONS** THAT ARE NO LONGER PERTINENT OR HAVE ALREADY BEEN IMPLEMENTED

policy

The historic district should be expanded to include the half-block to the south of Williams St. between Sycamore St. and Second St.

effect

Will promote unified development for both sides of Williams St.

general suggestions for implementation

* Ask the planning board to expand the existing historic district to include [1]:
  –all residences on Fourth St. south of Williams St.
  –all properties north of the alley that connects Sycamore St. with the Williams St. extension.
  –the lot south of Williams St. between Fifth St. and the Williams St. extension.
  –all property between Williams St. and Main St., Fifth St. and Fourth St. except lot on the southeast corner of the block.
  –Rea house and Hardee's Restaurant lots.
  –all property between Williams St. and Main St., Third St. and Second St. except lot on the southwest corner of the block and the lot directly east of that one.

* Ask the planning board to prepare, or have the MHA prepare, a report on the quality of the area to be added. [1]

* Encourage owners in area to attend public hearing(s) concerning the proposal to zone the area historic district. [1]

After historic district zoning has been secured:

* Encourage the planning board to adopt the practice of negotiating with developers, and other property owners, who request zoning changes on property outside the areas covered in this report. If the developer owns property within the Williams St.–South Side area, and will agree to grant to the MHA or to the town a preservation easement on this property–or will agree to relocate an obnoxious use to a site outside the area covered in this report--his zoning change application should be accepted. [1]

* Encourage the city to sell general obligation bonds to underwrite a rehabilitation loan and grant program for this area. If unsuccessful, encourage the city to recapture surplus revenues from increased property valuations in the central section of the historic district and reinvest in rehabilitation of residences on the south side of Williams St. [1]

* Encourage elderly owners of significant buildings and land to grant a preservation easement to the MHA in order to avoid the Federal Estate Tax. [1]

* Encourage the city to apply for federal monies for rehabilitation of the three residences on the west side of Fourth St.—south of Williams St. [1]

* Encourage rehabilitation of the commercial businesses on the Williams St. extension. [1]

specific suggestions for implementation

SUGGESTIONS PERTAINING TO SPECIFIC BUILDINGS—

* Encourage Rea house owner to keep building in present location. [1]

SUGGESTIONS PERTAINING TO LANDSCAPING—

* Encourage owner of lot between Fifth St. and Williams St. extension to store building materials elsewhere. [1] If successful, buy and use for relocating one (or two if small) significant building(s), or encourage the town to buy and use as public, open space. [2]
  EFFECT—open space will make more of the historic district visible from Main St. Building(s) would make corner busier—more conducive to commercial activity.

* Encourage owner of Daniel's Nursery to expand business. Encourage elimination of the storage buildings just south of the south property line of nursery. [1]
  EFFECT—will bring plant life further into the CBD. Plant life is desirable for separating different land uses.

* Under the powers granted to the Historic District Commission, deny a curb cut on Williams St. for Hardee's Restaurant. Encourage owner to plant a large barrier hedge. [1]

* Encourage Belk-Tyler to plant a large barrier hedge on Williams St. and Fifth St. between the sidewalk and the parking lot. [1]
  EFFECT—will help define the boundaries of the historic district.

Williams St.

* Encourage owner of commercial enterprise on the northwest corner of Second St. and Main St. to plant a large barrier hedge on Williams St. on the street side of existing fence. [2]

* Encourage the city to plant trees along the south side of Williams St. at irregular intervals. [2]

SUGGESTIONS THE MHA SHOULD MAKE TO OTHERS—

* Encourage owner of residence directly east of Belk-Tyler to keep building residential. [1]
  EFFECT—eases transition.

* Encourage owners of residence on the southeast corner of Third St. and Williams St. and residence directly east to keep buildings in residential use. [1]
  EFFECTS—keeps east end of Williams St. predominately residential. This increases the probability that the inhabitants of the residences on the north side of Williams St. will remain.

alternative plan 2

SCRATCH OUT ALL SUGGESTIONS THAT ARE NO LONGER PERTINENT OR HAVE ALREADY BEEN IMPLEMENTED

policy

The half-block to the south of Williams St., between Sycamore St. and Second St., should be zoned C-3 (Commercial-Office District)

effect

Will keep commercial development near the CBD. Will make the historic district a more convenient place to live.

general suggestions for implementation

* Design, setbacks and scale of all new commercial/public development should be in keeping with the design, setbacks and scale of the existing buildings within the historic district.

* Encourage relocation of significant buildings from outside of this area to within this area for commercial/public use. [1]

* Encourage demolition of all dilapidated (beyond repair) residences and development of commercial/public uses compatible with residential activities. [1]

* Specific uses of all new development should be compatible with residential activities.

* The MHA should compose a listing of all possible land uses compatible with historic district uses and activities. Encourage the zoning board to review this list and add it to the town's zoning ordinance as a guide to permitted and prohibited uses in the historic district. For example, highway commercial would not be a compatible land use. [1]

* Encourage the owners of all commercial properties adjacent to both Main St. and Williams St. to subdivide each property into two commercial lots. [1]

* Encourage owners of empty lots to develop with commercial/public uses compatible with residential activities. [1]

* Encourage rehabilitation of the commercial businesses on Williams St. extension. [1]

* If any more highway commercial, such as Hardee's Restaurant, or any other commercial/public development occurs in this area between Williams St. and Main St., Sycamore St. and Second St., the building should be oriented so that the rear, or service facilities, will not be visible from the historic district. [1]

* Discourage owners of any commercial/public development located between Williams St. and Main St., Sycamore St. and Second St., from placing parking lots adjacent to Williams St. [1]

specific suggestions for implementation

SUGGESTIONS PERTAINING TO SPECIFIC BUILDINGS—

* Consider the potential of the Rea house for commercial/public use. [2]

SUGGESTIONS THE MHA SHOULD MAKE TO OTHERS—

* Encourage owner of Daniel's Nursery to expand business. [1]

* Encourage Belk-Tyler to develop the north end of their parking lot for commercial /public use. Parking for both businesses could be located between the two. [1]

* Encourage Hardee's to orient the back of their building to the west, instead of north towards the historic district. [1] EFFECT—this will keep Hardees from having such an isolating effect on the historic district.

* Encourage owner of lot between Fifth St. and Williams St. extension to store building materials elsewhere. [1] If successful, the MHA should buy and relocate a significant building here for commercial/public use, or encourage a developer to buy and develop commercially/publicly. [2]

OTHERS—

* Investigate the possibility of a shopping center development within the two contiguous blocks bounded by Williams St. and Main St., Williams St. extension and Fourth St. [2]

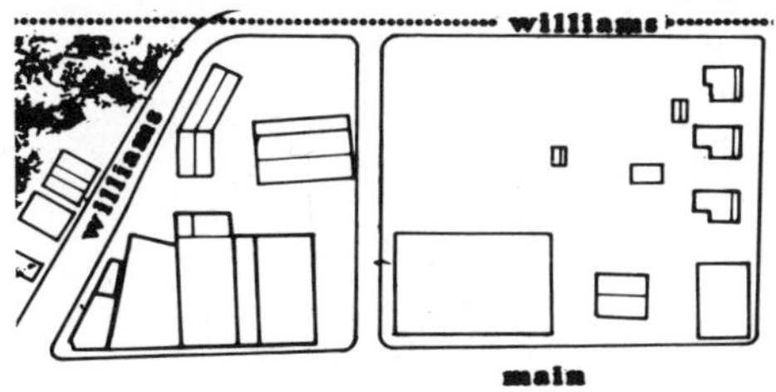

alternative plan 3

SCRATCH OUT ALL SUGGESTIONS THAT ARE NO LONGER PERTINENT OR HAVE ALREADY BEEN IMPLEMENTED

policy

All property adjacent to the south side of Williams St., between Sycamore St. and Second St., should be zoned R-12 (Residential District)

effect

Will keep Williams St. a predominately residential area where people will want to live.

general suggestions for implementation

* Ask the planning board to zone all properties adjacent to the south side of Williams St., between Sycamore St. and Second St., R-12 (Residential district). [1]

* Encourage owners of area to attend the public hearing(s) concerning the proposal to zone the area R-12 (Residential district). [1]

After area becomes residential:

* Encourage the planning board to adopt the practice of negotiating with developers, and other property owners, who request zoning changes on property outside the areas covered in this report. If the developer owns property within the Williams St.—South Side area, and will agree to grant to the MHA or to the town a preservation easement on this property—or will agree to relocate an obnoxious use to a site outside the area covered in this report—his zoning change application should be accepted. [1]

* Make a record of all buildings using the **Historic District and Buildings Survey**. [1]

* Encourage the owners of all commercial properties adjacent to both Main St. and Williams St. to subdivide each property into two lots, zoning the lot adjacent to Williams St. for residential use. [1]

* Encourage owners of empty lots on the south side of Williams St. to develop for low and moderate income residency. [1]

* Encourage the owners of all commercial buildings on Main St. which are visible from Williams St. to plant a large barrier hedge on the Williams St. side of the buildings. [1]

J

CENTRAL SECTION~ HISTORIC DISTRICT

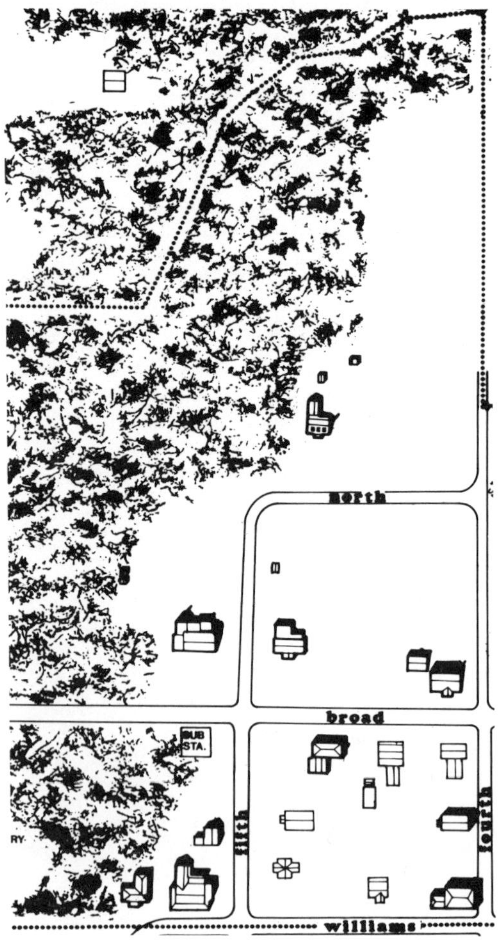

goals related to area

ADD & DELETE AS THE GENERAL GOALS OF THE **MHA** CHANGE (see **PROCEDURE FOR CHOOSING THE 'RIGHT' PLAN** in the previous chapter)

* Preserve architecturally and historically important buildings
* Boost the community interest and civic pride
* Control change in the historic district
* Use river and ravines to full potential
* Restore many homes to period authenticity
* Involve more people in the MHA's plans
* Improve business along Main St.
* Eliminate the non-historic homes from the historic district

____

____

____

____

____

____

____

____

____

____

____

____

qualities/importance of area

ADD & DELETE AS THE QUALITIES OF THE AREA CHANGE

* Highest density of historic buildings in Murfreesboro.
* Largest concentration of Federal style architecture.
* Oldest buildings in Murfreesboro.
* Rural character still evident.

Area north of Wheeler house

* Rural-town interface.
* Wide range of architectural styles represented.
* Largest concentration of MHA-owned land.
* Borders on largest ravine.
* Part of original town as laid out by William H. Murfree in 1787.
* Largest concentration of historic brick structures in Murfreesboro.
* Thompson house yard includes a large amount of open space, suitable for a wide range of uses.

* Many architecturally significant buildings.

Wheeler house

* Distinct residential character.
* One block from the CBD.
* Tourist attraction.
* Very little vehicular traffic through or bounding this area.
* Offers room for relocating buildings from elsewhere in town.
* Bounded by a lively residential area.
* One block from the City Hall and community center.
* Close to the Meherrin River.
* Large number of trees and shrubs.

Rear facade of Myrick house

* Includes an old cemetery.
* Large yard space per building.
* The significant buildings express the life styles of the early 1800's.

choosing the 'right' plan

EVERYTIME IMPLEMENTATION OF ANY **GENERAL** OR **SPECIFIC SUGGESTION** IS CONSIDERED UNDER ANY OF THE FOLLOWING **ALTERNATIVE PLANS**, GO THROUGH **PROCEDURE FOR CHOOSING THE 'RIGHT' PLAN** IN THE PREVIOUS CHAPTER

implementing the 'right' plan

IF THE **MHA** HAS DIFFERENT, BETTER, OR MORE EXACT **SUGGESTIONS** FOR IMPLEMENTING ANY OF THE FOLLOWING **ALTERNATIVE PLANS**, GO THROUGH THE **DESIGN GUIDELINES** IN THE FOLLOWING CHAPTER TO DETERMINE IF THE **SUGGESTIONS** ARE ACCEPTABLE

alternative plan 1

SCRATCH OUT ALL **SUGGESTIONS** THAT ARE NO LONGER PERTINENT OR HAVE ALREADY BEEN IMPLEMENTED

policy

A program promoting a distinct historic image for the area should be implemented.

effect

Area will have the largest density of significant structures in Murfreesboro

general suggestions for implementation

* Existing MHA funds should be used to rehabilitate/restore MHA-owned buildings presently located in the area, rather than for moving buildings into the area from other locations. Follow the priority list below for spending on rehabilitation/restoration of MHA-owned buildings in the area:
(1) restore the Ferguson house exterior [1]
(2) rehabilitate exterior of the building on the SW corner of Broad St. and Fourth St. [1]
(3) restore the Thompson house exterior [1]
(4) restore the exterior of the Winborne Law Office [1]
(5) rehabilitate interior and restore exterior of Peter Williams house [1]
(6) restore the interior of the Winborne Law Office [1]
(7) rehabilitate interior of the building on the SW corner of Broad St. and Fourth St. [1]
(8) restore exterior of Murfree Law Office [2]
(9) rehabilitate the interior of the Thompson house [2]

(10) restore the interior of the Murfree Law Office [2]

(11) rehabilitate the interior of the Ferguson house [3]

* Restore/rehabilitate all additions to the significant buildings owned by the MHA unless dilapidated or not needed for existing or intended use. [Follow the above spending priorities]

* Encourage the inhabitants of the remaining non-significant residences in this area to remain. If unsuccessful, encourage others to move in. [1] If unsuccessful again, buy the residences at reasonable cost, and rehabilitate for residential use, if possible. [2] If this cannot be done, and the lot is not needed for relocating a significant building, encourage a developer to build low or moderate income residences. [2] Such new development will be subject to the design criteria utilized by the Historic District Commission, because it will be located in the historic district. Additional criteria that should be followed by the Historic District Commission should include:
  –brick or block should not be utilized on new residential buildings.
  –garages visible from the street are not acceptable.
  –Imitations of Colonial designs and motifs should not be used.
  –long, low front facades are not acceptable.
  –see Chapter 7–**Design Guidelines**–for other criteria.

* When moving buildings into this area (or building new residential buildings), observe the following maximum densities:
  –Block between Broad St. and Williams St., Fifth St. and Fourth St.: 8 buildings
  –Block between North St. and Broad St.: 4 buildings
  –Block between Broad St. and Williams St., west of Fifth St.: 4 buildings
  –Block north of North St.: 2 buildings

* Follow these criteria for land use when relocating significant buildings (or for new construction):
  –Williams St.: commercial/public and residential mix
  –Broad St.: residential

Myrick house detail

* Encourage the city to reinvest all additional property taxes, accrued from increases in property values in the central section of the historic district, in the east end of the historic district for rehabilitation. [1]

* Encourage the city to put down a porous walkway (made of wood chips, packed sand, or gravel; no asphalt or concrete) along the west side of Fourth St. between Williams St. and Broad St. (going around the east side of the Rea Store by reducing the street width), along the east side of Fifth St. between Williams St. and Broad St., along the north side of Williams St. between Fourth St. and Fifth St., and along the south side of Broad St. between Fourth St. and Fifth St. [2]

* Encourage the city to leave Fifth St. and Fourth St., north of Broad St., and North St. unpaved. [1]

* Encourage the utility companies to put all telephone/electrical poles and wires underground. [1]

* Relocate buildings to block between North St. and Broad St. only as a last resort. Encourage preservation of the open space within this block. [1] If buildings must be relocated here, one should be located on the southwest corner of North St. and Fourth St. and the other located on the southeast corner of North St. and Fifth St. Both should face north. Such buildings should be of a design found in rural areas, not of a design found only in the town. A large structure (2 stories) would be most appropriate on the corner of Fourth St. and North St., and a small structure (1 story) most appropriate on the corner of Fifth St. and North St. Do not relocate a building on Broad St. between the Gingerbread house and the Wheeler for any reason. Do not plant any trees or hedges in this area between the two buildings.

* Discourage the city, Georgia-Pacific, and all other owners of land contiguous to the two ravines in the historic district from landfill or tree-cutting in the ravines. [1]

* Encourage continued archeological research throughout the central section of the historic district. [1]

* Encourage the town to install well-designed, unobtrusive markers on or near all significant buildings. If located on the building itself, the markers should be of an easily visible color (but not a greatly contrasting color) and should include only the street number as verbal information. All markers should be the same color, should be no more than 12 inches in any dimension, and the shape of the

marker must be simple (round, square, etc.). If located near the building, the markers should stand no farther than 12 inches above ground level, should be no more than 12 inches in any dimension, and should be located within 5 feet of a street. Again, the markers should be the same color, of a simple shape, and, in addition, of a durable material (metal, plastic, stone, etc.). [1]

specific suggestions for implementation

SUGGESTIONS PERTAINING TO SPECIFIC BUILDINGS–

* Peter Williams house–Rehabilitate as a rental efficiency apartment. Remove all additions except the earliest, to be used for kitchen/bathroom/closets. Remove most of the porch except the section between the two front windows. [1]
  EFFECT–Only minimal expenditures would be required to convert this building from residence to information center for the historic district if that ever becomes necessary.

Peter Williams house and Murfree Law Office

* Winborne Law Office–restore the first floor as a commercial museum. Encourage a private law firm to restore the second floor as a law office, utilizing all the original furniture. If unsuccessful, restore the second floor as a law office museum. Reconstruct the second floor porch on the north side, and the steps on the south side. All exterior trim should be rehabilitated/restored, even when the trim is not original to the building. [1]
  EFFECT–would be of great educational benefit since most of the original furnishings still exist.

* Building on the southwest corner of Broad St. and Fourth St.–rehabilitate as a low or moderate income residence, and either sell or rent when completed. Remove additions if not needed by prospective owner/tenant. [1]

P. S. White house

* Thompson house–Encourage the group responsible for this building to rehabilitate it as a moderate income residence and sell when completed. Rehabilitate/restore all original outbuildings. The building is not well-located for public use. If unsuccessful, encourage the group to sell to someone interested in rehabiliing it as a residence. In either case, rehabilitate/restore all exterior trim, even when the trim is not original to the building. [2]

* Murfree Law Office–encourage a private law firm to buy and restore as a law office and law library. If unsuccessful, encourage the city to buy and reuse as the town library. [2]
  EFFECT–would give the library room to enlarge their collection. Would put the law books in the same building with the public books.

* Ferguson house–Rehabiliate as a moderate or low income residence, and sell or rent when completed. Or, encourage someone to buy and rehabilitate as a moderate or low income residence. Keep addition if needed. Plant trees on the north side of building. [3]

* Encourage the utility company to remove substation at Fifth and Broad Sts. as soon as possible. If not possible, encourage the utility company to hide the substation from view with a large barrier hedge. When substation is gone, lot should be bought by MHA and utilized for relocating a large significant residential structure. Such a structure should be an example of design found in the town, not of design found only in rural Hertford County.[1]

* Captain Meredith house–Encourage owners to finish the restoration of the front facade, including the porch. [1]

* Rea Store–Erect a sign perpendicular to the street in front of the Rea Store, advertising the museum. Encourage commercial use of the one-story brick addition on the west side of the Rea Store. Encourage more public use of the existing museum. [1]

* Use the building on the northeast corner of Fifth St. and Williams St. for storage until such time when the lot is needed for relocating a large significant building (residential or commercial). Such a building should be of a design found in town, not of a rural design. Commercial use should be considered for such a building. If possible, do not remove any of the trees from this lot. [1]

SUGGESTIONS THE MHA SHOULD MAKE TO OTHERS–

* Encourage Belk-Tyler to allow visitors to the historic district to use the Belk-Tyler parking lot. Erect signs on Main St. indicating the location of the lot and directing approach from Main St. to the Fifth St. entrance. Encourage Belk-Tyler to close the Williams St. entrance to the lot. [2]
  EFFECT–this could be advantageous to Belk-Tyler's business and would keep autos off the streets of the historic district.
* Encourage owner of the Myrick house to finish restoration as soon as possible. [1]

SUGGESTIONS PERTAINING TO LANDSCAPING–

* Do not remove any trees, shrubs, or plants without first consulting the Historic District Commission and the MHA as a whole.
* Remove the hedge along North St. in front of the Thompson house. [1]
* Rea Store–Do not allow any parking between the Rea Store and the Winborne Law Office. An orchard, formal garden or informal garden are all acceptable alternatives for this area. Any plans for this area though should include a path from the west side of the Rea Store to the front of the Winborne Law Office.
* Wheeler house–If off-street parking must be provided, locate parking directly north of the main building, extending no farther west than the west facade of the main building.
* Continue using land directly east of the Thompson house for farming until such time as the land is needed for relocating a significant building. Such a building should be of a design found in rural areas, not of a design found only in the town.
* Encourage Georgia-Pacific to plant a visual barrier of fast-growing trees along the southern edge of the area where they are currently storing lumber on the Thompson property. [1]
* Discourage owner from cutting down trees on the lot in the historic district where the east ravine is adjacent to Fourth St. The trees provide an important visual barrier between the Georgia-Pacific plant and the historic district. [1]
* Clean up cemetery on the Thompson property. [2]
* If at any time the MHA owns all property in the block between Broad St. and Williams St., and Fifth St. and Fourth St. (except the Myrick house) consider putting a parking lot, if needed, in the center of the block behind all the buildings. Also, consider putting a path behind the buildings, connecting the corner of Broad St. and Fourth St., with the Rea Store. Properties within this block sold to private owners after installing such a path and/or parking lot would be sold with the added stipulation that the path and/or parking lot shall remain.
* Encourage the town to clear an 80 ft.-wide area along the north side of Broad St. in the west ravine, removing all but the very large trees or rare plants. Encourage the owners of the Captain Meredith house and the owners of the community center to support this recommendation. [2]
  EFFECT–will visually open up the central section of the historic district to the west end of the historic district. Will encourage visitors to the City Hall and community center to go into the central section of the historic district, and will encourage recreational use of the ravine.

alternative plan 2

SCRATCH OUT ALL **SUGGESTIONS** THAT ARE NO LONGER PERTINENT OR HAVE ALREADY BEEN IMPLEMENTED

policy

A program promoting a distinct small-town residential image for the area should be implemented.

effect

Will upgrade area to building code standards.

general suggestions for implementation

* All MHA-owned buildings in this area, whether significant or not, should be rehabilitated. When interior rehabilitation of a building has been completed, the building should be rented or sold for residential use. Follow this priority:
  (1) exterior of Ferguson house [1]
  (2) interior and exterior of Peter Williams house [1]
  (3) interior and exterior of the building on the SW corner of Broad St. and Fourth St. [1]
  (4) exterior of Thompson house [1]
  (5) interior of Ferguson house [1]
  (6) interior of Thompson house [2]
  (7) exterior of Winborne Law Office [2]
  (8) interior and exterior of Murfree Law Office [2]
  (9) interior of Winborne Law Office [3]
  EFFECT–will quickly increase the stock of one and two bedroom houses in Murfreesboro.

* Rehabilitate all additions to all buildings, significant or not, owned by the MHA unless dilapidated or not needed for existing or intended use. [Follow the above spending priorities]
* Discourage the city from putting down sidewalks of any kind in the central section of the historic district. [1]
* Encourage the city to pave Fifth St. and Fourth St. north of Broad St. (curbs unnecessary). [2]
* Encourage new low and moderate income residential development in vacant lots. [1] When building new residential buildings in this area, observe the following maximum densities:
  –Block between North St. and Broad St.: 7 buildings.
  –Block between Broad St. and Williams St., Fifth St. and Fourth St.: 8 buildings
  –Block between Broad St. and Williams St., west of Fifth St.: 4 buildings
  –Block north of North St.: 3 buildings
* Discourage the town, Georgia-Pacific and all other owners of land contiguous to the two ravines in the historic district from land-fill and tree-cutting in the ravines. [1]
* Encourage low and moderate income residential use for both Broad St. and Williams St. [1]
* Discourage use of signs or markers to identify significant buildings. [1]
* Set up a revolving fund, with the city's help, to rehabilitate residences within this area. [1]

specific suggestions for implementation

* Do not remove any trees, shrubs or plants without first consulting the Historic District Commission and the MHA as a whole.
* Encourage the utility company to remove the substation at Fifth and Broad Sts. as soon as possible. If not possible, encourage the utility company to hide substation from view with a large, high, barrier hedge. [1]
* Encourage Georgia-Pacific to plant a visual barrier of large, fast-growing trees along the southern edge of the area where they are currently storing lumber on the Thompson property. [1]
* Remove cemetery on the Thompson property. [1]
* Discourage owner from cutting down trees on the lot in the historic district where the east ravine is adjacent to Fourth St.

OTHERS–
* Move all antique furniture from rehabilitated significant buildings to the Rea Store Museum. [2]

K

PEDESTRIAN TRAFFIC~ HISTORIC DISTRICT

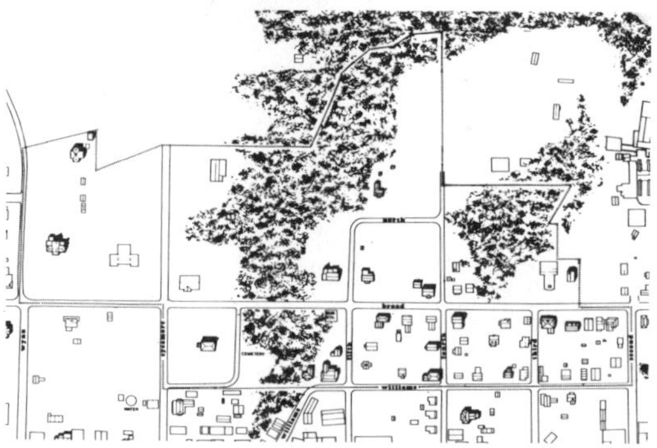

goals related to area

ADD & DELETE AS THE GENERAL GOALS OF THE MHA CHANGE (see PROCEDURE FOR CHOOSING THE 'RIGHT' PLAN in the previous chapter)

* Preserve the history of Murfreesboro
* Boost the community interest and civic pride
* Control change in the historic district
* Use the river and ravines to full potential
* Involve more people in the MHA's plans
* Improve business along Main Street
* Eliminate the non-historic homes from the historic district

qualities/importance of area

ADD & DELETE AS THE QUALITIES OF THE AREA CHANGE

* Visual connections between Main St. and the historic district exist on Williams St. extension and Fourth St.
* Presently there is a low level of vehicular traffic in the historic district with the exception of Broad St., Second St., and Wynn St., where the traffic is moderate at times.
* There is a considerably high level of pedestrian traffic on Second St. during the Georgia-Pacific lunch hour.

Broad St.–east end of the historic district

* The highest concentration of historic buildings in Murfreesboro is in the central section of the historic district.

choosing the `right´ plan

EVERYTIME IMPLEMENTATION OF ANY **GENERAL** OR **SPECIFIC SUGGESTION** IS CONSIDERED UNDER ANY OF THE FOLLOWING **ALTERNATIVE PLANS**, GO THROUGH **PROCEDURE FOR CHOOSING THE 'RIGHT' PLAN** IN THE PREVIOUS CHAPTER

implementing the `right´ plan

IF THE **MHA** HAS DIFFERENT, BETTER, OR MORE EXACT **SUGGESTIONS** FOR IMPLEMENTING ANY OF THE FOLLOWING **ALTERNATIVE PLANS**, GO THROUGH THE **DESIGN GUIDELINES** IN THE FOLLOWING CHAPTER TO DETERMINE IF THE **SUGGESTIONS** ARE ACCEPTABLE

alternative plan 1

SCRATCH OUT ALL **SUGGESTIONS** THAT ARE NO LONGER PERTINENT OR HAVE ALREADY BEEN IMPLEMENTED

policy

Clearly identify pedestrian entrances and walkways within the historic district, particularly in the central portion where most of the historic buildings are located in close proximity.

effect

This will organize and direct flow of people, separating them from vehicular traffic and making it more desirable to walk through the historic district.

general suggestions for implementation

* Initiate clean-up programs for the Williams St. extension area. [1]

* Encourage the owners of businesses along Williams St. extension to repair or restore structures to standard conditions. [1]

Buildings on the west side of the Williams St. extension

* Rehabilitate/restore the MHA-owned buildings visible from Main St., particularly the Murfree Law Office and the Peter Williams house. [2]

Williams St. extension–looking northeast

* In the central section of the historic district, make walkways which lead to all significant buildings owned by the MHA. Encourage other owners to do the same. [1]

* The sidewalk system should eventually be completed to include the whole historic district if the district is further developed similar to the central section. [2]

* Encourage the town to provide distinctive pavement and sidewalks, by using a different material such as brick, to indicate areas of historic buildings. [2]

* Encourage the city to regulate the amount and type of vehicular traffic along Broad Street. [1]

* Encourage the city or owners to cut back foliage enough to make it possible to see significant buildings from a distance when looking from one side of the district to the other: from in front of the Hertford Academy looking east, one could see the Myrick house, or the Murfree Law Office, and vise-versa. Be sure to save all valuable vegetation in the ravine. [1]

West ravine–Southall cemetery in foreground, Capt. Meredith house in background

* Begin planting programs, using donations and free labor as much as possible. The planting should emphasize connections between build ings, and between areas, while providing some continuity within small areas or street facade sequences. [2]

* Construct a footbridge across the west ravine from the Williams St. extension to the south boundary of the Hertford Academy (see **Corner of Williams St. and Fifth St.** in this chapter). [2]

* Place markers at the intersections of Sycamore St. and Main St., Williams St. extension and Main St., and Fourth St. and Main St., designating entrances to the historic district at these points. [1]

specific suggestions for implementation

SUGGESTIONS PERTAINING TO SPECIFIC BUILDINGS–

* Encourage conversion of the buildings on Williams St. into commercial use. Specific business types should be regulated to relate to the character of the historic district. [1] For further suggestions, see the area **Williams Street–South Side** in this chapter.

* Ensure visual connections between key corner structures–between Rea Store and Wheeler house, between Peter Williams house and Rea Store, between Peter Williams house and the Gingerbread and Capt. Meredith houses, etc. [1]

Broad St. looking east (Myrick house on the right)

SUGGESTIONS PERTAINING TO LANDSCAPING–

* Initiate 'planting day' in which businesses and civic groups donate money, labor and materials to improve landscaping. [2]

alternative plan 2

SCRATCH OUT ALL **SUGGESTIONS** THAT ARE NO LONGER PERTINENT OR HAVE ALREADY BEEN IMPLEMENTED

policy

A pedestrian connection between the CBD, the central section of the historic district, and any future public use area in the west ravine, should be emphasized.

effect

Will increase pedestrian activity in and around the CBD, since there will be more to do there than just shop.

general suggestions for implementation

* Initiate clean-up programs along Williams St. extension. [1]

* Encourage the owners of the buildings along Williams St. extension to rehabilitate buildings to standard conditions. [1]

* Strongly encourage any new commercial businesses to locate in vacant buildings within the CBD, or on vacant land adjacent to the CBD. [1] (see the area **Main Street Commercial–CBD & E. Main** in this chapter)

* Encourage development of the west ravine for public use, with nature trails, picnic areas, etc. [1]

* Encourage location of a public entrance to the west ravine at the southern end, near the rear of the CBD businesses. [1]

Alley behind the CBD businesses between Sycamore St. and Williams St. extension–looking west

* Use the Peter Williams house for public/commercial use to attract visitors. Emphasize an informal entrance to the historic district at this point. [1]

* Strongly encourage businesses on Main St., between Sycamore St. and Fifth St., to clean-up the rear of their buildings and properties. Encourage them to provide rear entrances to the stores in anticipation of pedestrian walkways and future development of the CBD. [2]

* Encourage the town to place markers along Main St. where it intersects with Williams St. extension and Fourth St., designating entrances to the historic district at these points. [1]

L

WEST END~ HISTORIC DISTRICT

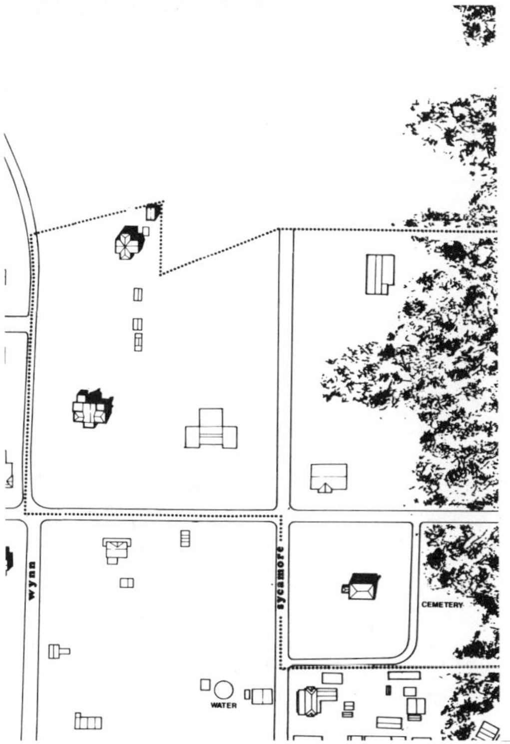

goals related to area

ADD & DELETE AS THE GENERAL GOALS OF THE **MHA** CHANGE (see **PROCEDURE FOR CHOOSING THE 'RIGHT' PLAN** in the previous chapter)

* Control change in the historic district

* Preserve the history of Murfreesboro

* Preserve architecturally and historically important buildings

* Boost the community interest and civic pride

* Increase business along Main St.

qualities/importance of area

ADD & DELETE AS THE QUALITIES OF THE AREA CHANGE

* The area has a large concentration of buildings with public activities

* There is a low building density in the area

* There is no uniformity of architectural styles in the area

* Only one building in the area is of historical value

Hertford Academy

* Hertford Academy is a reminder of the many academic institutions which have existed in Murfreesboro

* The community center is located within a new building

* The new City Hall is the center of local governmental activities

* There are large open spaces, especially behind the city hall, within the area

* All of the buildings within the area are under-used

Town Hall

* This area is located only one block away from the central commercial district

* The ravine on the east side of the Hertford Academy and the community center has potential for recreational use

* The vegetation growth in the ravine both visually and physically divides this area from the eastern part of the historic district

West ravine and Southall cemetery (Captain Meredith house in background)

* The city hall generates a continuous flow of traffic including police, fire and town maintenance vehicles

* There is a moderate amount of traffic on Broad Street and only a light amount of traffic on Sycamore Street

* There is little pedestrian activity in the area

____

choosing the 'right' plan

EVERYTIME IMPLEMENTATION OF ANY **GENERAL** OR **SPECIFIC SUGGESTION** IS CONSIDERED UNDER ANY OF THE FOLLOWING **ALTERNATIVE PLANS**, GO THROUGH **PROCEDURE FOR CHOOSING THE 'RIGHT' PLAN** IN THE PREVIOUS CHAPTER

implementing the 'right' plan

IF THE **MHA** HAS DIFFERENT, BETTER, OR MORE EXACT **SUGGESTIONS** FOR IMPLEMENTING ANY OF THE FOLLOWING **ALTERNATIVE PLANS**, GO THROUGH THE **DESIGN GUIDELINES** IN THE FOLLOWING CHAPTER TO DETERMINE IF THE **SUGGESTIONS** ARE ACCEPTABLE

alternative plan 1

SCRATCH OUT ALL **SUGGESTIONS** THAT ARE NO LONGER PERTINENT OR HAVE ALREADY BEEN IMPLEMENTED

policy

The amount of civic club activities within the Community Center and the Hertford Academy should be increased.

effect

The increased club activities will bring more people to this under-used portion of the historic district. This will help to expose the west end of the historic district to a greater number of people, and also help business along Main Street.

general suggestions for implementation

* Encourage the civic clubs which are currently using both the Community Center and the Hertford Academy to invite other community clubs to use these facilities at times when they are not presently being used. [1]

* Encourage any community club which is now using an historic building for their meeting place to relocate to one of these two structures [2]

* Consider the possibility of using the Community Center for daytime senior citizen activities. [2]

* Consider the possibility of using the Community Center for more daytime youth activities such as day care, summer day camps, or short subject courses (art, reading, crafts, etc.). [2]

* Increase the recreational activities both inside and outside the Community Center. Consider providing for small-team sports in the rear portion of the lot. [2]

* Develop nature trails in the adjacent ravine and use it as part of the expanded daytime youth program. [3]

alternative plan 2

SCRATCH OUT ALL SUGGESTIONS THAT ARE NO LONGER PERTINENT OR HAVE ALREADY BEEN IMPLEMENTED

policy

The Hertford Academy should be used for small scale commercial activities.

effect

This new commercial activity will bring more people to this underused area. This will also tend to help business along Main Street.

general suggestions for implementation

* Encourage the Murfreesboro Women's Club to lease the Hertford Academy to potential small scale commercial interests who will maintain the historical character of the building. [1]

* Have the Women's Club relocate their meeting place to the Roberts-Vaughan center, or another historical structure within the community. [2]

* Find one or more small commercial activities to locate within the Hertford Academy. A dress shop, a fabric shop, or a children's clothing store are a few of the possible activities which could draw people into the area and still maintain the historic character of the structure. [2]

Hertford Academy

* Consider the possibility of a professional–such as a lawyer, realtor, or dentist–using the Hertford Academy. This will also produce more activity in the historic district while maintaining the historical character of the structure. [2]

alternative plan 3

SCRATCH OUT ALL SUGGESTIONS THAT ARE NO LONGER PERTINENT OR HAVE ALREADY BEEN IMPLEMENTED

policy

The area should be used for residential activities.

effect

This will increase the activity within the area and help to link the western end of the historic district with the central and eastern sections.

general suggestions for implementation

* Encourage the Murfreesboro Women's Club to rehabilitate and lease the Hertford Academy as a single family residence. [2]

* Encourage the Town of Murfreesboro to relocate the existing maintenance shed on Sycramore Street to a site outside the historic district. [1]

* Consider the possibility of moving four historically or architecturally significant buildings to Sycramore Street (north of the City Hall) if the town agrees to relocate the maintenance shed to outside the historic district. Rehabilitate the buildings as single-family residences and either lease or sell them. [2]

* Consider the possibility of moving two historically or architecturally significant buildings to Sycramore Street (north of the City Hall) if the town does not agree to relocate the maintenance shed. Rehabilitate the buildings as single-family residences and either lease or sell them. [2]

M

MAIN STREET COMMERCIAL~ CBD & E. MAIN

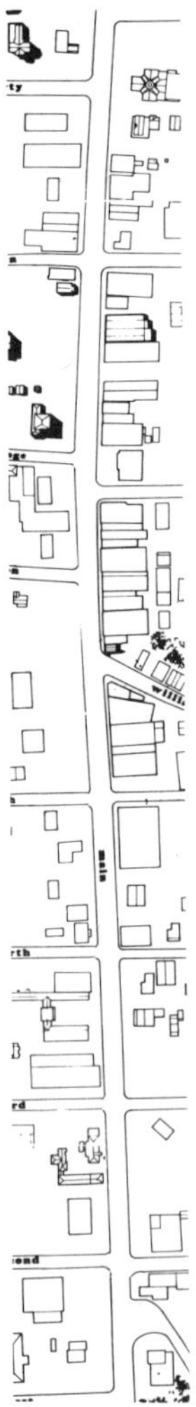

goals related to area

ADD & DELETE AS THE GENERAL GOALS OF THE MHA CHANGE (see **PROCEDURE FOR CHOOSING THE 'RIGHT' PLAN** in the previous chapter)

* Preserve the history of Murfreesboro
* Boost community interest and civic pride
* Use river and ravines to full potential
* Influence restoration in other parts of town
* Improve business along Main St.
* Involve more people in the MHA's plans
* Preserve architecturally and historically important buildings

qualities/importance of area

ADD & DELETE AS THE QUALITIES OF THE AREA CHANGE

* The area borders the southern edge of the historic district.
* The Central Business District (CBD) encourages concentrated pedestrian activity.
* The area reflects Murfreesboro's 'on-going' history.
* The area has adequate space available for new construction as well as rentable existing space.

Empty lot in the CBD

Empty building in the CBD

* There exist potential connections to any recreation area development in the west ravine.
* The area is within easy walking distance of most residential areas within Murfreesboro.
* The area is within easy walking distance of existing public activities—city hall, the community center, Hertford Academy, etc.
* The CBD is accessible to both pedestrian and vehicular traffic.

choosing the 'right' plan

EVERYTIME IMPLEMENTATION OF ANY **GENERAL** OR **SPECIFIC SUGGESTION** IS CONSIDERED UNDER ANY OF THE FOLLOWING **ALTERNATIVE PLANS**, GO THROUGH **PROCEDURE FOR CHOOSING THE 'RIGHT' PLAN** IN THE PREVIOUS CHAPTER

implementing the 'right' plan

IF THE **MHA** HAS DIFFERENT, BETTER, OR MORE EXACT **SUGGESTIONS** FOR IMPLEMENTING ANY OF THE FOLLOWING **ALTERNATIVE PLANS**, GO THROUGH THE **DESIGN GUIDELINES** IN THE FOLLOWING CHAPTER TO DETERMINE IF THE **SUGGESTIONS** ARE ACCEPTABLE

alternative plan 1

SCRATCH OUT ALL **SUGGESTIONS** THAT ARE NO LONGER PERTINENT OR HAVE ALREADY BEEN IMPLEMENTED

policy

CBD revitalization and new commercial construction between Winder St. and First St. should be promoted.

effect

Will keep commercial activities concentrated in one area. This will keep residential areas on other parts of Main St. from being destroyed by commercial development.

general suggestions for implementation

* Encourage building owners to clean up building facades. [1]
* Encourage any repair work to be done in the style authentic to the building–no imitation Colonial fronts. [1]
  Develop and distribute guidelines suggesting do's and don'ts for remodeling and repairs of the exteriors of commercial buildings (see **Remodeling Old Houses Without Destroying Their Character** by George Stephen for a good example). [1]

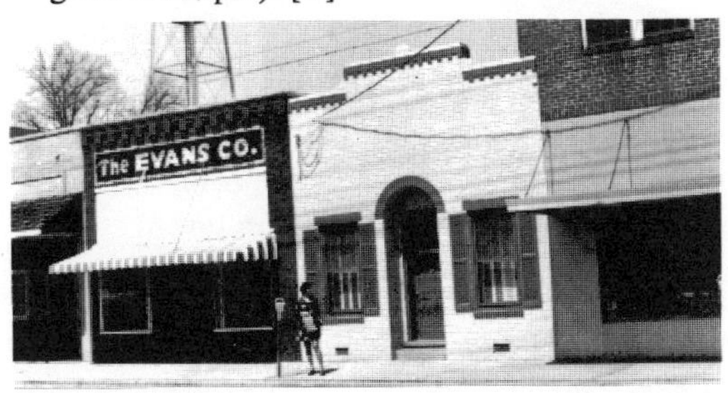

CBD colonial front

CBD–metal facade over brick

* Encourage all owner/tenant occupants of all commercial buildings to remain. If unsuccessful, encourage other businesses to move in. [1]
* Encourage those people seeking commercial space to utilize the presently vacant buildings in the CBD. [1]
* Encourage the owners of all empty lots to develop with new commercial buildings. [1]
* Encourage all owners of proposed new commercial construction to build using only designs that are:
  (1) compatible with the other buildings in the CBD, if it is to be located there, or on E. Main St., if it is to be located there.
  (2) compatible with the buildings directly adjacent to it.
  (see Chapter 7—**Design Guidelines**). [1]

CBD–inconsistent setback

* Encourage all owners of proposed new commercial construction to locate parking areas behind the building. [1]

CBD–parking in front of Fast-Fare

* Encourage the city to grant new commercial building permits only for property on Main St. between Winder St. and First St. [1] EFFECT–will support existing businesses there by increasing the number of people in that area, as opposed to spreading them down Main St.

* Encourage the city to change Article VII–Section 6.33 (C-1 central business district–Yard space) of the General Ordinances of the Town of Murfreesboro to read as follows:
  –Maximum front yard depth shall be 10 feet from the street right-of-way line; if the rear of the plot abuts a residential district, there should be a 20 foot rear yard; and side yards are not permitted.

  [1] Encourage the city to delete Section 6.34 and add same to Section 7.33–Article VII. [1]

* Encourage owners to clean up/repair the backs of their buildings as well as the facades. [1] EFFECT–this will encourage recreational use of the west ravine and encourage people to walk to/from the historic district.

* Encourage the city to utilize revenue sharing monies for revitalizing the CBD. [2]

* Encourage the city to adopt Article VIII (Conditional Uses Permitted)–Section 5 of the General Ordinance of the Town of Murfreesboro for all signs for permitted and conditional uses within the area. [1]

CBD–signs along the sidewalk

specific suggestions for implementation

SUGGESTIONS PERTAINING TO SPECIFIC BUILDINGS–

* Encourage the owner of the property from the Roberts-Vaughan center west to Wynn St. to do the following:
  –encourage the tenant currently using the brick building on the southeast corner of Main St. and Wynn St. to remain. [1]
  –donate the law office next to the brick building to the MHA. [1]
  –build (or sell lots and have someone else build) new commercial building(s) connecting to the east side of the brick building at the same height and setback as the brick building (and extending as close to the Roberts-Vaughan property as necessary).[2]
  EFFECT–would hide the Roberts-Vaughan parking lot from view, and would increase pedestrian activity at the corner of Wynn St. and Main St. and along Main St.

CBD–buildings directly west of Roberts-Vaughan house

  –sell or donate parking lot behind the brick building for Roberts-Vaughan use. [1]

* Acquire some property on Williams St. (on either the north or south side) and move the above mentioned law office, if donated to the MHA, to that location. [2] Encourage someone to rehabilitate/restore the building and use it commercially, publicly or residentially. [2] If unsuccessful, the MHA should rehabilitate/restore for public use, or to sell or rent for residential use. [3]

* Encourage demolition of the storage buildings on the north side of the alley between Sycamore St. and Williams St. extension. [2]

* Encourage owner of the Winborne house in Duncan Park to move and rehabilitate/restore the outbuildings still on Main St. If unsuccessful, encourage others interested in the outbuildings to move them. [1]

* Encourage owner of Beaman's Store at the corner of Williams St. and Main St. to rehabilitate/restore as active commercial. [2]

SUGGESTIONS PERTAINING TO LANDSCAPING–

* If any new shopping center development occurs in the block across from the City Hall, encourage inclusion of a pedestrian connection to Main St. in the middle of the block. [1]

SUGGESTIONS THE MHA SHOULD MAKE TO OTHERS–

* Encourage the CBD business owners to organize CBD-wide sidewalk sales now and then. [1]

* Encourage the town council to:
  (1) retain all parking spaces on Main St. in the CBD.
  (2) remove parking meters and install "one hour parking limit" signs.
  [1]

* Encourage the city to delete the following from the General Ordinance of the Town of Murfreesboro–Chapter 3–Section 3-1–"No person shall advertise any article of any kind for sale by crying out the same on the streets or sidewalks. No person shall advertise any article of any event or any kind by crying out the same...." [1]
  EFFECT–would make the CBD a more interesting and enjoyable place for people to congregate.

* Encourage owners of residences on Main St. between Fifth St. and Fourth St., between Fourth St. and Third St., and between Wynn St. and Winder St. to keep buildings in residential use. [1]

E. Main–Belk-Tyler and residence directly east

alternative plan 2

SCRATCH OUT ALL **SUGGESTIONS** THAT ARE NO LONGER PERTINENT OR HAVE ALREADY BEEN IMPLEMENTED

policy

All new construction or renovation should be designed in Greek Revival style or character.

effect

The Main St. commercial area will fit in with the predominant design of the remaining significant residential buildings throughout the town.

general suggestions for implementation

* Encourage building owners to clean up building facades. [1]

* Encourage any repair work to be done in Greek Revival style. [1] Develop and distribute guidelines suggesting do's and don'ts for remodeling and repairs in Greek Revival style. [1]

How a Greek Revival facade would look in the CBD

* Encourage owners to clean up/repair the back of their buildings as well. [1]

* Encourage all owner/tenant occupants of all commercial buildings to remain. If unsuccessful, encourage other businesses to move in. [1]

* Encourage those people seeking commercial space to utilize the presently vacant buildings in the CBD. [1]

* Encourage owners of empty lots to develop with new commercial buildings. [1]

* Encourage all owners of proposed new commercial construction to build using only designs that are:
  (1) compatible with the Greek Revival exteriors on the other buildings in the CBD (if it is to be located there) or on E. Main St. (if it is to be located there).
  (2) compatible with the Greek Revival exteriors of the buildings directly adjacent to it.
  (see Chapter 7–**Design Guidelines**). [1]

* Encourage all owners of proposed new commercial construction to locate parking areas behind the building. [1]

* Encourage the city to grant new commercial building permits only for property on Main St. between Winder St. and First St. [1]
  EFFECT–will support existing businesses there by increasing the number of people in that area, as opposed to spreading them down Main St.

* Encourage the city to adopt Article VIII (Conditional Uses Permitted)–Section 5 of the General Ordinances of the Town of Murfreesboro for all signs for permitted and conditional uses within the area. [1] Encourage use of signs with Greek Revival design. [1]

* Encourage the city to utilize revenue sharing funds for revitalizing the CBD in Greek Revival style. [2]

N
VANCE STREET

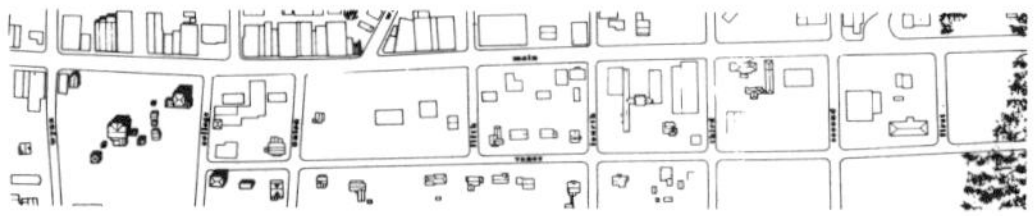

Sequence of four views down Vance St. looking east

goals related to area

ADD & DELETE AS THE GENERAL GOALS OF THE MHA CHANGE (see **PROCEDURE FOR CHOOSING THE 'RIGHT' PLAN** in the previous chapter)

* Preserve architecturally and historically important buildings
* Boost the community interest and civic pride
* Influence restoration in other parts of town
* Involve more people in the MHA's plans
* Improve business along Main St.

qualities/importance of area

ADD & DELETE AS THE QUALITIES OF THE AREA CHANGE

* Numerous turn-of-the-century buildings
* Southern boundary of original town as laid out by William H. Murfree in 1787.
* Location one block from Main St. offers easy access to goods and services.
* Originally was the main entrance to the town from the east.
* Residential area on north side of Vance St. is important to ease transition from commercial to residential areas.

Vance Street--north side

* Little vehicular traffic
* The medium setback of the buildings to Vance St. is conducive to residential activities
* The small scale of the residential structures is conducive to neighborly interaction.
* Numerous empty lots offer potential for new residential development

* Most existing residences are in standard structural condition.

* Two blocks to local historic district

* Western end of Vance St. is at a major north-south route—the route that runs directly south to Chowan College and north to the city hall.

* Bordered on the south by the High St. residential area

* Includes two pre-Civil War buildings between College St. and Union St., on the south side

Masonic Hall

Hope House

* Most residences are well-landscaped.

* Buildings less dense than those in east end of historic district.

choosing the 'right' plan

EVERYTIME IMPLEMENTATION OF ANY **GENERAL** OR **SPECIFIC SUGGESTION** IS CONSIDERED UNDER ANY OF THE FOLLOWING **ALTERNATIVE PLANS**, GO THROUGH **PROCEDURE FOR CHOOSING THE 'RIGHT' PLAN** IN THE PREVIOUS CHAPTER

implementing the 'right' plan

IF THE **MHA** HAS DIFFERENT, BETTER, OR MORE EXACT **SUGGESTIONS** FOR IMPLEMENTING ANY OF THE FOLLOWING **ALTERNATIVE PLANS**, GO THROUGH THE **DESIGN GUIDELINES** IN THE FOLLOWING CHAPTER TO DETERMINE IF THE **SUGGESTIONS** ARE ACCEPTABLE

alternative plan 1

SCRATCH OUT ALL **SUGGESTIONS** THAT ARE NO LONGER PERTINENT OR HAVE ALREADY BEEN IMPLEMENTED

policy

All properties adjacent to Vance St. between College St. and the ravine east of First St. should be zoned R-12 (Residential District).

effect

Will keep Vance St. a residential area where people will want to live.

general suggestions for implementation

* Ask the Planning Board to zone all properties touching Vance St. between College St. and the ravine east of First St. R-12 (Residential District). [1]

* Encourage owners in area to attend the public hearing(s) concerning the proposal to zone the area R-12 (Residential District). [1]

* Encourage the owners of all commercial properties adjacent to both Main St. and Vance St. to subdivide each property into two lots, zoning the lot adjacent to Vance St. residential. [1]

* Encourage owners of empty lots on Vance St. to develop with moderate income residences. [1]

Vance Street—north side

* Make a record of all buildings using the **Historic District and Buildings Survey**. [1]

* Encourage the owner(s) of empty lots on the south side of Main St. between Union St. and Fifth St. to develop only for commercial use, with the same setback, height, and scale as those commercial buildings on the north side of Main St. between Sycamore St. and Fifth St. [1]

* Develop and distribute guidelines suggesting do's and don'ts for remodeling and repairs of exteriors (see the publication **Revitalizing Older Houses in Charlestown** by George Stephen for a good example). [1]

* Encourage the planning board to adopt the practice of negotiating with developers, and other property owners, who request zoning changes on property outside the areas covered in this report. If the developer owns property within the Vance St. area, and will agree to grant to the MHA or to the town a preservation easement on this property--or will agree to relocate an obnoxious use to a site outside the area covered in this report--his zoning change application should be accepted. [1]

* Encourage the City to install unobtrusive, well designed markers on the corners of Vance St. and College St., Vance St. and Union St., and Vance St. and Fourth St. to indicate the southern boundary of the original town as laid out by William H. Murfre [3]

* The MHA should compose a listing of all possible land uses compatible with residential use and activity. Encourage the zoning board to review this list and add it to the town's zoning ordinance as a guide to permitted and prohibited uses in a residential area. For instance, highway commercial would not be a compatible use. [1]

specific suggestions for implementation

SUGGESTIONS PERTAINING TO SPECIFIC BUILDINGS—

* Encourage owner of the Winborne house in Duncan Park to move and rehabilitate/restore outbuildings still on Main St. If unsuccessful, encourage others interested in the outbuildings to move them. [1]

SUGGESTIONS PERTAINING TO LANDSCAPING—

* Encourage the owners of all commercial buildings visible from Vance St. to plant a large barrier hedge on the Vance St. side of the buildings. [2]
* Encourage the City to plant trees along Vance St. at irregular intervals. [3]

OTHERS—

* Consider the effects of a pedestrian path along the south property line of the Rose Bower house, connecting to the Roberts-Vaughan ty. property. [3]

alternative plan 2

SCRATCH OUT ALL **SUGGESTIONS** THAT ARE NO LONGER PERTINENT OR HAVE ALREADY BEEN IMPLEMENTED

policy

All properties adjacent to Vance St. between College St. and the ravine east of First St. should be zoned historic district.

effect

Will protect the existing character of Vance St.

general suggestions for implementation

* Ask the Planning Board to prepare, or have the MHA prepare, a report on the quality/importance of the area. [1]

* Ask the planning board to zone as historic district all properties adjacent to Vance St. between College St. and the ravine east of First St.

* Encourage owners in area to attend the public hearing(s) concerning the proposal to zone the area historic district. [1]

* Ask the Planning Board to accept proposals from owners of property adjacent to both Main St. and Vance St. to subdivide their properties. This would allow the owners to commercially develop the part adjacent only to Main St. without historic district restriction. [1]

* Make a record of all buildings using the **Historic District and Buildings Survey**. [1]

* Consider buying empty lots for relocation of significant buildings. [2]

O
RAVINES/RIVER

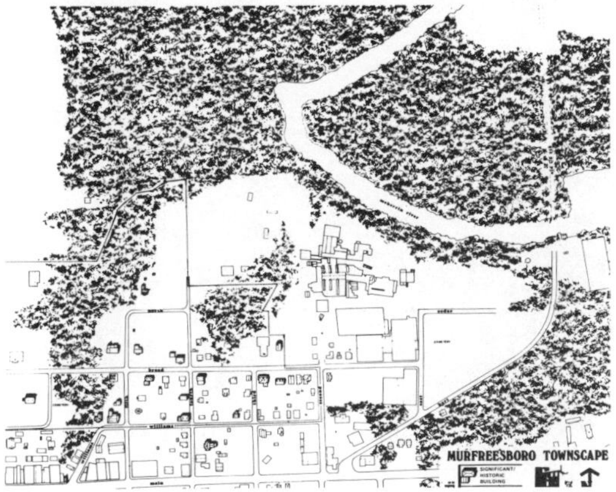

goals related to area

ADD & DELETE AS THE GENERAL GOALS OF THE MHA CHANGE (see **PROCEDURE FOR CHOOSING THE 'RIGHT' PLAN** in the previous chapter)

* Control change in the historic district
* Preserve the history of Murfreesboro
* Boost the community interest and civic pride
* Preserve architecturally and historically important buildings
* Influence restoration in other parts of town
* Use river and ravines to full potential
* Improve business along Main St.
* Involve more people in the MHA's plans

qualities/importance of area

ADD & DELETE AS THE QUALITIES OF THE AREA CHANGE

* The Meherrin River is the dominant physical feature of the area.

Meherrin River

Meherrin River–looking west (Georgia-Pacific in background)

* The River has a strong potential for recreational use.
* The River is a valuable site for archeological research, near its banks and underwater.
* The River is close to the historic district, so the potential exists for a physical and visual connection.

* The River was a key factor in the history and development of Murfreesboro, beginning with the establishment of the King's Landing in 1740.

Adams Floating Theater at Murfreesboro's Wharf in 1920

Old Murfreesboro Landing

Meherrin River covered bridge at Murfreesboro in 1870

* The Georgia-Pacific industrial plant is not visible from the River, due to the thick trees on the river bank.

* The potential exists for reconstruction or recreational use of the King's Landing.

King's Landing

* The ravines visually block Georgia-Pacific from view in the central section and west end of the historic district.

* The west ravine (between the Hertford Academy and the Murfree Law Office) offers unobstructed access to the River.

* Undeveloped historic landing sites further up the River from the King's Landing offer an excellent opportunity for archeological research, restoration/reconstruction programs, and/or recreational use.

* The west ravine extends from the central part of town (the CBD) to the River.

* The two ravines in the historic district had a strong influence on the physical development of Murfreesboro.

* The ravines contain a number of unusual and rare plants.

* The west ravine is adjacent to existing public activities located in the Town Hall, Hertford Academy, and the community center. The likelihood of a shopping center one block away and adjacent public use of the Murfree Law Office and Peter Williams house puts the west ravine itself in an excellent location for public use.

* The west ravine is also adjacent to existing commercial activities, located on Main St.

* The west ravine divides the historic district into two parts.

* The ravines are very deep, and are quite an unusual topographic feature for eastern N. C.

Old Broad Street bridge in 1890

* The two ravines give the area between them north of Broad St. its distinct rural character.

choosing the `right´ plan

EVERYTIME IMPLEMENTATION OF ANY **GENERAL** OR **SPECIFIC SUGGESTION** IS CONSIDERED UNDER ANY OF THE FOLLOWING **ALTERNATIVE PLANS**, GO THROUGH **PROCEDURE FOR CHOOSING THE 'RIGHT' PLAN** IN THE PREVIOUS CHAPTER

implementing the `right´ plan

IF THE **MHA** HAS DIFFERENT, BETTER, OR MORE EXACT **SUGGESTIONS** FOR IMPLEMENTING ANY OF THE FOLLOWING **ALTERNATIVE PLANS**, GO THROUGH THE **DESIGN GUIDELINES** IN THE FOLLOWING CHAPTER TO DETERMINE IF THE **SUGGESTIONS** ARE ACCEPTABLE

alternative plan 1

SCRATCH OUT ALL **SUGGESTIONS** THAT ARE NO LONGER PERTINENT OR HAVE ALREADY BEEN IMPLEMENTED

policy

The west ravine should be cleaned up and used for recreational purposes, in conjunction with recreational use of the Meherrin River.

effect

Will offer Murfreesboro's residents an opportunity to experience a wilderness environment right in the middle of town; will bring more people from outside of town in order to use the river.

general suggestions for implementation

* Determine the amount of land required for recreational use, from the beginning point of the west ravine to the River; then ask that owners of this land donate it to the town. If the town is unable to secure donations, encourage them to acquire partial use rights or fee-simple ownership. [1]

* Encourage the various community organizations, such as the Women's Club, the Women's League, the Garden Club, etc., to support the MHA's efforts. [1]

* Encourage the city and/or community organizations to clean up the west ravine from its beginning point to the river. Antique bottles or other items that are found could be sold to finance the cleaning up; or in the case of volunteer community organizations, to raise money. [1]

* Encourage the city, and/or community organizations, to plot and clear a walking trail from the beginning of the ravine to the River. [1] The trail should be laid out with the following in mind:
  - all rare or unusual vegetation should be protected.
  - no large trees should be cut to make way for the trail.
  - the trail should follow that path that offers the most interesting experience to the hiker.
  - the trail should be above that elevation where the stream level peaks during flooding.
  - the natural habitats of wild animals should be protected from intrusion.
  - the trail should follow that path that offers the least danger to humans.
  - the trail should be a dirt path, and should be well-marked to minimize the danger of getting lost (for example, place small coded posts at convenient intervals, and at turns.
  - vistas of the River should be established at certain points along the trail.
  - the trail should cross back and forth from one side of the stream to the other with small foot bridges.
  - the trail should pass by any historic building or structure in the west ravine.
  - depending on how far the town and community organizations wish to develop the trail, clearings with logs for sitting might be desirable.
  - the trail should end where the stream enters the River, at a point which offers a vista of both straight stretches of the River; or alternatively, at the location of an historic landing site.
  - the trail should follow a path which allows no view of the Georgia-Pacific plant.
  - access paths to the trail should be at the following points—at the southern end (from the alley), next to the Murfree Law Office (from both sides of the ravine), and next to the community center on Broad St. (from the west side of the ravine).

West ravine near the Meherrin River

* Encourage the town to look into the feasibility and desirability of establishing the west ravine as a town park, once the required land has been acquired. [2]

* Encourage the town, and/or private owners, to consider various recreational facilities at or near the point where the trail meets the Meherrin River. Possibilities include a marina, picnic area, campground, etc. Recreational facilities could also be considered for the King's Landing. [1]

* The MHA should research the location and use of the historic river landings up the river from the King's Landing. The necessary ground and underwater archeological investigations should be considered in order to better understand what existed at these landings, as well as at the King's Landing. Restoration/rehabilitation and adaptive use of any existing buildings located at these landings should be considered. The assistance of Chowan College and the N. C. Department of Cultural Resources should be secured for both the historic and archeological research. [1]

* Encourage the city to construct an east-west pedestrian bridge (wood construction) over the west ravine to connect Williams St. with the dirt road between Hertford Academy and Daniel's Nursery. Its exact placement regarding the existing vegetation and the trail to the River must be considered before construction. The cemetary next to the Hertford Academy must be protected from intensive use relating to the bridge. The bridge offers an excellent opportunity to connect commercial activities located on Williams St. with: a shopping mall in the water tower block, the businesses in the CBD, and the public activities located in the town hall and community center. A different and more intensive use of the Hertford Academy and Daniel's Nursery would make such a bridge even more valuable. The bridge should be as unobtrusive as possible from the ravine trail. [1]

Pedestrian bridge across the west ravine

* Encourage the town to clear an 80 ft. wide area along the north side of Broad St. in the west ravine, removing all but the very large trees or rare plants. Encourage the owners of the Captain Meredith House and the owners of the community center to support this recommendation. Consider the same for the south side of Broad St., once the electric substation is removed. [2]

* Encourage the city and/or owners to thin out the west ravine south of Broad St., so that Hertford Academy and the rest of the west end of the historic district is visible from the Murfree Law Office and other parts of the central section, and vice versa. [1]

West ravine (Hertford Academy and cemetery in background)

* Discourage the town, Georgia-Pacific, and others from land-fill or tree-cutting in either ravine in the historic district. This should apply to the entire length of the ravines, from their beginning point to the River. [1]

specific suggestions for implementation

SUGGESTIONS PERTAINING TO SPECIFIC BUILDINGS—

* Encourage the owners of the businesses on the west side of the Williams St. extension to fix up the rear facades of their buildings (the facades facing the west ravine). [2]

SUGGESTIONS PERTAINING TO LANDSCAPING—

* Encourage the city and/or owners to clean up the alley at the south end of the west ravine. Encourage the city to construct a walkway, either concrete or wood, to connect the corner of Main St. and Williams St. extension with the southern end of the ravine. The walkway should be located between the Beaman's Store and the building directly northeast. [1] Encourage the owners of the businesses located on Main St. between Sycamore St. and Williams St. extension to fix up the rear facades of their buildings. [1] The storage buildings directly north of the alley behind these businesses should be demolished and replaced with either a tree lined walk or expansion of Daniel's Nursery. [2]

* Discourage Georgia-Pacific from cutting down any more vegetation along the River next to their industrial plant. [1]

* Especially discourage clearing of the trees in the east ravine where they extend to Fourth St. near the corner of Fourth St. and North St. [1]

* Encourage the town to install a culvert underneath Broad St. where it crosses the west ravine. This culvert should be large enough for a person to walk through, so one can follow the trail in the ravine without having to walk up, over, and down the Broad St. fill. [1] Broad St. should have been constructed on a bridge over the west ravine.

West Ravine–looking east on Broad Street

SUGGESTIONS THE MHA SHOULD MAKE TO OTHERS–

* Encourage the community center to utilize the educational resources of the west ravine in its programs. [1]

OTHERS–

* Any future by-pass around Murfreesboro to the north should be located an adequate distance from the point where the west ravine stream empties into the Meherrin River, so that bypass traffic cannot be seen or heard at that point.

alternative plan 2

SCRATCH OUT ALL **SUGGESTIONS** THAT ARE NO LONGER PERTINENT OR HAVE ALREADY BEEN IMPLEMENTED

policy

The west ravine should be cleaned up and physical and visual connections made from one side to the other; the historic river landings should be researched as to their location and use.

effect

Will decrease the isolation of the central section and east end of the historic district from the west end.

general suggestions for implementation

* Encourage the city and/or community organizations to clean up the west ravine, from its beginning point at the alley behind the CBD buildings to a point directly west of North St. Antique bottles or other items that are found could be sold to help finance the clean-up; or in the case of volunteer community organizations, to raise money. [1]
* Encourage the various community organizations, such as the Women's Club, Women's League, Garden Club, etc., to support the MHA's efforts. [1]
* Encourage the city to construct an east-west pedestrian bridge (wood construction) to connect Williams St. with the dirt road between Hertford Academy and Daniel's Nursery. Its exact placement regarding protection of the existing vegetation, and the protection of the cemetary next to the Hertford Academy must be considered before construction. The bridge offers an excellent opportunity to connect commercial activities located on Williams St. with: a future shopping mall in the water tower block, the businesses in the CBD, and the public activities located in the town hall and community center. A different and more intensive use of the Hertford Academy and Daniel's Nursery would make such a bridge even more valuable. The bridge should be as unobtrusive as possible. [1]
* Encourage the town to clear an 80 ft. wide area along the north side of Broad St. in the west ravine, removing all but the very large trees and rare plants. Encourage the owners of the Captain Meredith house and the owners of the community center to support this recommendation. Consider the same for the south side of Broad St. once the electric substation is removed. [1]
* Encourage the city and/or owners to thin out the west ravine south of Broad St., so that Hertford Academy and the rest of the west end of the historic district is visible from the Murfree Law Office and other parts of the central section and vice versa. [1]
* Discourage the city, Georgia-Pacific, and anyothers from land-fill or tree-cutting in either ravine in the historic district. This should apply to the entire length of the ravines, from their beginning point to the River. [1]
* The MHA should research the location and use of the historic river landings up the river from the King's Landing. The necessary ground and underwater archeological investigations should be considered in order to better understand what existed at these landings, as well as at the King's Landing. Restoration/rehabilitation and adaptive use of and existing buildings located at these landings should be considered.

maintaining visual qualities

An important consideration in the development of an area is its overall visual cohesiveness. The way in which a street facade sequence, a block, a micro-neighborhood, or an intersection reads is entirely dependent on the way its components fit together. Structures in close proximity to each other should be compatible with respect to scale, proportion, use, and so on, and should be situated on their sites in a way that will contribute to an area's unified appearance. By providing for continuity of building types, setbacks and sideyards, landscaping features, architectural detail, etc., an area's character and identity can be maintained or enhanced. It is critical, then, that the MHA have some control over future development plans in order to ensure desirable relationships between existing structures and landscapes and those planned for the future.

Williams Street—east end of historic district

The purpose of this section is to offer a set of guidelines that can be used in evaluating the degree to which a relocated, or newly constructed building will 'fit' into a specific site. The guidelines should be used in conjuction with the **general** or **specific suggestions** for any of the target areas/issues described in the previous chapter. They can also be used for deciding the value of new suggestions not covered in the plan, and as a guide for implementing suggestions which do not conflict with the guidelines or with the general intent of the policy being followed.

how to use the guidelines

The guidelines are divided into two general areas of concern, 1) the potential 'fit' of a building to a specific area, and 2) the placement of the building on its future site. In the first step, the MHA should evaluate each structure according to how well it will fit in with existing structures in the area. The building to be moved into, or built within, the historic district should be openly discussed with regard to each of the criteria given. Determine by consensus which features of the structure in question are similar to those of existing structures adjacent to the intended site. It is necessary that the building exhibit a majority of similar features. If this is not the case, some of the options might be:

- choose a new site and reevaluate the same building for it
- alter the building to fit the original site
- choose another building for the site in question
- concentrate resources elsewhere, and wait for a more suitable building to become available
- alter the existing character of the area to be more appropriate for the new building.

If a majority of the criteria outlined can be met, the actual location of the structure on its future site can then be determined using the second set of guidelines. It should be noted that there are a variety of factors which may act in determining the relationship between setbacks/sideyards, and building itself. The final location can be determined in one of two ways: based on **the average** of all factors that pertain to the particular building and site in question; or reflection of one factor that is seen as an overriding element. A general 'rule of thumb' method for determining a proper fit can be based on the existing setbacks and scale of neighboring buildings. The building should line up with the average setback of its neighbors, and should be consistent with the existing proportion of setback to building size relationships.

Williams Street–central section of historic district

It is very important to remember, however, that these are not the only determinants for locating a building on its site. Other factors may play key roles in this decision, and should be taken into equal consideration.

determining building appropriateness

Measure the building's features against each of the following characteristics:

HEIGHT: buildings should be within 10% of the average height of existing buildings.

PROPORTION OF BUILDING: the relationship between the width and height of a building's front facade should be consistent with those of existing structures.

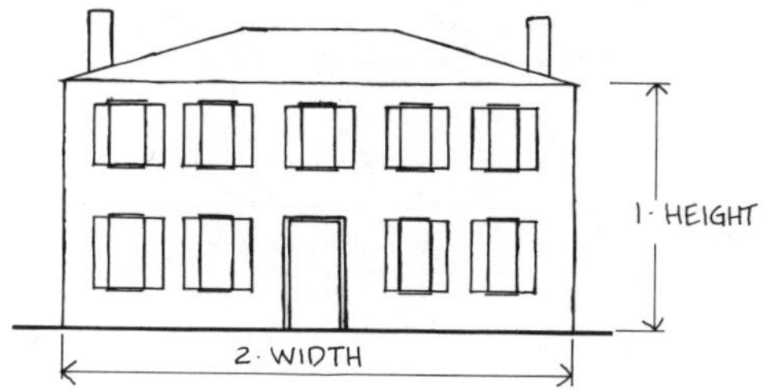

PROPORTION OF FACADE OPENINGS: The relationship of width to height for windows and doors on the building should be consistent with those of adjacent buildings.

RHYTHM OF BUILDING'S FRONT FACADE: the ordered, recurrent alternation of the masses and openings on the front elevation of the building should be consistent with those already existing.

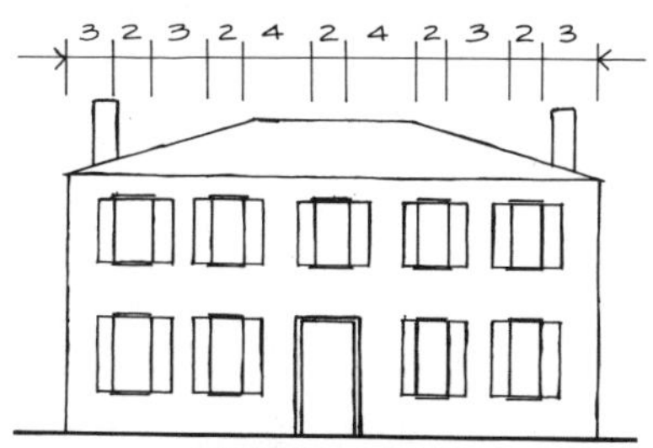

RHYTHM: 3-2-3-2-4

RHYTHM OF SPACING OF BUILDINGS: Moving past a sequence of buildings, one experiences a rhythm of recurrent building masses to spaces between them.

ROOF SHAPES: buildings should have roof shapes compatible with those already existing.

MATERIALS/TEXTURE/COLOR/DETAIL: buildings should be compatible with those characteristics predominant in the area.

Materials might include wood siding, brick, stucco, concrete.

Textures range from smooth (stucco) to rough (brick with tooled joints), to horizontal wood siding, etc.

Colors range from the materials' natural colors to the accenting or blending of painted surfaces.

Details may include cornices, lintels, other trim, chimneys, etc.

DIRECTIONAL EXPRESSION: structural shape, placement of openings and architectural details may give a predominantly vertical, horizontal, or non-directional character to the building's front facade.

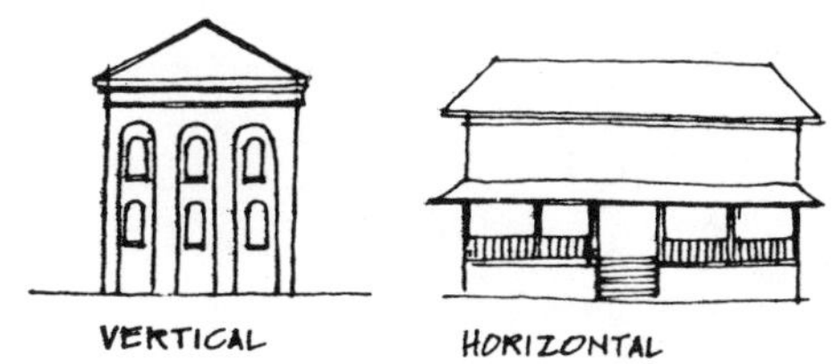

SCALE: scale is the relationship between the size of the units of construction and architectural detail to human size. Scale is equally dependent on the relationship between the size of a structure and the amount of space around it.

WALLS OF CONTINUITY: Physical ingredients such as brick walls, wrought iron fences, landscaping masses, building facades, or combinations of these, form continuous, cohesive walls of enclosure along the street.

The building characteristic guidelines above were derived from **Revitalizing Older Houses in Charlestown** by George Stephen.

setback/sideyard guidelines

These guidelines are divided into a set of criteria to determine 1) setback, and 2) sideyards. The final location should be determined by an **average** of the following, or by one **overriding** factor.

Ferguson house

determine setback of building by...

USE:

Commercial: structures should have a setback of 15 - 20 feet.

Residential: structures should have setbacks of 75 - 85 feet for large residences; 35 - 45 feet for medium size residences; 25 - 35 feet for small residences. (Measure to porch for major porches, and to building facade plane for minor porches.)

Residential (rural): should have a setback of 75 - 85 feet.

Public: structures should have a setback of 100 - 150 feet.

SCALE OF STRUCTURE:

Two-story, wood frame with major porch: should have a setback of 35 - 45 feet.

Two story, wood frame with grand facade: should have a setback of 115 feet or more.

Two story, wood frame with minor porch: should have a setback of 25 - 30 feet.

Two story, brick with major porch: should have a setback of 25 - 30 feet.

Two story, brick with minor porch: should have a setback of 25 - 30 feet.

Two story, brick with no porch: should have a setback of 10 - 15 feet.

One story, brick or wood frame: should have a setback of 25 - 30 feet.

STREET ACTIVITY:

Structures on major commercial streets: should have a setback of 65 feet for two story wood frame with major porch, 110 feet for two story wood frame with grand facade.

Structures on minor commercial streets: should have a setback of 10 - 20 feet.

Structures on major residential streets: should have a setback of 35 - 40 feet.

Structures on minor residential streets: should have a setback of 25 - 30 feet.

determine sideyards relative to...

SETBACK:

General rule: total width of sideyards, from building to building, is equal to half the setback.

Exception: building-to-building sideyard width for buildings on Main St. should be twice the setback.

LENGTH OF FRONT FACADE:

General rule: building-to-building sideyard width should be equal to the length of the front facade of the building.

Exception: building-to-building sideyard width should be equal to twice the length of the front facade of grand scale structures.

USE:

Residential: on Main St., the building-to-building sideyard width should be half the setback; elsewhere, sideyard width should be twice the setback.

Commercial: for two or more commercial structures next to each other, the sideyard width, from building to building, should be equal to the setback; elsewhere, sideyard width should be twice the setback.

Public: building-to-building sideyard width should be equal to the setback.

CURRENT GOALS OF THE MHA

1) PRESERVE THE HISTORY OF MURFREESBORO
2) PRESERVE ARCHITECTURALLY AND HISTORICALLY IMPORTANT BUILDINGS
3) BOOST THE COMMUNITY INTEREST AND CIVIC PRIDE
4) CONTROL CHANGE IN THE HISTORIC DISTRICT
5) USE RIVER AND RAVINES TO FULL POTENTIAL
6) RESTORE MANY HOMES TO PERIOD AUTHENTICITY
7) INFLUENCE RESTORATION IN OTHER PARTS OF TOWN
8) INVOLVE MORE PEOPLE IN THE MHA'S PLANS
9) IMPROVE BUSINESS ALONG MAIN ST.
10) ELIMINATE THE NON–HISTORIC HOMES FROM THE HISTORIC DISTRICT

4

Options for a Central Business District

Stephen Frary
and
David Kammeraad

Community Development Group
School of Design
North Carolina State University at Raleigh

In August of 1976 the Murfreesboro Historical Association requested that the Community Development Group of North Carolina State University continue the planning process in the historical district, begun the previous year, and begin planning in the central business district. Since Murfreesboro is a small town it is recognized that any decisions concerning one area, such as the central business district, will ultimately affect other parts of the town. Considering this interrelationship, it is important that when one group takes action to achieve its goals others are aware of these actions and their consequences. Each group should be aware of what the other is doing to eliminate conflicts and to coordinate plans so that each has the best chance to achieve their goals. This is necessary so that time, effort, and resources may be conserved, and that conflicts can be identified and eliminated.

GOALS AND STRATEGIES

With these purposes in mind, the CDG met with members of the downtown community to propose a list of goals for their consideration and revision. Also proposed was a list of possible strategies to act as a stimulus for producing suggestions from the community concerning ways to achieve the goals. In addition, it is important to realize that although the primary result from implementing the strategies should be the achievement of the goal, other results may occur. To become aware of these possibilities, the members of the downtown community were asked to predict the consequences of their strategies, to identify possible "side effects."

The method for making these choices was a group process procedure containing the following stages:

Proposed Goals

Preserve existing uses in the downtown buildings.
Promote the establishment of new businesses in the downtown.
Connect the downtown with the historic district.
Promote the establishment of a new shopping center west of town.
Divert auto traffic away from downtown.
Encourage people to walk instead of drive.
Reduce the noise level downtown.
Promote the establishment of new residences downtown.
Increase tourist related business.
Develop historic attractions downtown.
Encourage highway commercial development to the edge of town.
Control new development downtown.
Develop weekend and nighttime uses for the downtown.
Separate the central business district from the historical and residential districts.

Proposed Strategies

Widen sidewalks.
Construct new buildings on vacant lots to lease.
Develop open lots as parks.
Put parking lots behind buildings.
Declare Main Street an historical district.
Develop historical buildings behind Main Street for reuse.
Remove buildings behind Main Street stores.
Landscape Main Street.
Rent buildings in the historic district as stores.
Build new buildings behind Main Street.
Remodel existing buildings.
Develop in-town shopping center.
Develop outdoor market area.
Shield parking lots from view.
Develop rear entrances to Main Street stores facing the historical district.
Put outdoor furniture on Main Street.
Remove parking from Main Street.
Request speed-up of the by-pass project.
Open and close stores at the same time.
Control size and nature of signs downtown.
Prohibit stores in any location except downtown.
Develop off-street parking lots.
Reuse buildings downtown for retail.
Have periodic street fairs.

Instructions

1) Select three goals that you feel are important to the downtown of Murfreesboro. You may also propose your own goals. Individual sets of goals are then pooled and the four most important goals are selected by the group. You are encouraged to defend your choices until a mutually agreeable set of goals is obtained.

2) The group then chooses the three best strategies to achieve each goal. A strategy may be used more than once and you are again encouraged to propose your own strategy. One goal should be completed before the next is started.

3) After all the strategies are noted, you must try to anticipate the consequences of your actions. Hopefully, the result of the strategies will be the accomplishment of the goal. However side effects may occur. Making the downtown available for new housing, for example, may increase the traffic, or perhaps result in an increase in business. You must find two side effects for each set of strategies. The side effects may be either harmful or helpful. The list of side effects is then pooled. The group must decide which are most likely and delete the rest. Any side effect that conflicts with a goal is then noted.

Illustrations

Each strategy was illustrated to show possible before and after appearances to aid in the prediction of side effects and to define them more clearly.

Develop historical buildings on Main Street for reuse

Remove parking from Main Street

Control size and nature of signs downtown

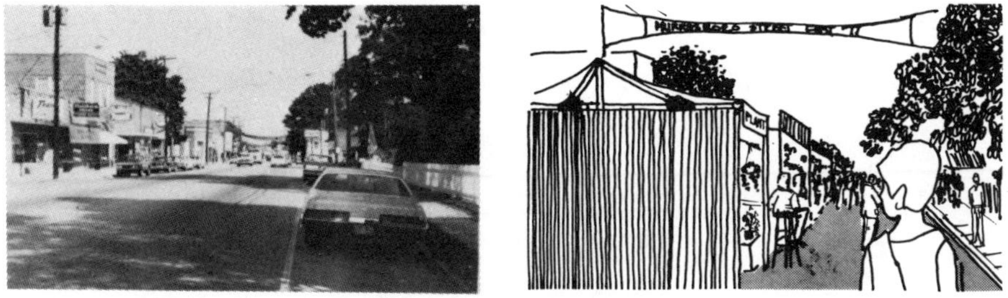

Have periodic street fairs

Rent buildings in the historical district as stores

Construct new buildings on vacant lots to lease

Shield parking lots from view

Develop area for outdoor market

Open and close stores at the same time

Reuse buildings downtown for retail

Widen sidewalks

Landscape Main Street

Build new buildings behind Main Street stores

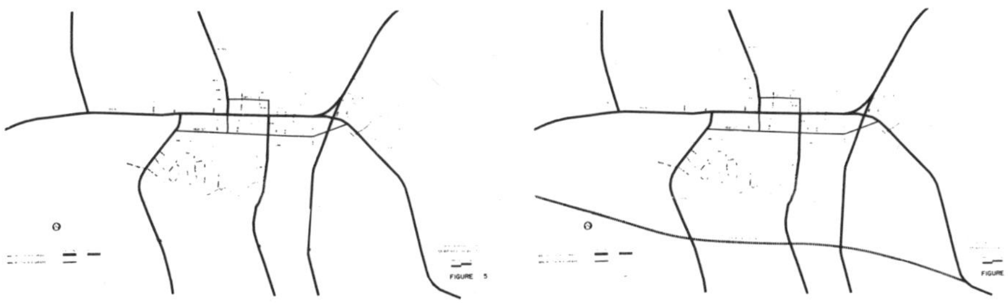

Request speed-up of bypass project

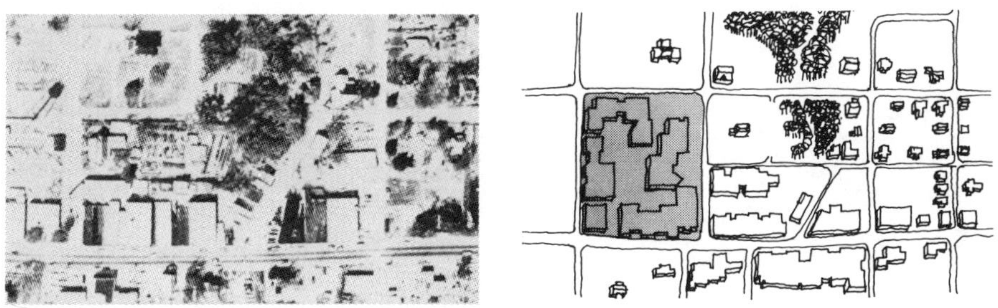

Develop in-town shopping center

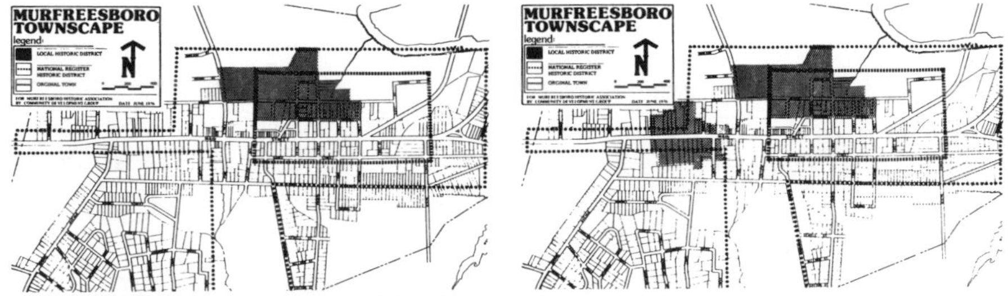

Declare Main Street an historical district

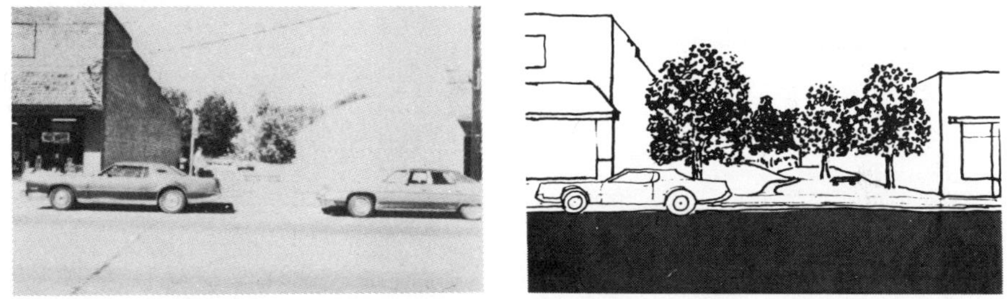

Develop open lots as parks

Put parking lots behind buildings

Develop rear entrances to Main Street stores facing historical district

The goals, strategies, and side effects most often mentioned by the group that used the process are listed below.

GOALS	STRATEGIES	SIDE EFFECTS
1 PROMOTE ESTABLISHMENT OF NEW BUSINESSES DOWNTOWN	LANDSCAPE MAIN ST DEVELOP OFF-STREET PARKING CONSTRUCT NEW BUILDINGS	LACK OF OFF STREET PARKING
2 DIVERT AUTO TRAFFIC AWAY FROM THE DOWNTOWN	REQUEST SPEED-UP OF BY-PASS DEVELOP OUTDOOR MARKETS & FAIRS DEVELOP OFF-STREET PARKING	
3 PRESERVE EXISTING USES IN DOWNTOWN BUILDINGS	DEVELOP OFF-STREET PARKING DEVELOP IN-TOWN SHOPPING CENTER LANDSCAPE MAIN ST.	COST OF ACQUIRING PROPERTY
4 CONNECT THE DOWNTOWN WITH THE HISTORIC DISTRICT	RENT BUILDINGS IN THE HISTORIC DISTRICT AS STORES RESTORE HISTORIC BUILDINGS FOR RE-USE; REMOVE UNUSABLE BLGS. PROVIDE REAR ENTRANCES TO STORES FACING HISTORIC DISTRICT	BUSINESS MAY NOT CONFORM TO HISTORIC DISTRICTS

PLANNING PROCESS

To illustrate an approach to solving a problem, we will select one goal and the strategies chosen to achieve that goal to show how a plan for action can be developed. The goal to be investigated here is "Connect the downtown with the historical district."

Area

First we identified the geographical areas affected by the goal. The selected areas, shown in the photograph, are bounded by Main Street on the south, Fourth Street on the east, Broad Street on the north, and Sycamore Street on the west. This area seems to be the most important in developing a viable connection because:

1. The area between Broad and Main Streets is where the historical and business areas presently meet.

2. The area bounded by Williams, Fourth and Sycamore Streets was identified by the MHA as the central section of the historical district with:

   A. The largest concentration of MHA land.
   B. Highest density of historic buildings.
   C. One block from the CBD.

3. This area includes the corner of Fifth and Williams, an area identified by the CDG study for the MHA as:

   A. Containing important neighborhood businesses.
   B. One block from Main Street, offering easy access to goods and services.
   C. A good informal vehicular and pedestrian entrance to the historical district from Main Street.
   D. Having great potential within the area for commercial reuse of buildings.

The area between Sycamore and Wynn Streets was not included, although it is part of the CBD, because:

1. This block has the greatest amount of developable land.

2. Connecting the downtown and the historical district is a goal to be met by this area as it will be developed, rather than as it is now.

Diagnosis

After describing the geographical area, a diagnosis must be made. A diagnosis is an evaluation of the area that identifies factors that will effect the achievement of the goal. These factors may be physical barriers to the goal or special situations that help or constrain the progress toward achievement.

The preliminary diagnosis suggests several problems that prevent the downtown and the historical district from working together. There is no attraction for

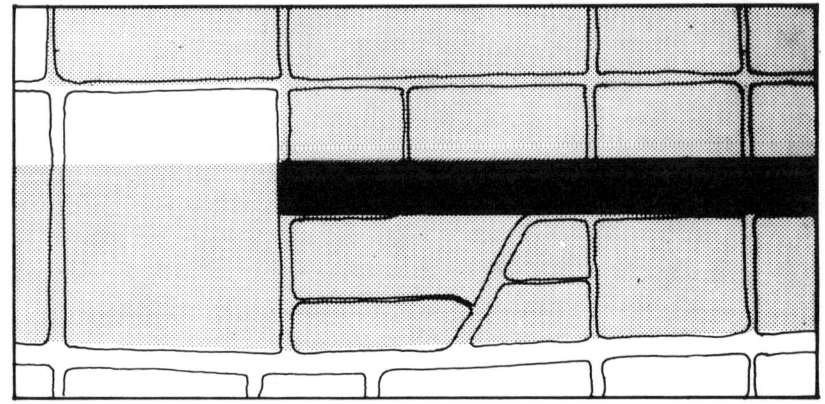

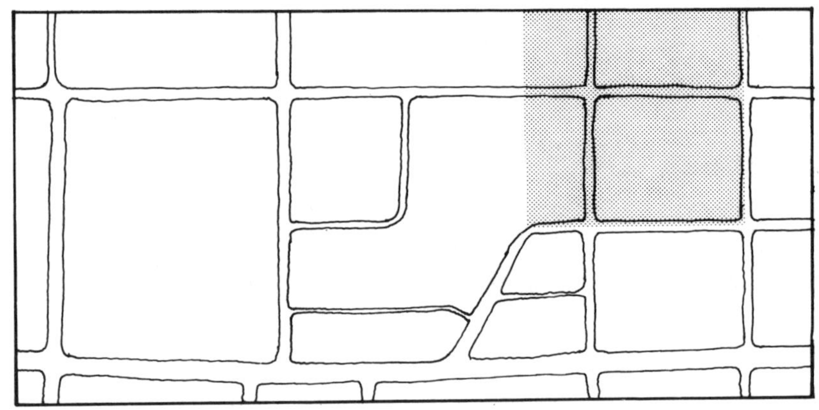

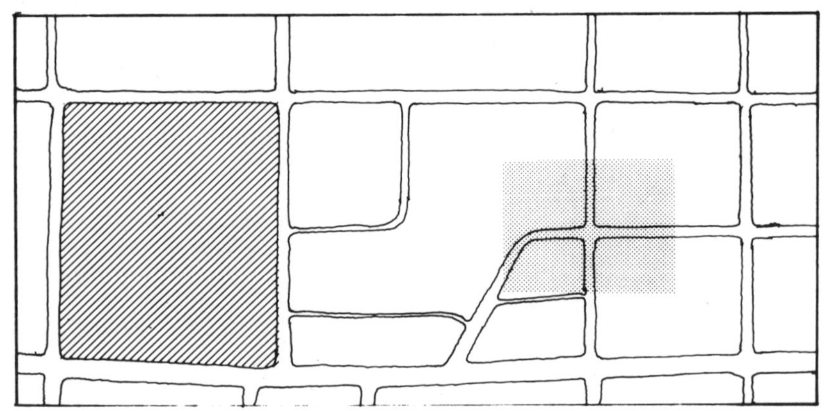

people in the downtown to enter the historical district. The historic area is used for residences and the downtown for shopping. In order for the historic area to be connected with the CBD there must first be activities in the historical district that those who are downtown may take part in, and second, there must be a pleasant way to move between the two areas. The diagnosis shown on the map is a notation of factors that either prevent the above from occurring or that could possibly aid in bringing these conditions into being.

Plan

Following the diagnosis is a plan of strategies. Each is a course of action that will either correct a problem identified in the diagnosis, or that tends to enhance a potential benefit to the area. The plan is only one of many possible combinations of strategies. When one strategy is deemed unacceptable it may be replaced with one that deals with the same issue of the diagnosis. The set of strategies is listed according to priority. Those of highest priority should be undertaken first. In establishing priorities such questions as the following are considered:

1. Will the situation deteriorate if the strategy is not implemented?
2. Can a strategy help achieve more than one goal?
3. Are there opportunities that must be taken advantage of immediately?

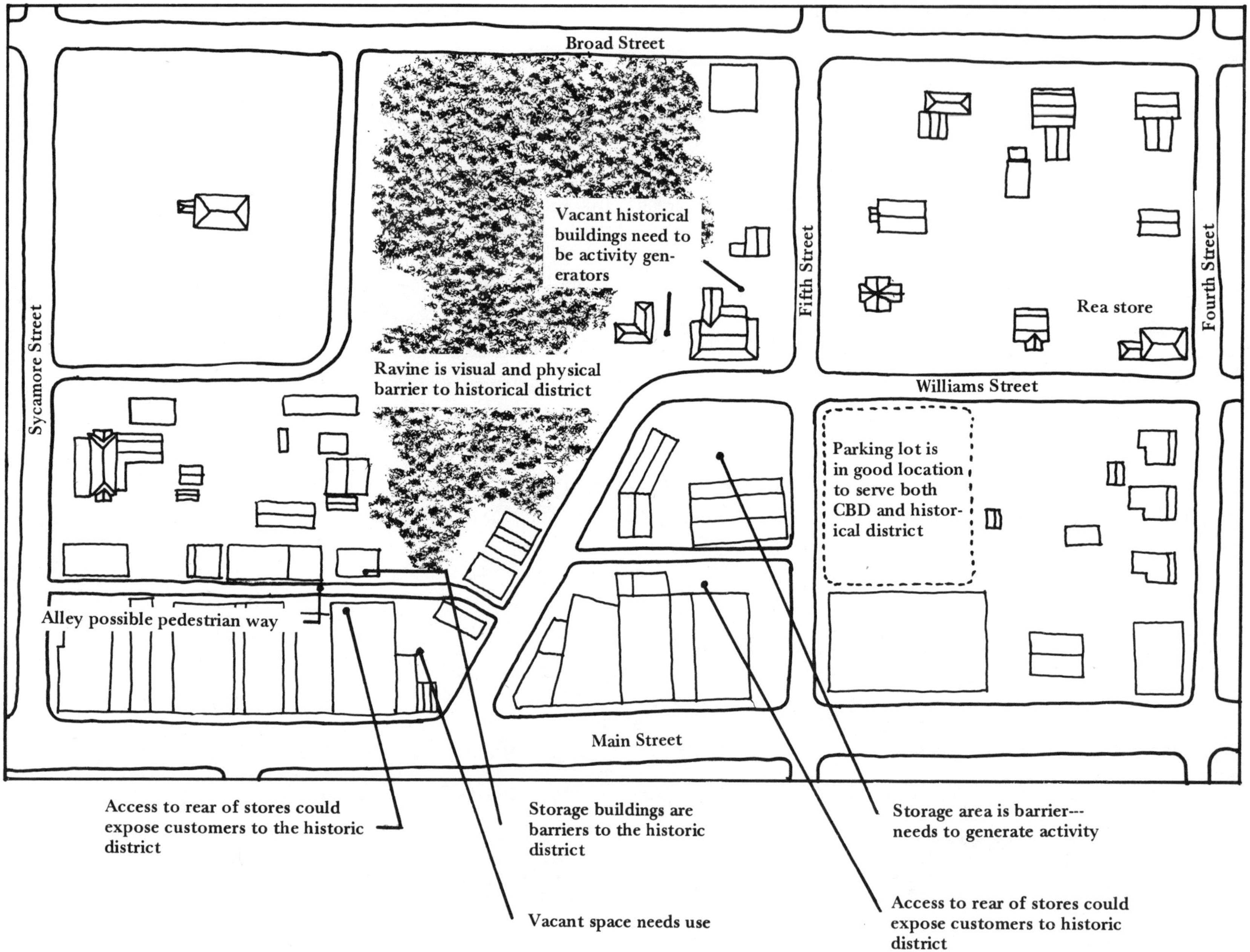

Broad Street
Sycamore Street
Fifth Street
Fourth Street
Williams Street
Main Street
Rea store
Vacant historical buildings need to be activity generators
Ravine is visual and physical barrier to historical district
Parking lot is in good location to serve both CBD and histor-ical district
Alley possible pedestrian way
Access to rear of stores could expose customers to the historic district
Storage buildings are barriers to the historic district
Storage area is barrier--- needs to generate activity
Vacant space needs use
Access to rear of stores could expose customers to historic district

Strategy

1. Use historical buildings for retail activity.

Reasons

A. The buildings at Fifth and Williams are in an important location, midway between the historical district and the CBD. If they can be reused as stores, they can attract people from the downtown closer to the historical district.

B. Reuse of these buildings will deter their physical and visual decay, and insure their maintenance.

C. When these buildings are given or sold to private individuals they will be re-entering the tax role and contribute to the community.

D. The historic appearance of the structures can be insured through deed restrictions.

Alternative

If no retail use can be found for the buildings, they should be given or sold to anyone willing to reuse them as residences or offices.

Agent

Murfreesboro Historical Association

Strategy

2. Use the building supply storage area as an outdoor produce or flea market.

Reasons

A. This location, too, is critical. At present there is no activity that can draw persons closer to the historical district.

B. Some retail activity should be found that can utilize both buildings and the space around them. The open space should not be used for parking.

C. If no retail use can be found at this time the building supplies should be stored out of sight and the weeds trimmed.

Alternative

If no retail use can be found and the owner wishes to dispose of the property, it should be purchased as a sight for historic buildings.

Agent

Murfreesboro Chamber of Commerce

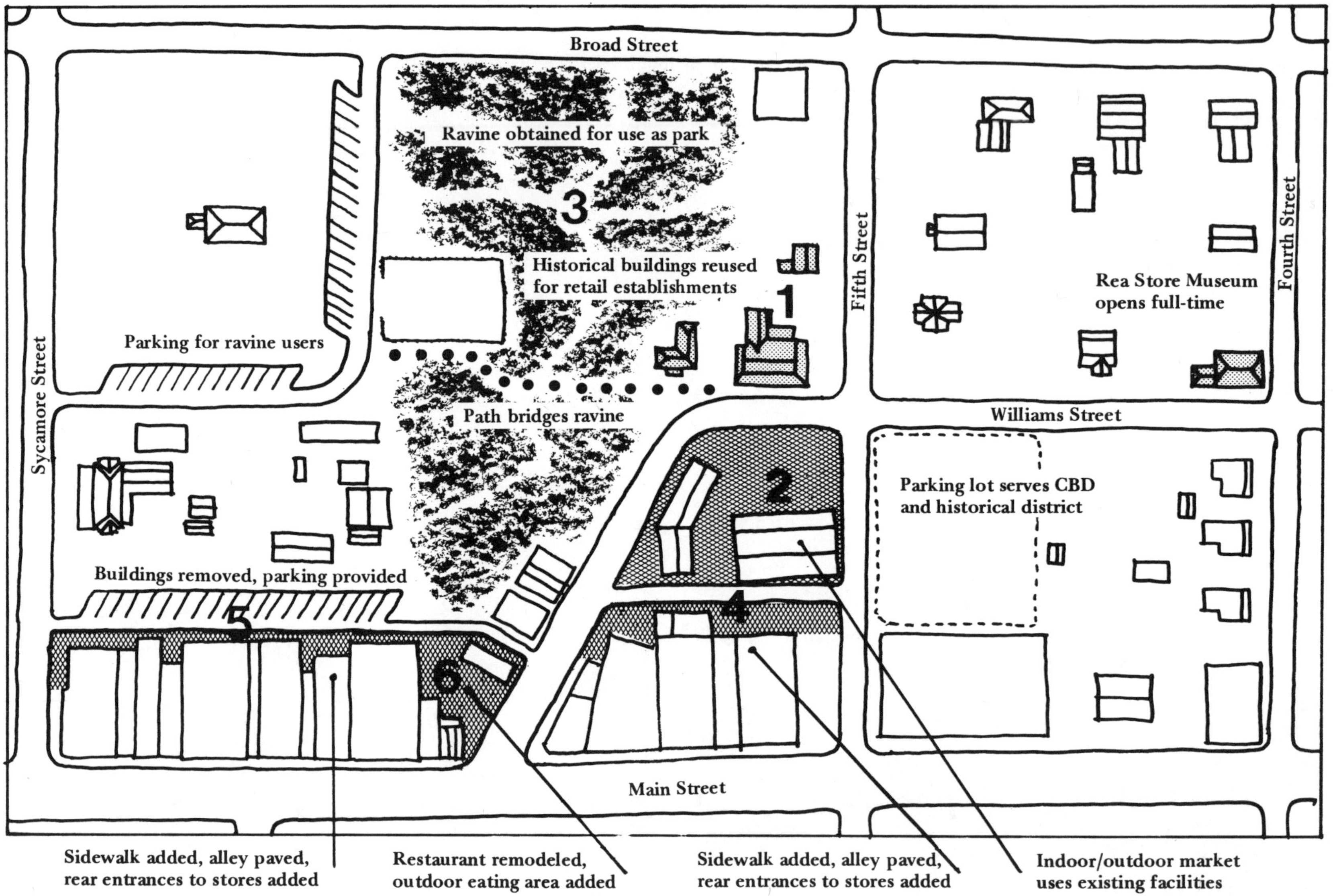
Broad Street
Ravine obtained for use as park
3
Historical buildings reused
for retail establishments
1
Fifth Street
Fourth Street
Rea Store Museum
opens full-time
Sycamore Street
Parking for ravine users
Path bridges ravine
Williams Street
2
Parking lot serves CBD
and historical district
Buildings removed, parking provided
5
4
6
Main Street
Sidewalk added, alley paved,
rear entrances to stores added
Restaurant remodeled,
outdoor eating area added
Sidewalk added, alley paved,
rear entrances to stores added
Indoor/outdoor market
uses existing facilities

Strategy

3. Obtain the ravine for use as a park, provide parking.

Reasons

A. It will provide another activity in the historical district attracting people to the area.

B. It will connect the east and west historical districts visually, and, when the ravine is bridged, physically.

C. It will preserve this natural area from development.

Alternative

If the ravine cannot be obtained, get permission to clear it and clean the litter.

Agent

City of Murfreesboro and civic clubs.

Strategy

4. Pave alleys, add sidewalks, provide rear entrances to Main street stores, from Williams Street to Fifth Street.

Reasons

A. Pedestrian traffic here will bring people closer to the historical district.

B. Pedestrian traffic here will make the Williams Street area a more lucrative business location.

C. Activity here will expose people visually to the historic area.

Alternative

Clean up the rear of buildings and add rear entrances.

Agents

Owners, tenants, and the City of Murfreesboro.

Strategy

5. Add sidewalks, pave alley, provide rear entrances to stores, remove storage buildings, add parking, from Williams Street to Sycamore Street.

Reasons

A. Removing the storage buildings will improve the appearance of the alley and remove a visual barrier to the historical district.

B. Parking here will expose people coming from the CBD to the historical district.

C. Rear entrances to stores will encourage the clean up of the alley and will be needed to serve the parking.

D. Sidewalks and paved alleys will be needed to keep the area clean and to serve pedestrians.

Alternative

If the Buildings cannot be removed, proceed with the clean up of the rear of the buildings until some activity can be found for this area.

Agent

Rear entrances—owners/tenants
Remove buildings—owners
Paving and sidewalks—City of Murfreesboro

Strategy

6. Remodel restaurant, add outdoor eating area.

Reasons

A. A restaurant in this outdoor area will bring activity "around the corner" exposing people to the historical district.

B. Outdoor amenities will improve the appearance of the corner.

C. The corner activity will draw people toward the alley activity area.

Alternative

Obtain the lot for a mini-park.

Agent

Owner

Part II

Environmental Awareness Techniques

Effective lay participation often requires some understanding of the concepts and methods by which productive decisions can be made. Although problems of housing, preservation, and day care are persistent throughout the nation, there are sufficient idiosyncratic differences in all communities to suggest the need for methods to focus on the design and planning components of these problems. Environmental awareness techniques propose to stimulate interaction between community participants from lay or professional positions as well as from residents of different communities.

Until people become conscious of the surrounding environment and understand it, thinking creatively about using or changing it is not possible. Awareness exercises encourage thinking about environmental objectives: the feelings an observer gets from looking at different kinds of space (parks, shopping centers, slums, etc.); the consideration of design criteria that might be used to decide what style of buildings citizens of downtown SERGORP (mythically a backward town) should use to fill space in a turn-of-the century residential block.

There are many approaches to sensitizing people to particular environmental issues. In the past, designers believed that they were endowed with certain abilities that permitted them to make proper aesthetic judgments that would guide society. This approach has not been successful since it required a certain amount of faith on the part of an uninformed public and more importantly it created an elite group separate from the general population. To further confound the situation, there was insufficient political support in the form of enabling legislation on guiding policies to support designers' recommendations about environmental quality.

Today, as the emphasis on citizen participation increases and becomes mandatory in many urban development programs it becomes necessary to find ways to integrate effectively user input into the design and planning process as well as ensure that user groups are not asked to participate in situations that are beyond their competence.

This section describes methods of sensitizing lay individuals and groups through the use of instructional materials designed to facilitate informed decision making. The instructional methods are oriented towards short intensive work sessions or workshops with small citizen groups that require some advance preparation in the understanding of design and planning concepts prior to their participation in local problems.

5

Workshops: An Experience in Learning

Henry Sanoff
North Carolina State University at Raleigh

The term "workshop" is used to emphasize that participants engage in those experiences that provide the material for human relations learning. The assumption is that learning is most functional when it grows out of personally involving experiences that require the processes of reflecting, and the developing and testing of new insights and approaches to problem solving. These processes come into focus when participants are obliged to reconcile their differences in pursuit of a common goal.

Workshops are an appropriate setting to achieve a high level of interaction between people sharing a common purpose. A workshop is a planned event whereby participants can learn from each other as they explore issues.

An important component in the development of a workshop is that of building cohesion. Opportunities should be provided for groups to get so involved with each other that they begin to see each other as persons, and therefore invest interest in each other.

It is important to use activities that are so powerful that they get off the cognitive level of relationships and onto the feeling level. Questions like "What are you feeling and perceiving?" and "What do you have concerns about?" help groups build a cohesive feeling.

The workshop format is the embodiment of experiential learning sometimes described as deep learning or real learning, the synthesis of raw experience and intellectual structuring. As our available structures become more complex and seemingly removed from reality, the need for appropriate experiences increases.

THE STRUCTURED EXPERIENCE

The theory of experiential learning, marked by the abiding impact of training, is a testimony to a growing maturity in the field of human relations training.

The term "structured experience" is a label for those activities that (a) focus the group process, (b)

STRUCTURED EXPERIENCE CATEGORIES

Getting Acquainted:	In order for participants to overcome initial barriers, it is valuable for them to reveal their first impressions—to present themselves as individuals.
Interpersonal Communication:	The successful functioning in groups depends upon the development of trust, learning how to listen and develop various types of relationships.
Group Problem Solving:	The ability for groups to engage in a situation requiring their ability to resolve a conflict and to provide a solution through consensus.
Organizational Development:	The ability for participants to organize themselves into work groups for production and planning as well as team building.
Awareness Expansion:	The development of sensory awareness is linked to broadening the individual's self-perception and enlarging personal boundaries.
Personal Feedback:	When participants interact, there is a giving and receiving of information through informal means as well as formal instruments.
Dyads:	Often it is necessary to subdivide a large group into pairs, partnerships, or two-person transactions for effective communications.
Competition and Cooperation:	In order to experience winning and losing, competitive and cooperative strategies are necessary for participants to engage in.

control extraneous variables, and (c) increase the probability that certain learning will occur for the participants. Although the learning goals of the structured experience can be specified and the outcomes somewhat predictable, the structured experience does not dictate what a participant should learn. Its intent is to facilitate learning that would be otherwise haphazard and diffuse.

The structured experience is a function of the objectives of the experience, the content, and the techniques employed to focus learning.

The objectives or learning goals of a structured experience may include cognitive, affective, and skill building aspects. The cognitive aspect of the experience may include an awareness of the content, incorporation, and the use of content, generalizing from the particular, and conceptual organization.

The participant's self-awareness, insight, and empathy illustrate the affective aspect of learning objectives. Development and implementation of skills such as listening and problem solving are the skill-building aspects of the objectives.

The content areas refer to individual and group events. They include methods of getting acquainted, interpersonal communication, group problem solving, sensory awareness, giving and receiving feedback, team building, and work groups.

Techniques employed to direct learning include activities such as making or building something, discussions, summarizations, board games, interviews, inventories or check lists, role playing, and tasks.

An appropriate combination of objectives, content, and techniques will produce a structure that will generate an opportunity for learning.

GETTING ACQUAINTED

In order to provide opportunities to become acquainted with other members of the group, to promote feedback and self-disclosure regarding initial perceptions, a getting acquainted activity is important.

One approach that can be employed is the interview. This method requires the interviewer to choose five nontraditional questions to ask. (Some examples are provided in figure 5-1.) Following the first interview, roles are reversed. After both interviews, each individual introduces his/her partner to the group.

> How do you feel right now?
> What turns you on the most?
> What is your favorite object?
> Where do you go to be alone?
> What do you think is very exotic?
> What value is most important to you?
> When do you feel most comfortable?
> What do you expect of me?
> When do you feel uncomfortable?
> What is difficult for you to do?
> How would you define friendship?
> If you weren't what you are, what would you be?

Figure 5-1
Examples of typical questions for an interview.

DESCRIPTIVE WORDS

Awareness, expansion, personal feedback

An initial environmental awareness task can focus on the place within which the workshop occurs and the use of verbal descriptions to expand each participant's spatial vocabulary. Since every place we inhabit can be described by its physical features as well as by personal feelings, a word list was developed to identify particular denotative and connotative spatial qualities. Participants can select five adjectives from the 200 word pairs that best describe the workshop setting. Familiarity with the setting might elicit spatial connotations related to previous activities occuring at the space while a lack of familiarity might generate judgments about the performance of the space for the intended activity.

An open discussion of the concepts can focus attention on spatial conditions that people often adapt to, as well as provide a useful vocabulary for continuous environmental exercises.

DESCRIPTIVE WORDS

How we behave in a situation depends upon what that situation means to us. Ordinarily, if we want to find out what something **means** to a person, we ask that person to tell us. Sometimes we understand the meaning but find it difficult to put into words, so we describe what **it** is like, you know what I mean.

Meaning can be very precise and descriptive; for example, Mr. Webster states that a **house** is a building to live in. Meaning can also be associative since Mr. Webster's house can be roomy, old, livable, urban and beautiful.

We believe that the environment has an important meaning for each of us although our associations about the environment may be different. Sometimes it is possible to understand an environment better if we **free-associate** or generate as many descriptive words that we can identify.

We have collected a large list of words and their opposites that are good environmental descriptors. Each pair of words can be used to describe your classroom, your house, your place of worship or any other environment you may choose. This new vocabulary can help you see and understand subtle and varied qualities about your **built environment.**

Active-Passive
ADEQUATE SIZE-INADEQUATE SIZE
Affected-Unaffected
Alive-Dead
APPEALING-UNAPPEALING
Ascending Color-Receding Color
ATTRACTIVE-UNATTRACTIVE
BEAUTIFUL-UGLY
BRIGHT-DULL
BRIGHT COLORS-MUTED COLORS
Busy-Calm
Calming-Upsetting
Changeable-Unchangeable
CHEERFUL-GLOOMY
CLEAN-DIRTY
Coarse-Smooth
COLORFUL-DRAE
COMFORTABLE-UNCOMFORTABLE
COMFORTABLE TEMPERATURE-UNCOMFORTABLE TEMPERATURE
Complete-Incomplete
COMPLEX-SIMPLE
Confused-Clear
Consonant-Dissonant
CONTEMPORARY-TRADITIONAL
Content-Discontent
CONVENIENT-INCONVENIENT
Coordinated-Uncoordinated
Cozy-Monumental
Cultured-Uncultured
Dated-Timeless
Decorated-Stark
Deep-Shallow
Defined Space-Undefined Space
Definite Volume-Indefinite Volume
Depressing-Exhilarating
DIFFUSE LIGHTING-DIRECT LIGHTING
Dignified-Undignified
Directed-Undirected
DISTINCTIVE-ORDINARY
Downward Scale-Upward Scale
DRAFTY-STUFFY
Dry-Humid
Dynamic Space-Static Space
EFFICIENT-INEFFICIENT
ELEGANT-UNADORNED
EMPTY-FULL
Encouraging-Discouraging
Euphonious-Diseuphonious
Even Texture-Uneven Texture
Exciting-Unexciting
EXPENSIVE-CHEAP
Expressive-Unexpressive
Familiar-Unfamiliar
FASHIONABLE-UNFASHIONABLE
Fatiguing-Invigorating
Feminine-Masculine
Finished-Unfinished
FLASHY COLORS-SUBDUED COLORS
Flexible-Rigid
Formal-Informal
Formed-Formless
Fragile-Sturdy
FREE SPACE-RESTRICTED SPACE
FRESH ODOR-STALE ODOR
Friendly-Unfriendly
Frilly-Tailored
FUNCTIONAL-NONFUNCTIONAL
GAY-DREARY
Gentle-Brutal
Glaring-Unglaring
Good-Bad
GOOD ACOUSTICS-POOR ACOUSTICS
GOOD COLORS-BAD COLORS
GOOD LIGHTING-POOR LIGHTING
GOOD LINES-BAD LINES
Good Odor-Bad Odor
GOOD TEMPERATURE-BAD TEMPERATURE
GOOD VENTILATION-POOR VENTILATION
Graceful-Clumsy
Hard-Soft
Hard Texture-Soft Texture

Harmonious-Discordant
Healthy-Unhealthy
Heavy-Light
Heterogeneous-Homogeneous
High-Low
Honest-Dishonest
Horizontal Volume-Vertical Volume
Hospitable-Inhospitable
Hot-Cold
HUGE-TINY
Human Scale-Inhuman Scale
Imaginative-Unimaginative
Impersonal-Personal
IMPRESSIVE-UNIMPRESSIVE
Inner-directed-Outer-directed
Inspiring-Discouraging
Interesting-Uninteresting
INVITING-REPELLING
LARGE-SMALL
Lazy-Energetic
LIGHT-DARK
Livable-Unlivable
Lively-Dull
Long-Short
Meaningful-Meaningless
Mechanical Space-Nonmechanical Space
MODERN-OLD FASHIONED
MULTIPLE PURPOSE-SINGLE PURPOSE
Mystic-Nonmystic
Natural-Artificial
NEAT-MESSY
NEW-OLD
Nice-Awful
No Odor-Strong Odor
Open-Closed
ODERLY-CHAOTIC
ORGANIZED-DISORGANIZED
ORNATE-PLAIN
Orthodox-Unorthodox
PLEASANT-UNPLEASANT
PLEASANT ODOR-UNPLEASANT
Pleasing-Annoying
Plush-Austere
Polished-Unpolished
Popular-Unpopular
Positive-Negative
Pretentious-Unpretentious
PRIVATE-PUBLIC
Progressive-Conservative
Proportional-Unproportional
QUIET-NOISY
Real-Phony
Rectilinear-Curvilinear
Refined-Unrefined
Refreshing-Wearying
Regular-Irregular
Related-Unrelated
Relaxed-Tense
Reputable-Disreputable
Reserved-Uninhibited
Resonant-Flat
Restful-Disturbing
Restrained-Unrestrained
Restricted-Unrestricted
Reverent-Irreverent
Rhythmic-Unrhythmic
Rich-Poor
Rickety-Stable
Romantic-Unromantic
ROOMY-CRAMPED
Scenic-Unscenic
Sectionalized Space-Undifferentiated Space
Secure-Insecure
Sedate-Flamboyant
Sensitive-Insensitive
Sensual-Prim
Serene-Disturbed
Serious-Humorous
Shaped-Shapeless
Sharp-Blunt
Sincere-Insincere
Sociable-Unsociable
SOFT LIGHTING-HARSH LIGHTING
Soothing-Distracting
Sophisticated-Unsophisticated
SPARKLING-DINGY
Spiritual-Nonspiritual
Stereotyped-Unstereotyped
Sterile-Filthy

6

Environmental Awareness

Henry Sanoff
Gregory Centeno
North Carolina State University at Raleigh

Sensitivity to the visual environment requires the development of a level of perceptual awareness that is relatively uncommon. Since a small portion of our everyday environment is consciously designed, the level of consumer awareness necessary to effectuate any changes is relatively low. Although it can be argued that the development of environmental awareness is the province of early education, requiring continual reinforcement, it is also necessary to recognize that decisions effecting the visual environment are made daily.

Research into techniques used to produce change have been identified as coercion, reason, emotion, and education. Change behavior involving environmental sensitivity must include a rational educational approach as well as directly engaging the personality.

The methods developed to nurture this sensitivity are included in this section. Each method has a clear educational component; a variable that contributes to visual satisfaction. In order to further nurture the change process, the variable is introduced into a small group setting where each participant would respond differently. The difference in response would be due to a difference in background or value orientation. Indirectly a conflict situation arises where each participant in the group is obliged to clarify his/her intentions and reach a consensus in seeking the best solution. This highly interactive group process relies in part upon coercion and reason for consensus and usually engages the emotions. The learning is often about other group members' perceptions of the problem as well as a heightened awareness to the initial variable under study.

WHAT WORDS DESCRIBE THESE PLACES?

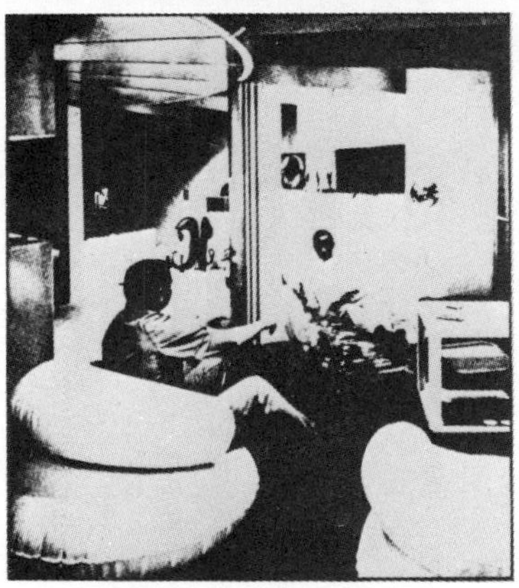

☐	RESTRICTED SPACE ● FREE SPACE	☐
☐	UNPLEASANT ● PLEASANT	☐
☐	OBSCURE ● CLEAR	☐
☐	SINGLE PURPOSE ● MULTI-PURPOSE	☐
☐	UNFINISHED ● FINISHED	☐
☐	COMMON ● UNIQUE	☐
☐	UNIMAGINATIVE ● IMAGINATIVE	☐
☐	BORING ● INTERESTING	☐
☐	HAPPY ● SAD	☐
☐	UNFRIENDLY ● FRIENDLY	☐
☐	ORDINARY ● DISTINCTIVE	☐
☐	SIMPLE ● COMPLEX	☐
☐	USUAL ● UNUSUAL	☐

KEEPS~A GAME TO OPEN UP DISCUSSION ABOUT PRESERVATION~CONSERVATION ISSUES

KEEPS is a game designed to provide groups, interested in preserving the many environmental **qualities** unique to older neighborhoods, districts and towns, with an understanding of the **strategies** open to them. Organizing and planning for the preservation of the **qualities** your group has identified as important, relies upon the consideration of:

The environmental **qualities** your group seeks to develop.

Your **goals**.

The type of **strategies** your group can realistically use to accomplish your **goals**.

To begin, each player selects from the **goal** list provided, no more than four **goals** that seem to be important in developing the environmental **qualities** your group has decided upon. When making your initial selections, brief notes should be made justifying each choice. When all the members of your group have made their **goal** choices, the individual lists are pooled.

Through negotiation the group must choose a total of four **goals**, with the additional constraint-that the four statements must be incorporated into a unified conservation program. Players are urged to forcefully support their individual choices, even if other members of the group differ. Discussion should continue until group members persuade or are persuaded to include four **goals** that reflect the groups priorities. This may require considerable discussion.

When consensus is reached the group should enter its choices on the record sheet.

Next, using the **strategies** list, each player should individually select no more than four implementation **strategies** that can be used to effectively accomplish each of the **goal** choices. Work through each **goal** completely before starting a new one. And, keep in mind that some **strategies** may relate to more than one of your **goal** choices. After all members of your group have made their **strategy** selections, pool your lists and negotiate on your final group selections. As before, players are urged to persuade the total group to include their own particular selections.

Your completed record sheet now contains the framework of a collaboratively generated conservation program. Combine the results of all the working groups and use these as a framework for future discussions and actions. Remember, the future is up to you.....and KEEPS is what we are playing for.

Goals

Preserve historically significant sites, landmarks, objects, and buildings.

More public and private involvement in decisions which could alter the character of the area.

Heightened public awareness to the area's unique physical character.

Optimal use, or re-use of sites in the area.

Preservation of the neighborhood's visual characteristics.

Influence public and private investment for the good of the area.

Neighborhood development which is compatable with the long range objectives for town development.

Public awareness to the area's historic resources.

Preservation of neighborhood social cohesiveness.

Maintenance and upgrading of properties.

Increased public participation in the development of the area.

Influence neighborhood improvement programs in other parts of town.

Control of growth and development in the area.

Strategies

Encourage property owners to increase property maintenence.

Encourage civic organizations to clean up, or maintain sites.

Offer preliminary architectural services to businesses and individuals interested in developing sites.

Encourage private planting programs.

Move some historically significant building to infill a key unoccupied site.

Encourage pedestrian activities in key areas by petitioning for walkway improvement programs.

Contact other organizations that have initiated similar projects for advice.

Have an area wide 'planting day .

Develop detailed design guidelines to maintain a consistent area image.

Organize for bulk purchase of materials.

Acquire public agency support.

Encourage the demolition of buildings that are hopelessly beyon repair.

Use local media sources to obtain issue visibility.

Look into the possibility of federal and state grants.

Lobby for zoning changes which can insure the implementation of your goals.

Purchase and restore key buildings and sites to 'period authenticity'.

Sponsor continuing area wide 'clean up day' programs.

Identify and evaluate historically significant buildings and sites.

Purchase, rehabilitate, and adaptively re-use significant buildings and sites.

Develop property easement programs and standards.

Put utilities underground.

Purchase, rehabilitate and sell.

Purchase, rehabilitate and rent.

Control of outdoor advertising.

Develop a revolving fund.

Tree planting and maintenence of publically owned property.

The above cartoon is a story about changes that have occured through the years in an imaginary town (not unlike your own). Organize into groups of three to five players. Then, as a group, discuss and list on your record sheet the qualities and characteristics that were lost through the process of change. Try to be as specific as possible; include only those qualities that your group agrees upon.

ENVIRONMENTAL QUALITIES:

RECORD SHEET

Goals	Strategies
1	
2	
3	
4	

7

House Choice

Henry Sanoff
and
Bruce Sanders

North Carolina State University at Raleigh

Current housing market mechanisms suggest that house types that sell the best continue to be produced in even greater quantity. The fallacy of this approach to the housing market is that choices must be made between available housing, and the choice of what is available may reflect a set of compromises made by consumers, because what is really desired is not reflected in the current alternatives. This housing market approach is supply oriented and rarely considers the demand side of the argument, or peoples needs.

It has often been stated that people generally do the same things but they may do them in different places. Also, housing choices often represent a compromise, and the trade-offs that people use vary considerably.

The house awareness game sets up various combinations of living and sleeping arrangements. Since each choice entails a variation in the amount of floor space required, there is a point score developed to reflect this difference. The planning variations are a result of the general categories of arrangements that were found in house plans developed by architects. Each arrangement is suggestive of a different style of family living. Although there is no attempt to identify life-style characteristics, players are involved in identifying their own life styles and preferences when they are involved in selecting an alternative.

In order to develop further the plan variations, alternatives were introduced that compare the number of floors and stylistic variations.

The game was developed for use with low-to-moderate income families considering home ownership. In all instances the players could not afford the services of an architect but did have a free planning service available to select the house of their choice to build.

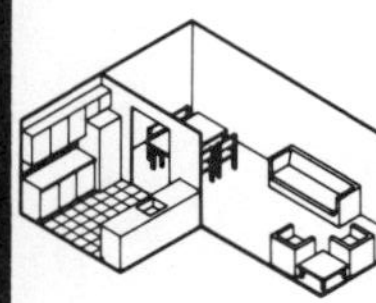

1a.
(Living-Dining) & (Kitchen) 31

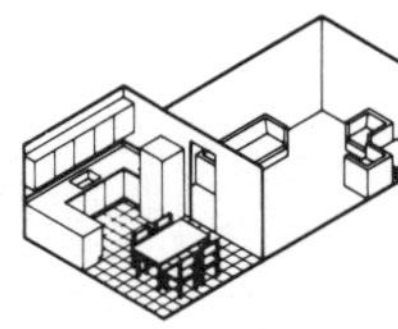

1b.
(Living) & (Dining-Kitchen) 34

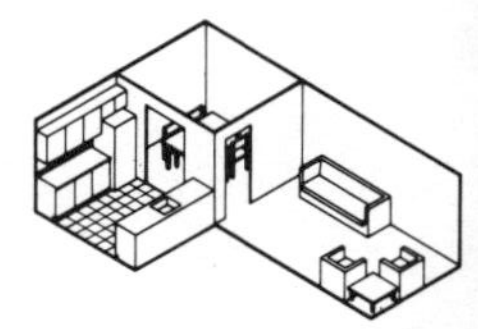

1c.
(Living) & (Dining) & (Kitchen) 36

living·dining·kitchen choices

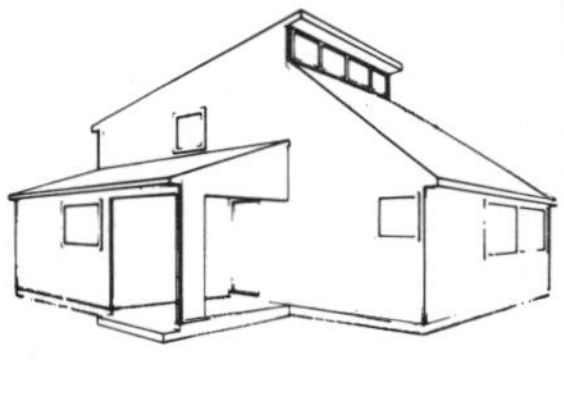

5d.
Two Story: (Exterior)

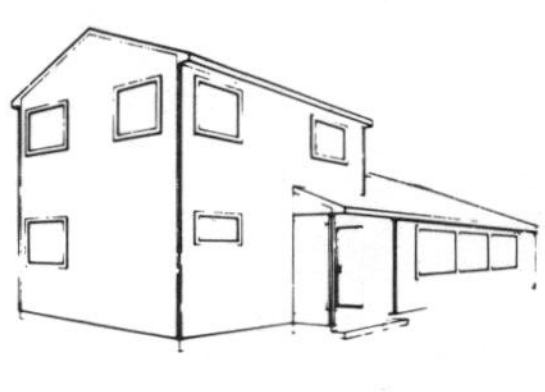

4c.
Two Story: (Exterior)

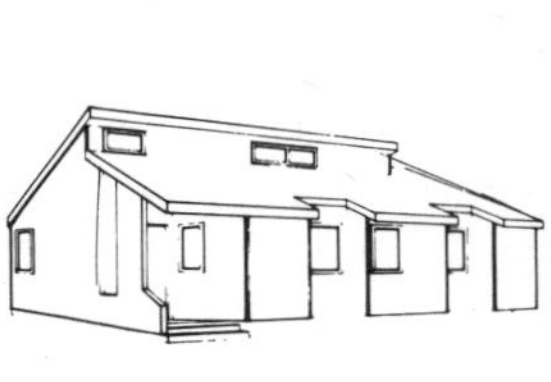

4d.
One Story: (Exterior)

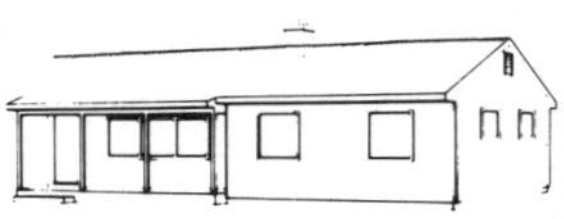

5c.
One Story: (Exterior)

record sheet

CARD	HOUSING COMPONENT	POINT VALUE TRIAL I	TRIAL II	TRIAL III	TRIAL IV
la	(Living-Dining) & (Kitchen)	___	___	___	___
lb	(Living) & (Dining-Kitchen)	___	___	___	___
lc	(Living) & (Dining) & (Kitchen)	___	___	___	___
2a	(Parent's Bedroom)	___	___	___	___
2b	(Parent's Large Bedroom)	___	___	___	___
3a	(Children's Bedroom)	___	___	___	___
3b	(Children's Bedroom) & (Playroom)	___	___	___	___
4a	(One Story) - Traditional	___	___	___	___
4b	(One Story) - Contemporary	___	___	___	___
5a	(Two Story) - Traditional	___	___	___	___
5b	(Two Story) - Contemporary	___	___	___	___
4c	(Two Story) - Traditional	___	___	___	___
4d	(One Story) - Contemporary	___	___	___	___
5c	(One Story) - Traditional	___	___	___	___
5d	(Two Story) - Contemporary	___	___	___	___
POINT TOTAL (Not to Exceed 75 Points)		___	___	___	___

house choice

GAME DESCRIPTION AND RULES

Contained in this packet is a Rationale Sheet, a Scoring Sheet, and a set of descriptive cards. Each card has a descriptive drawing of a housing component. Each component is identified by a number (such as 1a, 1b, etc.) and a subject title. Located in the lower right-hand corner is the point value, based on the relative cost of that component. There is no point value associated with style (exterior appearance), therefore, this is a free choice. The choice of style, however, must correspond with the choice of levels. For instance, if a two-story level preference is indicated, the style preference must be selected from the two-story options.

GAME PROCEDURE

(a) Group cards according to component subject.
(b) Select a level (one or two story) preference and record point value on score sheet.
(c) Select a style preference.
(d) Select a living-dining-kitchen arrangement and record point value on score sheet.
(e) Select a parent's-bedroom preference and record point value on score sheet.
(f) Select a children's-bedroom preference and record point value on score sheet.
(g) Total the point value of selected preferences. If the total exceeds 75 points, initiate trial two by beginning with step (a). Record the choices and point values under the next "trial" column.
(h) The game concludes when the game objectives have been satisfied.

RATIONALE

This game seeks to determine different preferences for housing components within the limits of an economic "budget". Each arranged component has an associated point value (related to its cost), and the player is given a limit on the number of points with which he can budget his choices. The possible choices are divided into the categories of "levels" (one or two story), living-dining-kitchen arrangements, and sleeping accomodations for adults and children.

The living-dining-kitchen choices differ according to size and the amount of separation between each activity area. The sleeping accomodations differ according to size and the possibility of a children's playspace. By limiting the points available to play the game, it is possible to encourage the player to make decisions based on the need for privacy between kitchen, living and dining, and the size of the sleeping areas with the possibility of a child playspace. An additional choice of style or exterior appearance (with no point values) is also offered.

GAME OBJECT

The objective of this game is to select, from an array of cards, your preference of a living-dining-kitchen arrangement, two sleeping components (one adult and one children's), a "level" (one or two story) component, and a style (appearance) component without exceeding a total of 75 points.

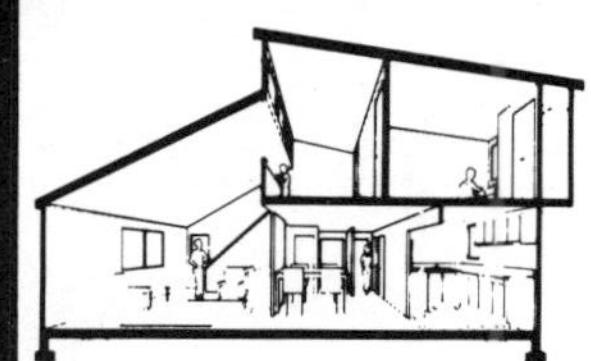

5b.
Two Story: (Interior) 8

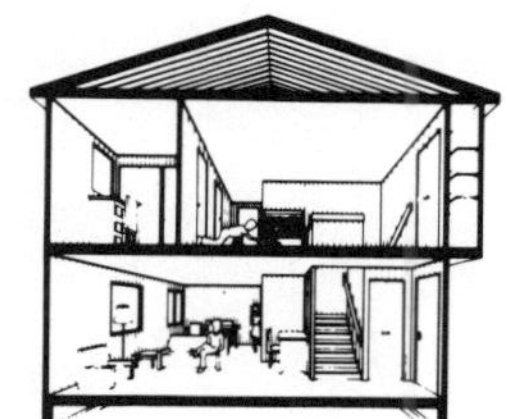

5a.
Two Story: (Interior) 8

4b.
One Story: (Interior) 10

4a.
One Story: (Interior) 10

sleeping choices

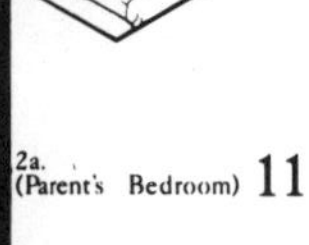

2a.
(Parent's Bedroom) 11

3a.
(Children's Bedroom) 14

2b.
(Parent's Large Bedroom) 14

3b.
(Children's Bedrooms) & (Playroom) 22

PLAN NO. 100

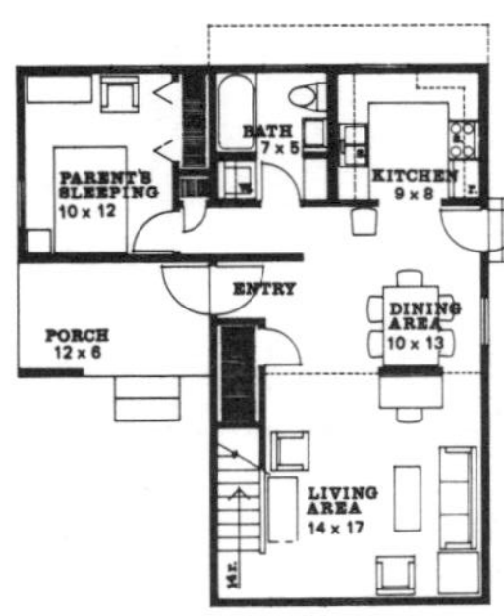

LOWER FLOOR

scale: 3/8" = 1'-0"

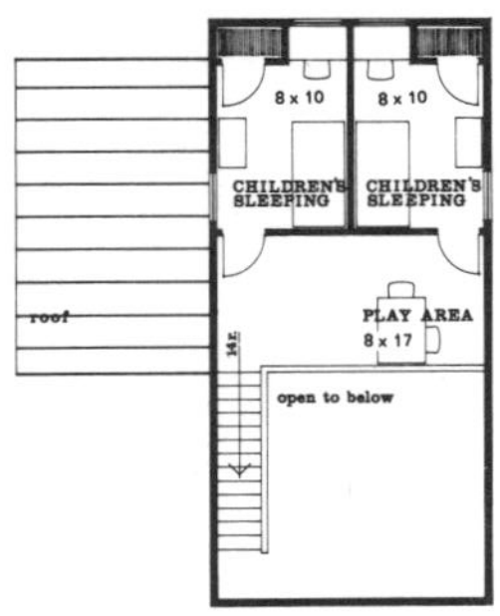

UPPER FLOOR

1090 sq ft

This plan is the first prepared by the Community Development Group at the School of Design, NCSU in conjunction with North Carolina Agricultural Extension Service.

The floor plan contains 1090 square feet of living space excluding the porch or optional carport. The house is designed in two levels with the children and their play area on the upper level with their parents' sleeping area on the lower level. With double deck beds, four children can easily be accommodated in this dwelling and given the privacy of living, play and sleeping that is separated from the adults' living area. All of the rooms are conveniently arranged so there is little wasted space.

This house was developed to provide good livability at a cost acceptable to low income families. The open plan arrangement permits more adaptable space for individual family needs.

An outdoor storage and carport unit can easily be added to the plan.

Blueprints of this plan are available free to residents of North Carolina by ordering through your county cooperative extension service or by writing to Extension Biological and Agricultural Engineering, North Carolina State University, P. O. Box 5906, Raleigh, North Carolina 27607.

PLAN NO. 102

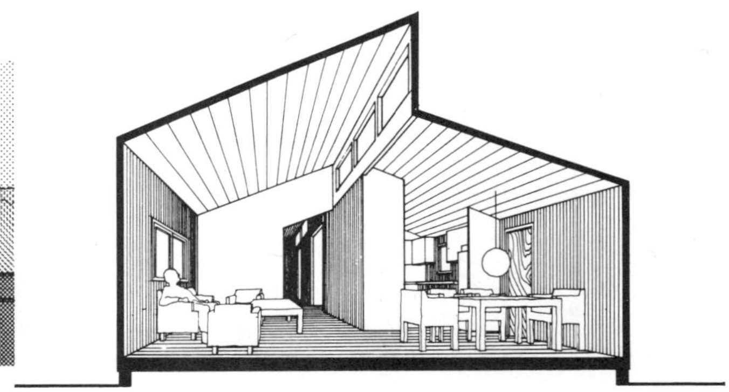

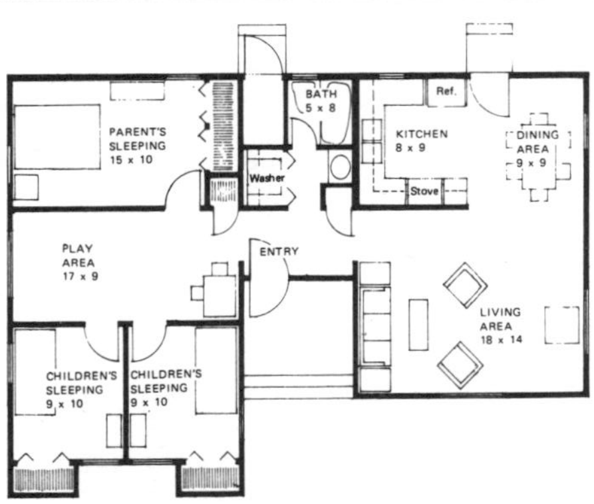

FLOOR PLAN 1100 sq. ft. SCALE: 1/8" = 1' – 0"

This plan is the third in a series of new designs prepared by the Community Development Group at the School of Design, NCSU, in conjunction with the North Carolina Agricultural Extension Service.

The floor plan is about 1100 square feet of living space excluding the front porch and optional carport. Each of the bedrooms are designed to provide cross ventilation in this one story plan. A major feature of the house is the childrens play area adjacent to the childrens bedrooms.

The house is planned in 2 zones: sleeping and living connected by the main entrance. Corridor space is eliminated in this design in order to provide more area in the living spaces. Each of the major areas within the house has high clerestory windows for ventilation as well as natural light.

Blueprints of this plan are available free to residents of North Carolina by ordering through your county cooperative extension service or by writing to Extension Biological and Agricultural Engineering, North Carolina State University, P. O. Box 5906, Raleigh, North Carolina 27607.

8

Behavior-Environment Relations

Henry Sanoff
and
Michael McNamara

North Carolina State University at Raleigh

The relationship between the activities people engage in and the places that accommodate those activities is the basis for designing. This notion is further enhanced by the need to understand the objectives that provide the stimulus for the human activities that fulfill the stated purposes. Developing an understanding of the importance of the environmental setting as the place where specific human activities occur is the purpose of the learning environments process. The context established to experience these relationships is a learning environment for young children. Participants in the process can be teachers, parents, or designers who would subsequently be involved in creating a place for children. Each member of the small group is involved in a sequential set of negotiations that begins with their selection of appropriate objectives for learning. After group consensus, individual participants select the activities that fulfill each goal and negotiate for consensus. Finally, settings are selected that are appropriate to contain the activities.

In planning for the effective achievement of your educational OBJECTIVES it is necessary to consider the following:

> The ACTIVITIES you wish to use to accomplish these OBJECTIVES.
> The type of physical SETTINGS you can use to properly accomodate these ACTIVITIES.

For each of these three catagories there is a list or deck of cards included in your packet:

OBJECTIVES
ACTIVITIES
SETTINGS

The game is planned to be played by a group of 3-5 people. To begin, each player individually selects, from the list provided, no more than 5 OBJECTIVES which seem to be the most important. Brief notes should be made justifying each choice. After each player has made his or her choices,the individual lists are pooled, and the corresponding OBJECTIVE cards are pulled from the deck. OBJECTIVE cards are arranged, face up, so that they can be seen by all players. Through negotiation the group must choose from these no more than 5 cards, with the additional constraint that these 5 must be able to be incorperated into a single, unified educational program. Players are urged to forcefully support their individual choices, even if other members of the group did not make the same choices. Continue negotiating until consensus is reached on the 5 OBJECTIVES your group feel most important. This may require conciderable discussion. Time should be limited to 20--30 minutes. After completing this phase group members should record their final choices on the record sheet provided.

Next, as a group examine each OBJECTIVE individually and select four ACTIVITIES which can be used to accomplish each OBJECTIVE. (You should work through each OBJECTIVE completely before starting a new one.) Keep in mind that some ACTIVITIES may relate to more than one OBJECTIVE.

Then, combining these two elements – OBJECTIVES and ACTIVITIES choose a physical SETTING which can be used to fulfill the requirements of each OBJECTIVE. Remember, SETTINGS should provide the right kind of props, materials and equipment to allow for the sucessful performance of the ACTIVITIES on your record sheet.

RECORD SHEET

objectives	activities	settings
1		
2		
3		
4		
5		

activities

Art Activities
Manipulatives
Water Play
Measuring
Construction
Animal Care
Cubbies
Table Games
Role Playing
Plant Care
Weighing
Listening
Looking through Books
Music
Block Play
Eating or Snacking
Outdoor Active Play
Cooking
Science Activities
Large Group Games
Sand Play
Resting
Quiet Activities

OBJECTIVES

Developing Language Fluency
Encouraging Student's Sense of Community Identity
Reinforcing Sense of Effectiveness of the Individual
Developing Cognitive Skills
Developing Motivation for Learning
Encouraging Self-Expression
Reinforcing Positive Self-Image
Developing a Sense of Confidence
Developing Persistence Toward A Goal
Developing Concentration
Developing Self-Regulation
Learning by Discrimination
Developing Communication Skills
Developing Concept Formation
Channeling Basic Biological Drives Constructively
Developing/Encouraging Resourcefulness
Developing Initiative and Spontaneity
Developing Introspective Skills
Developing Social Competence
Developing Tolerance of Differences
Developing A Sense of Responsibility
Learning by Conditioning
Encouraging Group Interaction
Learning Through Execution
Stimulating Curiosity and Imagination
Developing A Sense of Reality
Achieving Intra-Sensory Integration
Developing Motor Skills
Learning by Example
Developing Memory Skills
Developing Self-Actualization
Encouraging A Sense of Trust
Constructive Use of Fantasy
Developing Perceptual Acuity
Involving Parents In The Educational Experience
Developing Social Awareness

a
b
c
WOOD
d
e
f
g
h
i
j
k
41
l
read

9

Visual Quality Assessment

Henry Sanoff
Gregory Centeno

North Carolina State University at Raleigh

Many municipalities have adopted special regulatory controls over the architecture in an historic or central business district. Such controls may prevent exterior changes such as demolitions, alterations, and new construction without approval from a special commission or design review board. Review boards may set conditions regarding the design of the building, signs, and other visual aspects of the proposed action. Procedures such as those have potential for protecting historically important buildings, for preventing action that would cause harm to central business districts, and for coordinating the design of renovation work and new construction.

Often, however, when action is proposed in communities without the support of historic-district or CBD regulations, the visual impact of a proposed action does not lend itself readily to analysis, and recognized criteria are lacking to evaluate their significance and to weigh their beneficial and detrimental consequences. These problems have led to charges of insensitivity and even litigation against those responsible for urban development projects.

Attempts to control the alteration or destruction of valued buildings has resulted in design review ordinances. A typical ordinance as cited in *Downtown Improvement Manual* by Emanuel Berk (Illinois Department of Local Government Affairs) provides the following purposes:

> The purposes of the Architectural Board of Review are to protect property on which buildings are constructed or altered to maintain the high character of community development, and to protect real estate within this City from impairment or destruction of value, by regulating according to proper architectural principles the design, use of materials, finished grade lines and orientation of all new buildings, hereafter erected, and the moving, alteration, improvement, repair, adding to or razing in whole or in part of all existing building. . . ![1]

1. Codified Ordinances of the City of Cleveland Heights (Ohio), Section 137.05.

There is a considerable degree of variation in the specificity of design standards although most tend to be restrictive standards such as in Alexandria, Virginia, which requires that all buildings in the historic district be built in a pre-1846 style. For restoration purpose, restrictive standards may be appropriate; however, this approach is generally not practical or desirable. The important aspect of any control is to encourage new construction in a residential area or central business district that is compatible with the old in terms of the building's mass, scale, height, placement, and surface textures.

With this purpose in mind, the Best Fit Slide Rule is a tool for identifying those factors to be considered when contemplating new construction. The determination of compatibility of any proposed changes is facilitated by a visual guide for testing a range of options.

fold along this line

BEST FIT SLIDE RULE

fold along this line

DIRECTIONS:

This typical block face can be found in most communities where residential areas have remained unaltered for many decades. Today, neglect and rapid decay have placed many blocks of this type in need of replacement or repair. In order to insure that there is some continuity in the process of restoring older residential areas, it is often necessary to evaluate new building proposals very carefully. This is done by comparing each new proposal with specific criteria developed to insure the preservation of those unique qualities that make older buildings visually distinctive.

On your slide rule you will find a drawing of Front Street, a typical residential block in downtown SERGORP. One of the buildings has been condemned by SERGORP's building department because it is unsafe for occupancy. There are twelve proposals for reusing the site of the condemned building.

As an interested citizen it is your job to insure the selection of a proposal which will preserve Front Streets unique character. Using the criteria checklist provided, review each of the twelve design proposals and decide which proposal fits best within the available space. Only that proposal which satisfies all eight criteria can be considered acceptable.

CRITERIA CHECKLIST

Each proposal must be evaluated using the criteria below and must be similiar to the adjacent buildings regarding the following:

- alignment of window and door openings
- relative height
- roof silhouette
- proportion of window and door openings
- use of ornamentation
- building use
- surface variation
- relative width

ASSEMBLY INSTRUCTIONS

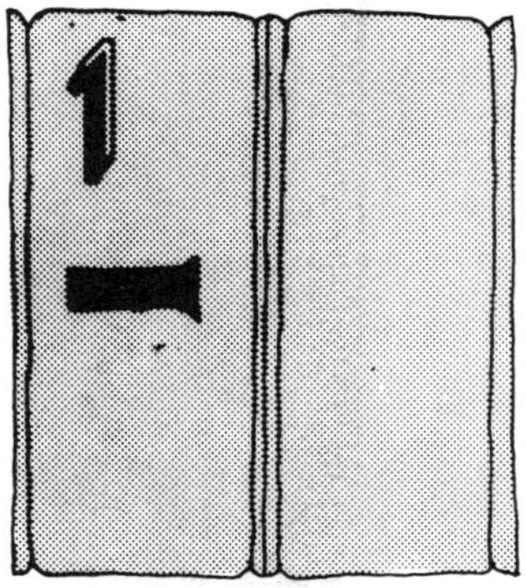

Remove the previous page from the workbook. Cut the slide rule out along the dotted lines which run along its outside edge.
Then, carefully cut out the black panel as indicated.

Using a ruler as a guide, fold along the dotted and dashed lines (—·—·—·—). Use this picture to determine the direction of your folds.

Then fold the front and back panels together, insuring that the tabs are sandwiched between the two panels.

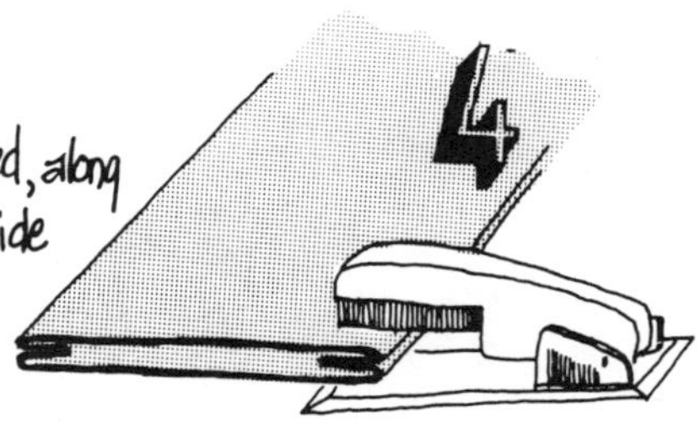

Staple where indicated, along both edges of your slide rule.

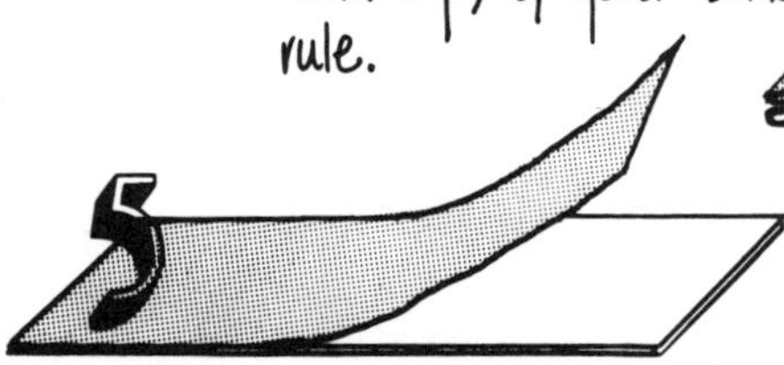

Cut out the strip (on this page) along the dotted lines and paste it to a piece of cardboard. Cut off any excess cardboard if necessary. And, insert this strip into the slot. Read the directions and your ready to go!

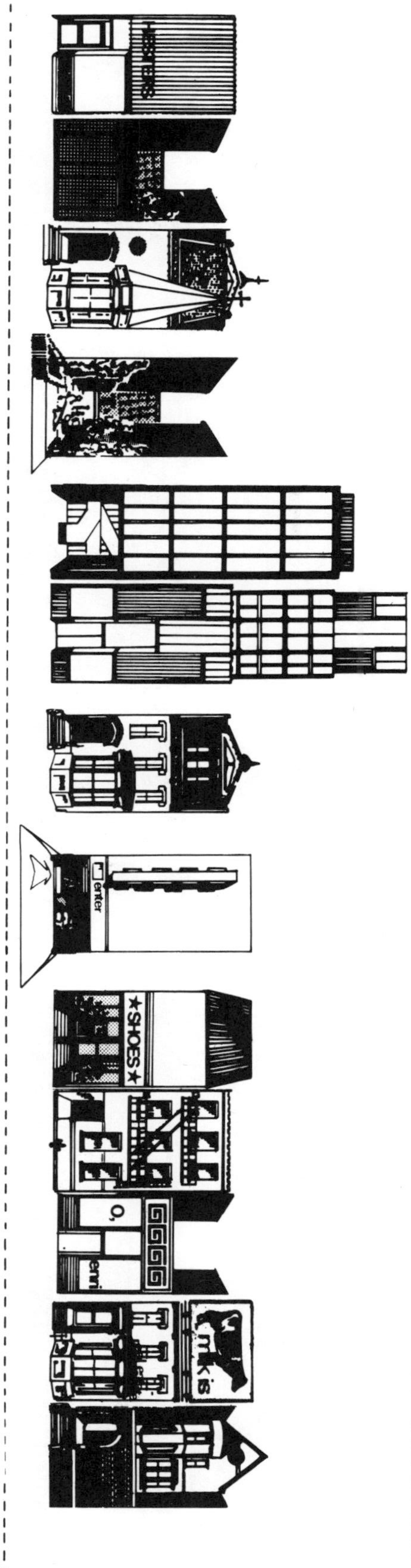

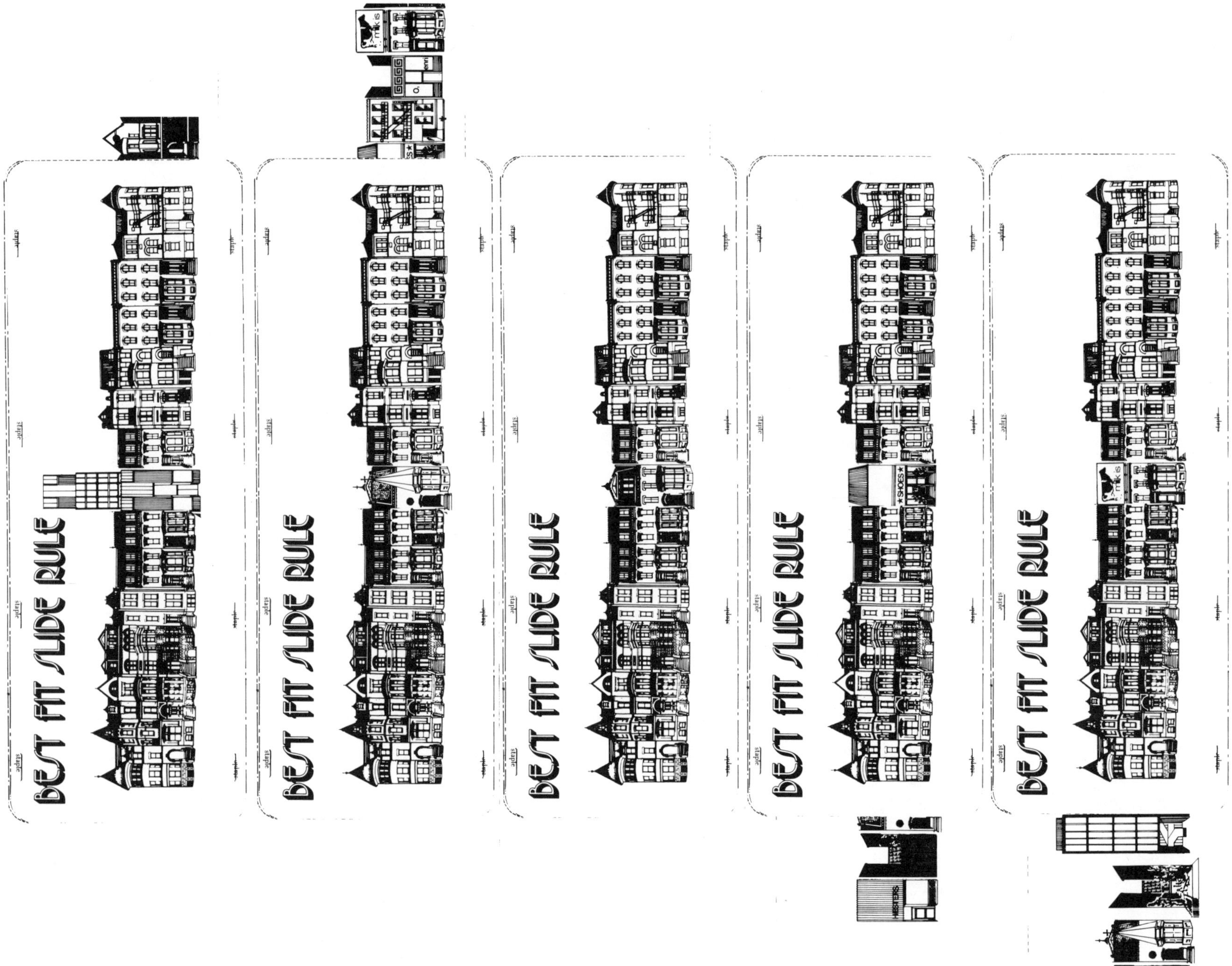
BEST FIT SLIDE RULE
BEST FIT SLIDE RULE
BEST FIT SLIDE RULE
BEST FIT SLIDE RULE
BEST FIT SLIDE RULE
staple
★SHOES★

COMMUNITY DEVELOPMENT GROUP
SCHOOL OF DESIGN
NORTH CAROLINA STATE UNIVERSITY

CBD CONSERVING BUILDINGS DOWNTOWN

A GAME FOR PREDICTING CONSEQUENCES

Developed by Jerry Seelen

RE-USE STICKERS

DESCRIPTIONS

City Hall.	Totally vacant, one unit up, one down.	Shoe store on first floor. One unit vacant on second.	Mens' wear store. Law offices on second floor.	Fabric store. Two units vacant, second and third floors.

RE-USE STICKERS

DESCRIPTIONS

Department store on ground floor. Two units vacant on second floor.	Totally vacant, one unit up, one down. Needs minor repairs.	Old hotel. Barber shop on ground floor, one unit vacant on ground floor, two units vacant on second floor.

DIRECTIONS

This booklet is a dramatization of the way the uses of individual buildings affect the quality of a whole town. The scenes you see on page one are pictures of a typical small downtown- some of the buildings are totally occupied, some are partially used, and some are completely vacant; but all are structurally sound. Each building is briefly described on the sheet above page one, with vacancy defined in available space units. There are a total of 12 vacant space units for the entire town, three on street level (down), and the remaining nine on upper floors (up).

Listed on the left-hand side of the use sheet are a number of possible facilities a town might need, with the amount and location of space each of these uses requires. Your job is to go through the list and select 12 units worth of uses that you think your town needs, list those uses in the space provided on the right-hand side of the page, and then find spaces for these activities. (Feel free to consider uses not included on the list.) Sound easy? It is. You have basically two choices: to use old spaces, or not to use them. After you have identified your town's needs, go through the inventory of available spaces in town. If you feel that you can't fit a particular activity in any vacant space in town, then you have the option of putting that activity in a new building somewhere outside of town, possibly in a new development or regional shopping center. If, on the other hand, you feel that you can use a vacant space in town, take the re-use sticker from the sheet above page one for the building you chose, and position it on the acetate overlay on top of the original building. Try to identify the immediate changes in and around the building that are brought on by re-use.

After you've allocated all the uses to spaces, either in town or out of it, you can then see what long-range changes might take place, for better or worse, in the entire town as a result of the decisions you made: count up the total re-uses of vacant space per block; if you have re-used more than ½ the vacant space in a block, lift the overlay, turn to page two, and reposition the overlay; if you have used less than ½ the vacant space in a block, lift the overlay and reposition it on page three; if you have re-used nothing on a block, remove overlay and turn to page four; and if you have occupied all space on both blocks, remove overlay and turn to page five. On each page identify the side effects that your decisions had on the whole town over the long run.

You can easily play this game by yourself, but there are many possible ways the vacant spaces can be re-used. So the more people playing and making decisions, the better, since then you can talk together about why you made the choices you did. Also you can talk about the side effects shown in the game, discuss what you think might have caused them, and try to think of good and bad side effects not shown that your special uses might cause.

USE SHEET

POSSIBLE USES: SPACE UNITS: LEVEL

1 . Day care: 1 unit; down
2 . Apartments: 1 unit per apt.; up or down
3 . Public offices: 1 or 2 units; up and down
4 . Movie house: 1 unit; down
5 . Elderly housing: 1 unit per apt.; up or down
6 . Health services: 2 units; down
7 . Library: 1 unit; down
8 . Craft center: 1 unit; down
9 . Office space: 1 unit per office; up or down
10. Youth center: 1 unit; up or down
11. Restaurant: 1 unit; up or down
12. Bingo parlor/Game room: 1 unit; down
13. Bar: 1 unit; down
14. Local history museum: 1 unit; down
15. Tourist info. center: 1 unit; down
16. Bank: 1 unit; down
17. Adult education center: 1 unit; down
18. Community center: 1 unit down
19. Thrift shop: 1 unit; up or down
20. Other retail: 1 unit per shop; up or down

SELECTED USES: SPACE UNITS; LEVEL

1 . ____________________
2 . ____________________
3 . ____________________
4 . ____________________
5 . ____________________
6 . ____________________
7 . ____________________
8 . ____________________
9 . ____________________
10. ____________________
11. ____________________
12. ____________________

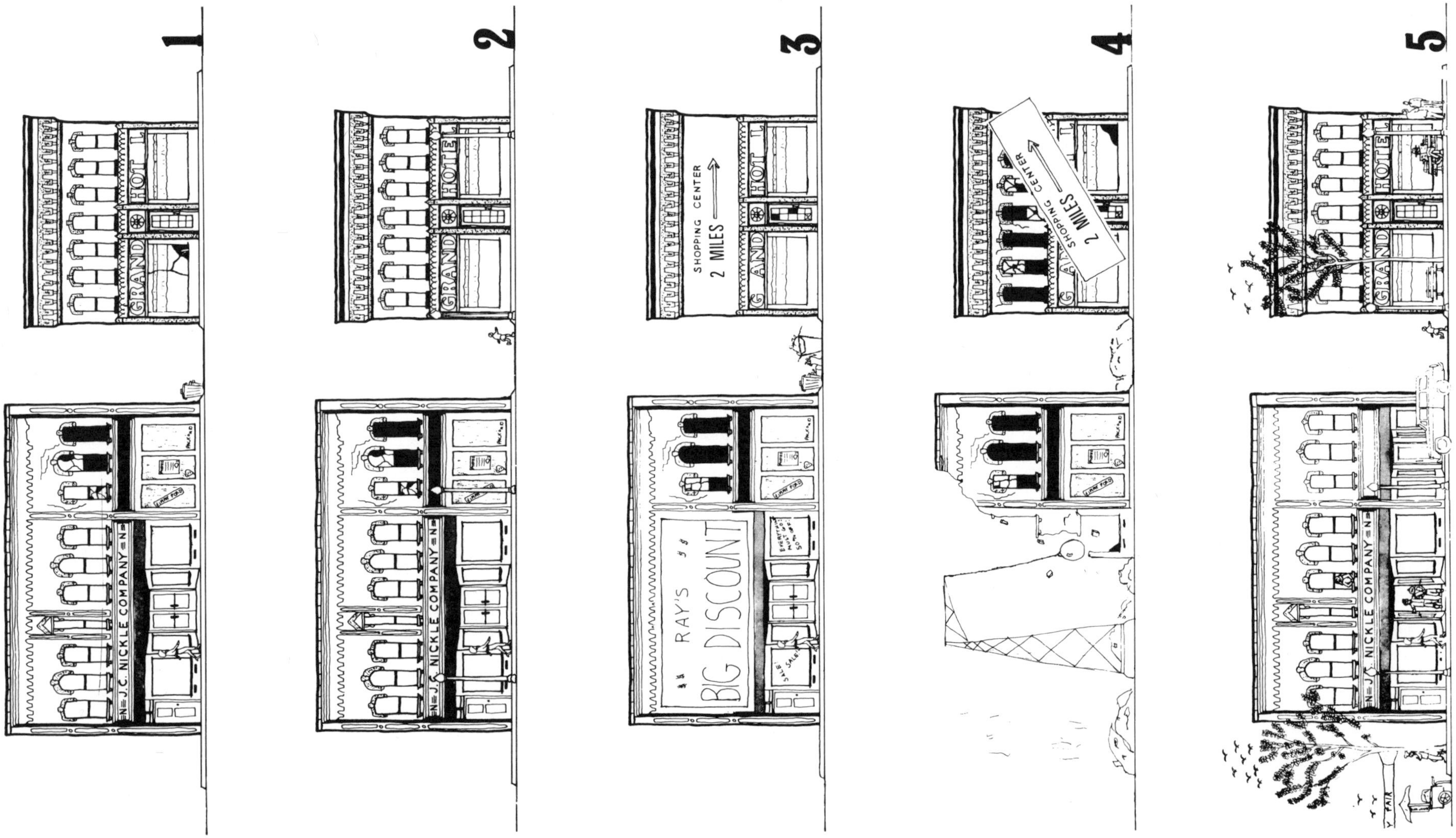
1
GRAND HOT
J.C. NICKLE COMPANY
2
GRAND HOTE
J.C. NICKLE COMPANY
3
SHOPPING CENTER
2 MILES
RAY'S
BIG DISCOUNT
4
SHOPPING CENTER
2 MILES
5
GRAND HOTE
J.C. NICKLE COMPANY

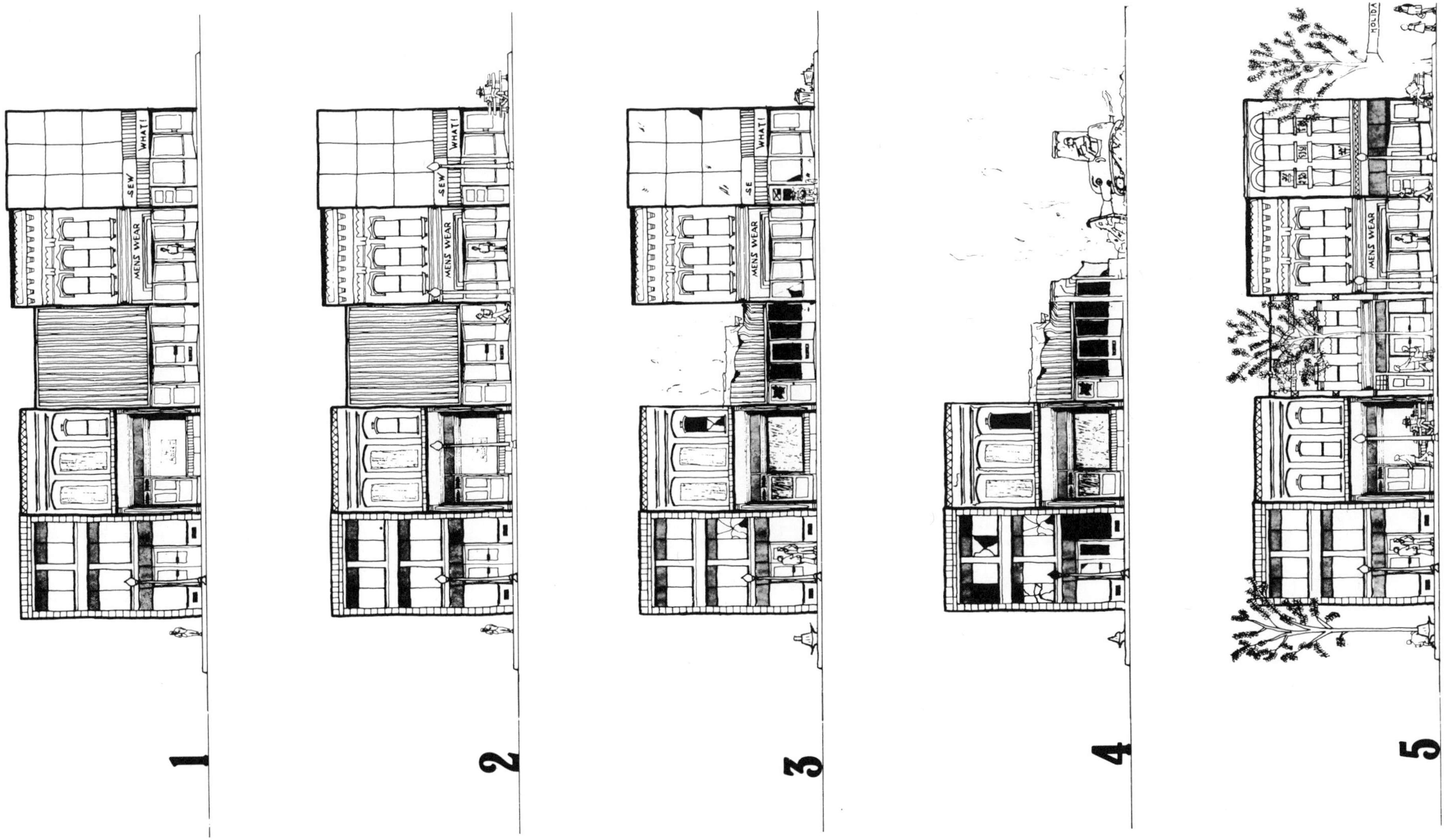
1
SEW WHAT!
MENS WEAR
2
SEW WHAT!
MENS WEAR
3
MENS WEAR
4
5
MENS WEAR

Part III

Field Approaches to Community Participation

This section represents the invited work of designers that share an interest in effective lay participation. The authors have developed and field tested methods for design participation, issue identification, and urban simulation. Each approach is unique to the areas they represent whether participation in housing design, responding to housing alternatives or community problems, or understanding the new communities planning process.

The methods cover an international spectrum and are oriented towards problems larger in scope than the previous sections. Pyatok and Weber seek to deinstitutionalize the delivery of housing and develop a design and organizational process that reflects the unique characteristics of the user population. Turner, Adams, and Beckerman of the San Francisco Design Center propose a planning ballot outlining alternative concepts, financial and administrative policies, and a selection of land uses for a controversial section of the community. Vonier and Scribner involve visitors in a policy-oriented exhibit to respond to topics of community interest. Using ballots, participants can respond to local issues and sponsoring organizations. Summers, Douglas, and Nasar propose a simulation game to facilitate predictability in planning for a new town. Each author proposes participating approaches that are practical and easily modified to suit a variety of similar situations.

This field-approach section includes alternative methods for achieving effective citizen participation. Through the utilization of exhibitions, games, voting, or direct involvement, lay participants can be involved in learning processes designed to produce informed judgments. Each strategy, however, differs depending upon the size of the affected population and the degree of involvement from the participants that would be necessary to yield satisfactory results. All the approaches are oriented towards developing more informed participants engaged in more effective and meaningful results.

10

Participation in Residental Design: A Method for Generating Choice and Its Ideological Implications

Michael Pyatok
Assistant Professor

Hanno Weber
Associate Professor

School of Architecture
Washington University

INTRODUCTION

The Production of Shelter—A Quest for Equity

One of the consequences of the last 200 years of industrialization is the now universal "housing problem" in both developed and developing countries. The large influx of populations in search of work opportunities from the abandoned rural countryside has and continues to beset cities unable to accept them. The corrosion of physical conditions resulting from social deprivations suffered and lived by millions for over two centuries, however, has been and continues to be addressed as an entirely physical problem of quantifiable health and building standards. Few solutions, if any, have defined the problem as an economic and political one of production systems characterized by limited resources that are inequitably distributed by centralized decision makers in both western capitalism and eastern socialism. In market economy systems, packaged housing is provided according to financial ability. Consequently large numbers of people are never a part of the market. Whether there is a shortage or an excess of housing, they are crowded into what the market lets them afford. In controlled economies, vast resources are allocated to housing by central planners. But the bureaucratic response is usually so expensive that there are never enough dwellings.

Well-meaning public legislation or philanthropy have sought to reduce overcrowding by providing more housing units; congestion and concentration have been countered with more open space and dispersal; and unsanitary conditions and squalor have been combated with sunlight, ventilation, and plumbing. Hence, most solutions have technocratically reduced the problem of the unequal distribution of resources into the quantitative "packaging" process of mass housing. This response transforms shelter into a utilitarian tool that hopes to preserve health, reform social behavior, accommodate economic inequity, save cities from decay, and modernize developing economies and rescue faltering developed ones.

But the "housing problem," like a Sisyphean ordeal remains to plague people. Given limited resources that are inequitably distributed, there are never enough dwellings that people can afford. In developing countries the problem is exacerbated. The centralization of resources is directed at crash industrialization and modernization. The efforts are channeled to the production of shelter as a marketable commodity within industrial centers only. This expensively produced supply can not be afforded by populations subjected to high unemployment and low wages who continue to migrate into cities seeking opportunities denied in the rural areas. In developed countries, the clearance of older housing stock not meeting the requirements of profitable commodities aggravates the problem for those with limited resources. Such removal simultaneously induces deterioration in other sectors that must absorb displaced populations. New housing stock, which is provided to meet higher building standards, becomes less affordable to an increasing number of people. In both contexts the exacerbated problem, to a large extent, centers on shelter being addressed primarily as packaged housing, designed by professionals, built by capital-intensive production, and owned and managed by institutions who do not use it.

The advocates of shelter as mass housing argue that people must be satisfied since most units provided by government and private production are occupied. Not only do they ignore that too many people still cannot afford the limited supply of mass housing, but even disregard the dissatisfaction of those who can occupy it and have no other recourse. Thus, some professionals have prodded bureaucracies and some developers to use planning and design methods that are more responsive to the needs of users of mass housing. These professionals believe that the "housing problem" can be solved by involving consumers in the process of defining the housing product. Admittedly, the advocates of housing as a process have expanded the technocratic response of housing as a product to include more than just *quantity*. Questionnaires, interviews, and gaming techniques involving users, however, limit people's attention to the improvement of the product. Such an approach still accepts the production of mass housing as an answer to the "housing problem" and ignores the inequitable distribution of limited resources that mass housing exacerbates. User involvement is narrowly circumscribed within decisions already made by others.

It would seem, then, that a more responsive approach to the production and distribution of shelter would advocate the involvement by dwellers in those major decisions about ownership and management of resources. Such an approach would maintain the concerns for achieving a shelter product that satisfies people's needs and aspirations. In addition, it would also seek to raise people's awareness of how existing institutional frameworks control the production and distribution of shelter. This understanding may enable people, if not to gain some control of the process, at least to loosen constricting bonds and widen the range of options. In undertaking such an approach to produce shelter, people would realize that they must cooperate and organize their efforts with others in order to achieve their own aims. By itself, this approach will not solve the "housing problem" that results from the inequitable distribution of limited resources. But, it could join other efforts seeking to gain equity for all facets of life.

Housing as a Consumer Product

When dwellings are provided either by public or private actions for the use of statistically described consumers, despite good intentions, shelter is reduced to a grossly simplified product for "everyman." "Shoe boxes" scattered in the landscape or stacked in "filing cabinets" may provide protection from the elements, safety, the ability to perform daily life, and at times even achieve symbolic identity. But commoditized shelter usually is supplied not to meet the needs of dwellers. It is used to satisfy other demands inherent to the production system. In the public sector large scale visible projects provide necessary political symbols; they meet the demands to stabilize large scale construction and material industries; they occupy growing bureaucracies and provide work for professionals; they raise consumer expectations beyond means; and they become testing grounds for new products and technologies. In the private sector, large speculative developments assisted by public subsidy minimize financial risk and insure profits; they induce speculation and settlement of undeveloped land; they foster the clearance of deteriorated shelter stock in order to create new demand; and they aid and abet the lucrative enterprise of banks and lending institutions.

Hence, satisfying the needs and preferences of dwellers and accommodating the cultural and social processes that characterize the complex wholeness of life are low priorities of market commodity shelter. After all, people are adaptable, particulars are unimportant, and it is easier to describe "average" people statistically. As long as the loose fit storage of people does not thwart or hinder their lives, and they can move if it does, the market product persists. Once shelter gets in the way, and standardized, oversimplified stereotypes of dwelling often do for those whose choices are limited, these naive and expedient prescriptions contribute to alienation and hostility toward environments and others. True, efforts have been made to generalize the consumers in order to tailor shelter to their needs and aspirations. These include surveys and interviews to establish user require-

Figure 10-1
Ixtacalco: Government sponsored housing in Mexico City (Photo by M. Pyatok)

Figure 10-2
El Roasrio: Government sponsored housing in Mexico City (Photo by M. Pyatok)

ments. But the stereotype of people through building standards based on general human performance criteria only aggravates simplification, leading either to trivial truths or exacerbating inaccuracies. Market commodity shelter not only tends to be socially and culturally unresponsive to a context and situation, but as a physical product, inflexible and unadaptable over time. Most of the criteria that generate "mass housing" are *a priori* conceptions about people's social-cultural preferences and priorities, building technology standards and performance, and financial and economic constraints. The resulting shelter product is so bounded by these imperatives that little room is left for its continual evolution and the future contributions by those who will use it. Unlike the mutable and anonymous shelter of the past that over time has accommodated variable use and transformation, unchangeable market commodities freeze and crystalize the present. When obsolescence has to be confronted, alteration and adaptation are not possible and too often the only answer is eradication.

The priorities of shelter as a product also inflict economic consequences on consumers. Since the standards, performance, technology, and supply are regulated to meet the values of the providers of shelter, the users only have a "take-it-or-leave-it" option. The high costs that result from inflated land and expensive building technologies, high interest rates over long amortization periods, limited supplies, and the encouragement to pursue conspicuous consumption, forces dwellers into one of two courses. Either they compromise health, amenity, and security in substandard environments they can afford while being able to allocate resources to other priorities; or they have to saddle themselves with excessive rents or mortgages, which become a burden to their economic viability. The latter path in many contexts even hinders the mobility of dwellers to seek and improve employment opportunities by tying them down to a place.

Finally, shelter as a packaged and finished product creates additional costs that rarely are exposed to the consumer. Among some of these hidden costs are both the direct and indirect impact of public and private "mass housing" on the social and economic context. Very often large scale projects hoard and create temporary artificial scarcities of material and labor to the consumer. Whenever commodity shelter dabbles with new technologies under the pretense of accelerating production and reducing costs, it generally creates unemployment, a cost eventually borne by everyone. Similar costs are created by urban periphery speculation that taxes the public sector with service and infrastructure costs, which in turn taxes the consumer. In addition land values adjacent to such large projects may experience excessive deflation or inflation. The resulting low property values or high taxes may encourage abandonment and/or speculation that lead to the instability of the social fabric. Perhaps the most disguised cost such projects incur is the political and social disruption and human suffering caused by forced land clearance and dislocation of households at the expense and in the name of the public good.

Housing as a Consumer Process

Some producers of housing attempt to involve the dweller as more than just a consumer. This is done to counter both the detrimental consequences of unresponsive packaging and the continual inability of dwellers to afford shelter as a completed market product. The basic intention of housing as a process is that people actively engage in the design, construction, management, and ownership of their dwellings. Such a process attempts to prevent incongruity and misfit, to satisfy people's needs and aspirations, to enable dwellings to adapt over time to changes, and to cope with limited family resources.

Figure 10-3
Ixtacalco: Government sponsored minimum core houses in Mexico City (Photo by M. Pyatok)

Figure 10-4
Ixtacalco: Core house expansion by dweller in court yard (Photo by M. Pyatok)

Efforts to make building into an adaptive mechanism rather than forcing dwellers to adapt, can even overcome some of the disguised by-products of commodity packaging. Encouraging dwellers to manage the building process distributes the demand for materials and labor. Smaller incremental actions tend to favor and maintain labor-intensive construction. They also allow by-passing the need to undertake long term extensive loans, thereby avoiding high interest costs. However, shelter as a process to accommodate personal and individual needs, without understanding the contradictions and constraining relations of the larger production system, leaves that process unguarded against new means of intervention and control by the institutions people were originally seeking to circumvent. This process of shelter is again seen as an end in itself because it does not address as a primary goal the alteration of the system of control over all facets of life, of which shelter is only one.

The involvement of dwellers in the design of their shelter may allow them to satisfy socio-cultural needs and aspirations. Priorities in the arrangement of dwellings, and the organization of neighborhoods, and their symbolic content can be determined by the occupants. Buildings liberated from *a priori* notions of completed objects can become adaptable frameworks for continuous change and contribution over time. But is there any appreciable difference between dweller as consumer before building and dweller as the consumer of a packaged product? If the constraining relationships between the dwelling process and the larger production system are maintained, the difference is minimal. Deciding what "style" of shelter or even how to arrange rooms or where to locate yards and parking are really inconsequential if those decisions are circumscribed by others who supply credit, insurance, and who influence those who enact and enforce regulations.

Similarly, involving dwellers in establishing their priorities over the amount of space or the extent of services based on what they can afford only replicates the existing market patterns. True, dwellers may not be saddled with excessive shelter costs and are able to allocate their limited resources to other priorities. But they still have to compromise comfort, and perhaps, even more critically, *time.* Incremental self-help building as a means to cope with shelter needs according to financial ability makes dwellers indentured servants to keeping a roof over their heads for extensive periods of time. It prevents time and energy from being directed to other endeavors, be they educational, recreational, or political. Subsequently, it conveniently diffuses dissatisfactions that exist about other inequities. Without any doubt there are benefits resulting from the involvement of dwellers in the design, construction, ownership, and management of shelter. All the advantages, including the existential value derived by dwellers in tailoring their environments, however, must be seen in a larger political and economic context. It seems, then, that dweller involvement and self-help shelter are expedient responses to ignore the public responsibility for altering the underlying structure that maintains inequity. They are not efforts committed to confront contradictions in a quest for political and economic equity.

What people must ask themselves is whether they just want to own their house, or their lives. An alternative process would seek to examine the limitations of and alternatives to the institutional framework that circumscribes the provision of shelter. Such an approach would expand an awareness that would enable people to consider seriously other options for gaining control over all facets of life, of which shelter is only one. It would engage people in an extensive discourse, not only about all the principal issues of shelter as a designed product, but also how all these issues relate to other facets of life and to the larger institutional framework. Designing a product, therefore, becomes both an end and a means. People would be providing themselves with the kind of shelter they want. More than that, they would develop skills to gain and maintain control over the relationships that link their shelter product to the larger context. Participating cooperatively in the inception, generation, and development of shelter would reintegrate the work of professionals and laymen in a design activity that could become a part of everyone's life. Unlike "housing," a commodity furnished by some to be consumed by others, this alternative transforms shelter into "dwelling"—a vital act of life, a culture-making process.

AN ALTERNATIVE

Dwelling as a Life Process

A self-defined group of people seeking to develop shelter for themselves must accept the basic assumption that the existing institutions, which presently produce commodity housing, fundamentally obstruct their desire to engage in dwelling as a process of life. In doing so, they must recognize they will embark upon a highly argumentative journey to understand critically, "things as they are" and "things as they would like them to be." Such a group must also realize that there is no single alternative universal solution to their dilemma. The group must accept that their argumentative efforts will devise a development strategy that is uniquely tailored to their particular needs as individuals and as a group, and to the institutions specific to their context.

To be involved in a well informed argument a group must be certain it will undertake the widest possible search of options for development. Such a search will unavoidably attempt to understand the residential development process in relation to transportation, recreation, education, employment, and the procurement of food, energy, and other necessities. A broadened search for possible action is accompanied by the obligation to pursue continuous critical assessment of previously selected strategies to insure that they reflect the changing realities internal to the group and in the surrounding institutional framework. Such a development process would require from participating professionals, not only attitudes supportive of the basic assumptions which necessitate a broadened search but also a commitment to provide continued assistance to a group operating within what may become an unreceptive climate.

More specifically, a group would need to identify: 1) the nature of its internal organization; 2) the types of preferred relationships to external institutions; 3) the different arenas in which it will take action including the rehabilitation and development of dwellings and other physical facilities; 4) a wide variety of options for taking action; and 5) the most appropriate combination of options that satisfy their desired ways of relating to each other and their context. This proposal for organizing a development strategy is illustrated in subsequent sections focusing only on issues related to residential planning and design. Any programs that a group would pursue concurrently, including residential development, would obviously have to be coordinated since they share limited resources. The following proposal can also serve as a model for generating and selecting options for programs in other arenas. This proposal is applicable to developing and developed countries, but the example demonstrates its use in a developing country—Mexico.

Internal Organization

If a group wishes to undertake a development project in an inclusive manner, it needs to engage at the start in a careful assessment of the resources desirable and available to them. They must engage in "economic design." However, unlike "economic design" undertaken by private and public developers, such a process should examine the potential social relations that may exist as a consequence of the obligations within the group for sharing and using resources. They should also examine the relations between the group and a variety of institutions in the larger context that may control some of those resources. For example, if each family in a group were to be economically responsible for only its own private property with no obligations to others in a development project, the autonomy of dwelling ownership by individual families could lead to uncontrolled speculation on the new dwellings, and uncertainty about the potential membership of the group. Self-help construction by each family working *in solo* could also lead to excessive consumption of materials and energy, resulting in higher construction costs, reducing funds available to a family to finance other activities of life. The complete economic self-sufficiency of all families in all activities of life may also create excessive and redundant expenses. As a consequence, some income levels in the group may be forced to maintain their substandard living conditions. Also, the elimination of inter-family dependencies in some contexts (such as shared kitchens, washing facilities, and other communal uses) may lead to an organizationally impotent group, which in turn would require members to expend more time, money, and labor instigating artificial activities to rekindle "community spirit" and collective concern. If each family is separately responsible to lending and insuring institutions, such institutions can wield significantly more power over the destiny of a group than if the group were organized as a cooperative with more legal, political, and economic influence.

The implications of economic and social linkages that may exist between groups of families in pursuit of residential development, therefore, must be considered in the early stages of planning and design. Definitive decisions need not be taken initially because people may change their minds after exploring several options for development. But a wide range of possibilities exists for internal relations. They can vary from a loosely affiliated collection of autonomous, nuclear families, through extended families and kinship networks, into legally contracted cooperatives and nonprofit and profit corporations. Perhaps they can even integrate into a pre-existing regional cooperative or corporation. The intention of any group to organize consciously into alternative forms rather than to unconsciously accept existing patterns of social relations for development clearly has economic motivations. However, to begin to design at this level has political implications as well. Depending on the type of organization a group chooses, as it increases in size and in economic leverage, its ability to affect decisions within its immediate context and region also increases. In fact, cooperatives may not only integrate themselves into regional networks but also become active members in larger political movements seeking to gain control over the economic infrastructure of their region, with the simultaneous evolution of more indigenous control over important financial, cultural, and educational institutions.

External Relations

The range of organizational forms into which a group may evolve to cement themselves internally can be accompanied by a range of attitudes they may take toward the institutional framework that is external to them as they undertake a development project. A group could attempt autonomous development in its extreme forms for economic, cultural, or even religious reasons. It may choose to avoid as much contact as possible with the existing system, supplying its own labor, materials, energy, food, etc. Such an attitude toward the context probably would delimit the range of internal relations possible in order to maintain solidarity against outside pressures. Extreme "autonomy," however, can shade into a mixture using the existing system when it suits the group's ensemble of economic, political, social, and cultural beliefs. For example, as a cooperative, it may wish to invest in its own energy generation facility and recycle its own wastes, but recognize the need to contract labor outside the cooperative membership to save time. Or it may choose to allow a portion of the development to be owned by members of the cooperative only, and other portions rented to outsiders in order to earn income. Finally, a group may choose a posture of "dependency" and encourage a profit motivated developer to implement a traditional project completely linked to all facets of the existing economic and technical infrastructure. Such a group would accept the ownership and management of dwellings by developers and pay their price for having borne the burden of development.

The range of internal organizational relations and the attitudes a group might take toward their institutional context are combined in simplified form in figure 10-5. A group can pursue any pure or hybrid combination of internal relations and external attitudes toward its context. As already demonstrated in the few examples cited above, to take action from any position, a group needs to address what seem to be three major sets of issues. First, the members of the group may wish to reassess how they undertake many of the major activities in their lives, from basic survival endeavors like the provision and maintenance of food supply and shelter, to educational and leisure time activities. What, if any, would be the benefits to them if they undertook these activities in the variety of ways possible by combining the range of internal and external relations? The answer to this first question becomes the focus of the next two. The second major question relates to labor. Who would undertake any work necessary to implement whatever is decided? Should it be accomplished by self-help labor, and if so, is it undertaken by individual families or by a cooperative? Or is the labor totally or partially contracted? A hybrid approach could include professional outside labor to build the infrastructure of roads and utilities while cooperative labor could build unfinished houses and individual families add their own finishings and additions as they need them over time. In similar manner a group may investigate the implications of labor for acquiring food. Should they grow it or buy it, and as individuals or as a cooperative? The third major question relates to ownership of any property, equipment, or capital that may accumulate as a result of the development project. Who should own and manage it? Should families be given complete control over the sale of their dwellings and land, or should property be governed by a cooperative? If a group builds its own dwellings as a cooperative, should it rely upon a mortgage from an outside lending institution, which in effect will own the project? Or should they initially build minimal dwellings through a minimum short term mortgage with gradual expansion accomplished by each family's own resources independent of outside credit?

	INTERNAL RELATIONS	FAMILIAL						LEGAL & CONTRACTUAL								
		Nuclear Family			Extended Family			Small Group Cooperative or Corporation			Neighborhood Cooperative or Corporation			Regional Cooperative or Corporation		
	EXTERNAL RELATIONS	Auto.	Inter.	Dep.	Auto.	Inter.	Dep.	Auto.	Inter.	Dep.	Auto.	Inter.	Dep.	Auto.	Inter.	Dep.
LABOR	Food															
	Shelter															
	Education															
	Health															
	Recreation															
	Transportation															
CONTROL	Food															
	Shelter															
	Education															
	Health															
	Recreation															
	Transportation															

(The shaded area indicates that activity of life which is developed in this text) Auto.–autonomy; Inter.–interdependence; Dep.–dependence.

Figure 10-5
Simplified form for determining development strategies based on internal and external factors.

Even from this brief sample, it can be seen that a group carefully reviewing the three basic questions outlined in figure 10-5 could create a wide variety of development strategies uniquely tailored to its special needs. Such an endeavor could free the group from preconceptions about the seemingly immutable established patterns of residential development that may, in fact, be counterproductive to their real desires. Equally important, then, is the educational experience that exposes the contradictions between what they would like to have on the one hand, and on the other hand, how the social and economic relations between themselves and their institutional context as presently organized often militate against them.

Defining Physical Issues

There can be a wide variety of means by which a group undertakes the task of deciding how it wishes to structure its physical environment. It is impossible to recommend a single design method that would apply to any context, or even one that could avoid significant alteration while in use by any one group or groups of similar social characteristics. Groups will differ in many ways, including their experience with such projects, their degree of internal disagreement, the amount of time available to their members, their familiarity with each other, the ways by which their individual members usually define and solve other problems of life, and many more. Some groups may have the propensity to fantasize, to imagine what their environment could be like if existing constraints were ignored. Others may be exceptionally reverent of techniques that carefully deliberate and assess all issues in a conscious way. However a group chooses to design its environment, it will eventually have decided about *what* it wants for its physical environment, *how much* of what it wants, and how all the things it wants in its environment should *relate* in space to each other. In addition, regardless of how any group makes these choices, they will all have made them at several scales of complexity. These include the overall area of development (a site), sectors of the development (clusters of dwellings), the individual places where families live (dwellings), and finally, small places and parts of a dwelling (components).

For example, about the *site* as a whole, any group would ascertain *what* it wants—park space, roads, parking, recreation facilities, communal areas, etc. *How much* of these would be necessary is another of the concerns, as well as how all these things *relate* to each other in an overall plan. On the second level of complexity, a group would need to determine *what* it would want to accompany a cluster of dwellings—children's play space, communal laundry facilities, parking, shared kitchens, etc. *How much* of these in size and numbers would be determined, as well as how they ought to be organized in space to *relate* to each other and to dwellings. On the next level of concern, a group would need to examine closely *what* kinds of dwelling units they want—individual detached homes, attached multi-family units, highly integrated communal living units, etc. *How much* space, equipment, outdoor area, etc., would be decided along with how all these things should be *related* in space to meet their needs. Finally, at the most intimate scales, members of the group would need to identify their needs and preferences about particulars of the dwellings—living and sleeping areas, sunlight, views, etc.—again answering what, how much, and in what relation.

To answer these three basic questions about topics related to the four scales of physical complexity, a group has to identify all the topics that are important to it at all scales. Thus far, this system for organizing decisions has required a method for partitioning reality that is generic enough to resist the usual criticism that such attempts may not relate to a group's special conception of reality. However, identifying topics of discussion at each scale may have special results for each group and may require special efforts to ascertain them. As an example, topics which often seem to be important from the viewpoint of architects can perhaps fit into three major categories: 1) places for activity (space uses); 2) paths of movement for people, goods, utilities between places (networks, e.g., roads, service lines, etc.); and 3) materials and methods for constructing the previous two categories of physical things (technics). Each topic is affected by necessities that insure comfort, such as ways for providing light, ventilation, view, privacy as well as amenities that consider significant cultural meaning (environmental requirements). The questions that should be answered for each general topic at each scale of complexity are graphically displayed in figure 10-6. One additional question is included as a composite summary of the previous three questions. Answers to it would be complete enough

TOPICS	QUESTIONS	WHAT ?	WHAT & HOW MUCH ?	WHAT & IN WHAT RELATION ?	WHAT , HOW MUCH & IN WHAT RELATION ?
SPACE USES	Site				12
	Cluster		6	9	13
	Dwelling	3			14
	Component				
NETWORKS	Site	4	7		
	Cluster			10	
	Dwelling				
	Component				
TECHNICS	Site				
	Cluster				
	Dwelling		8		
	Component	5		11	

(The shaded area indicates what topics are used to demonstrate options; the numbers refer to the figures which graphically display the options.)

Figure 10-6
Table indicating questions to be asked of architects before making design decisions.

to qualify as design proposals. The diagram indicates the large number of issues a group should encounter as they take on the task of designing their own proposal for a residential development project. What is most critical for a group to demand of professionals working with them is not only to guarantee that the group will participate in all these decisions, but that for each question of each topic at each scale, they *must* be shown a wide variety of options to discuss before deciding upon those that they most prefer.

Generating Alternative Answers

A group can be assisted in generating alternative answers to these four questions by many techniques. A technique for inventing answers can vary widely from a spontaneous, playful brainstorming, to a more structured reviewing of intuited possibilities; from an extremely rigorous itemizing of workable options from previous experience to a cautious and rational assembling of options by combining basic elements derived analytically. As noted earlier, the choice of design methods depends heavily on the composition of the participating group. On the one hand, a group more accustomed to and comfortable with a careful, rational review of clearly depicted options, should at times use brainstorming techniques to free itself from fixing on conventional options. On the other hand, a group that habitually enjoys wandering through a morass of possible answers, at times ought to use techniques that construct and sort out possibilities in an orderly way. It is important that a citizen group know that such a range of means for generating options exists, and that they require their assisting professionals to put them to use as they may be needed.

The following graphic displays, which exemplify one method for generating options, were used in residential development projects located in rural and urban areas of Mexico. These were simulated projects, undertaken by students and architects from Mexico and the U.S. to learn to generate many optional answers to the questions about the topics at the various scales of complexity to simulate working with citizen groups on residential projects. The particular method was chosen to generate options because it required the participants, many of whom were both still inexperienced in design and unfamiliar with the Mexican culture, to consider options carefully and systematically before assembling them into suggested design proposals. Finally, the graphics do not necessarily represent those that may be used with client groups, but were essentially prepared to facilitate communication among the participants of the exercise.

"What" options

Some possible answers to the question, *"What does your group want?"* are demonstrated for the topic of space use at the scale of rural dwelling in figure 10-7. The charts represent a variety of means by

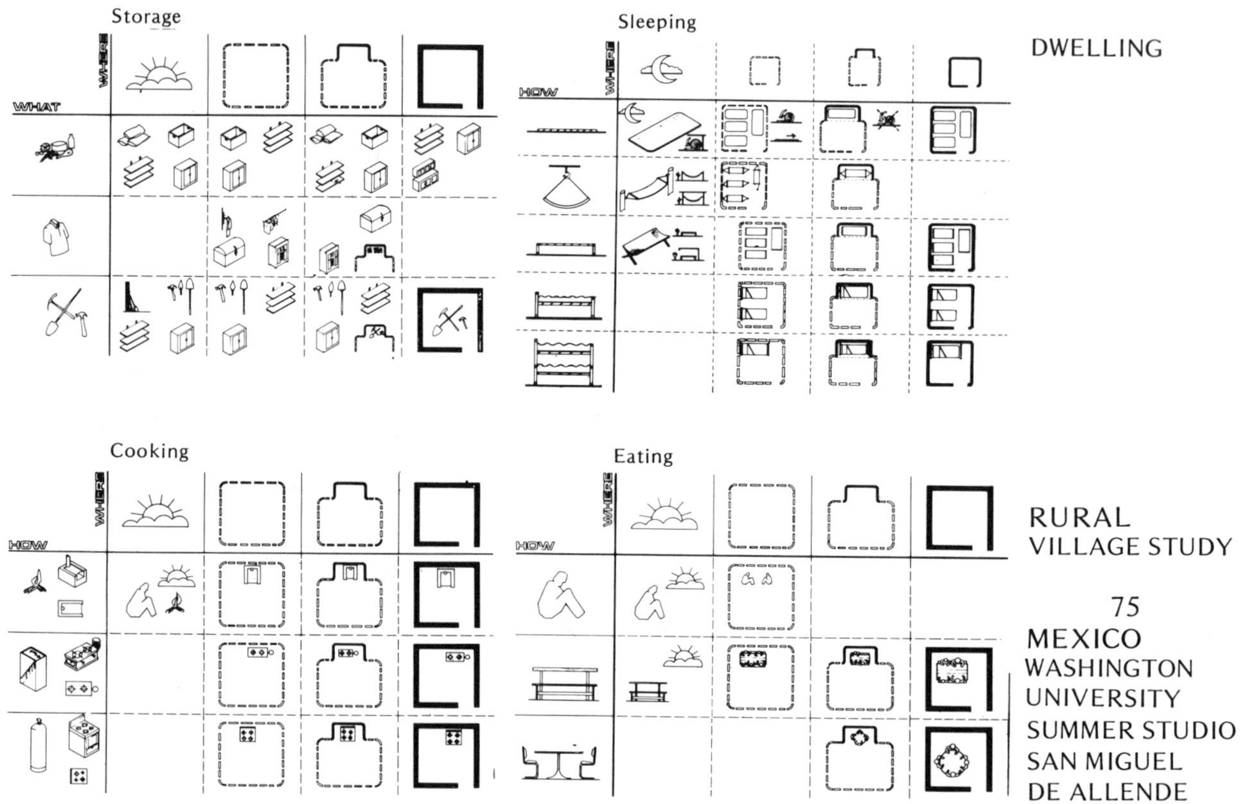

Figure 10-7
Chart representing ways families can select to support some of their activities.

which families may select to support some of the activities of their households, assuming that families wish to live independently of each other in their own dwellings. A simple device is utilized to generate the options. For each activity, two major properties of the supporting environment are identified and a spectrum of options is created for each of them. In one case, eating can be described as taking place in a space (where) and with some basic equipment (how). A spectrum of possible places includes: outside in the open air, inside a single multi-use space with other activities, in a special alcove adjacent to other activities, or in its own space. The same is true for equipment: from none, that is not relying on furnishings, to homemade and store bought furniture. Combining these two spectrums of where and how, generates a number of options for families to review before deciding what they need in order to live comfortably within their economic means. This can be considered for early stages of a self-help residential development, but also when incomes and capital savings permit expansion and alteration of dwellings.

The basic key to this technique then, is to identify at least two critical features that contribute to the basic essence of the piece of the environment under consideration and then to identify a variety of possibilities for each of them. Systematic combination produces a broad array of possibilities that spontaneous brainstorming or even searching historical precedents may often overlook. This is the basic technique that is demonstrated in all subsequent examples.

It is important to stress some caution when using this technique. When deciding what characteristics should be arranged on the horizontal and vertical axes before combining them, a group should carefully assess whether those that had been chosen will really contribute to generating useful options. The selection of unimportant characteristics can generate trivial options. Also, once they are convinced that those chosen will yield significant discussion, they should be careful to display a range of possibilities for each that has considerable variation, including contradictory possibilities. The wider the variety, the more informed the final choice.

Another example of how to generate optional answers to the question, "*What* does a group want?" is demonstrated for networks at the scale of the site in figure 10-8. In this case a group may wish to know how to organize the flow of vehicles and animals in their rural town, which already has a highway passing through the center. Across the top of the chart, three options are considered: a single network of roads combining vehicles and animals, two-road systems separating the two uses, or mixing them in the village but rerouting the higher speed through-traffic around the edge of the village. Along the vertical axis are arranged four possible patterns for organizing the roads in the town. One additional feature has been

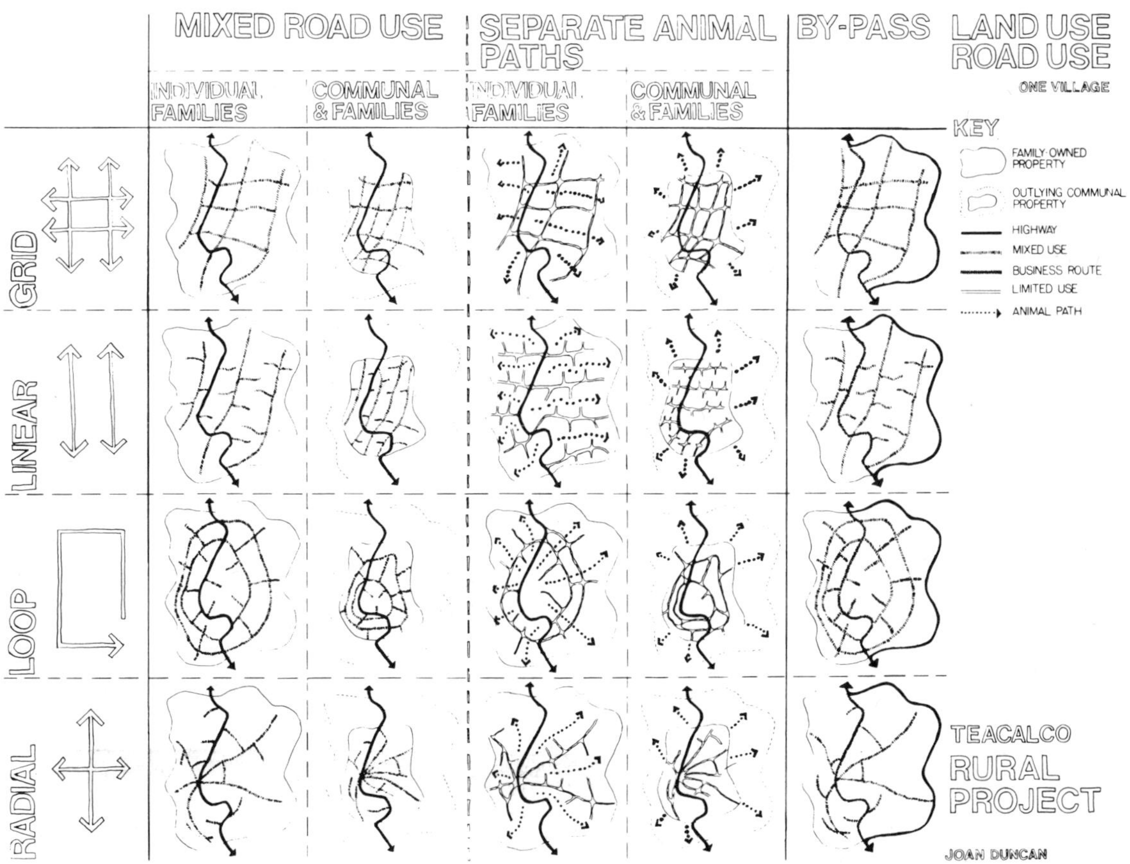

Figure 10-8
Chart representing alternative patterns for road systems.

added to indicate that if each individual family has its own animal corral, the town and its roads would be larger than if they agreed to a mixture of condensed communal corrals for larger animals along with the maintenance of smaller animals on the individual family lots. Thus, twenty possibilities for the road network can be generated for discussion purposes. Such a general discussion about *what* a town needs does not yet require accurate dimensions or clearly defined relationships with existing paths or the topography. The "what" types of question, then, are very useful at the start of a problem, because answers to them can set the general directions and limits for generating future options that also answer the questions "how much" and "in what relation."

A final example to demonstrate the nature of answers to a "what" question is provided for the topic of technics, this time at the scale of components in figure 10-9. The particular technical question asks, how can a village supply water for its people, animals, and crops? Eight ways of acquiring water are combined with two ways for distributing it, that is, communally or to each family, for the three uses. Also, two means for purifying it are combined with the ways of distribution. The possibilities resulting from combining the two sets of attributes offer some sketch ideas of what a village might do to solve its water problem. No dimensions or accurate, detailed descriptions are necessary yet to answer this "what" question. Only a nominal description is necessary to convey the basic intent of each option. In order to obtain a more detailed description of options, a group would go on to ask not only how much is necessary but how should any new possibilities relate to existing conditions and to each other.

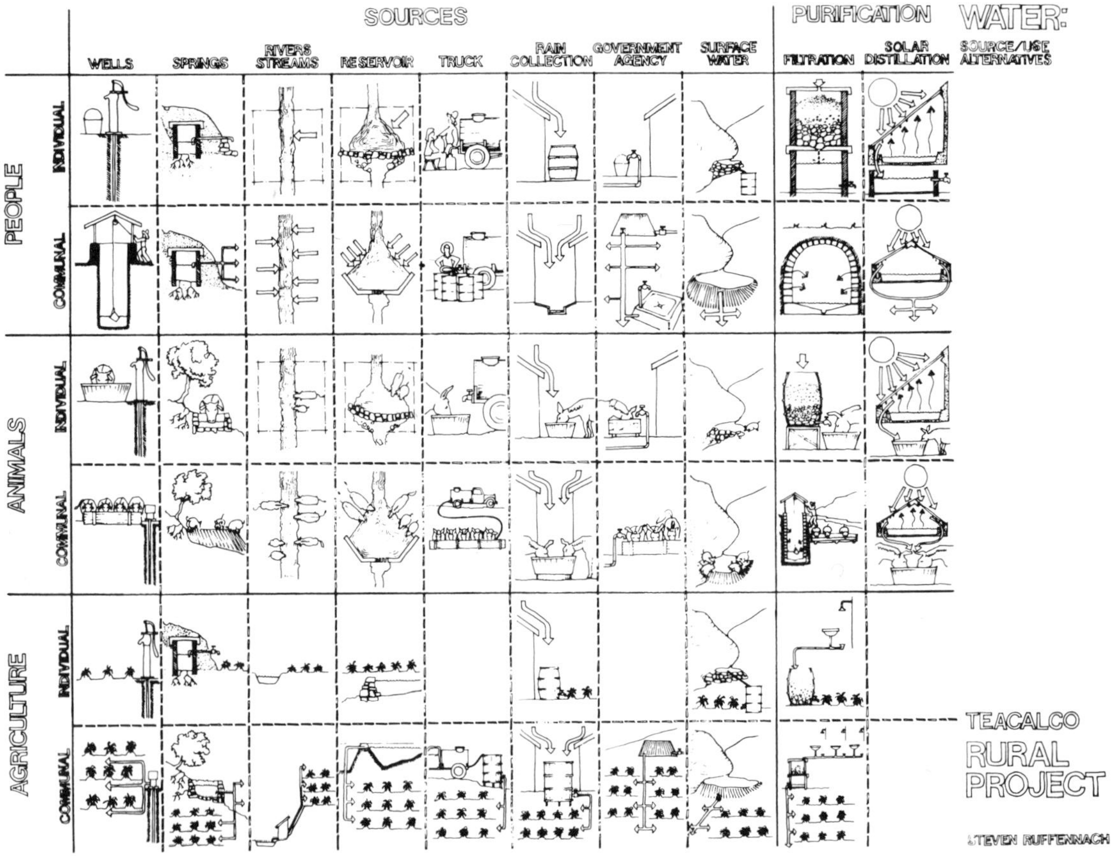

Figure 10–9
Chart representing options for water supply.

"What and how much" options

When the concern of "how much" is introduced and combined with "what" issues, a group addresses questions of relative size, area, height, and/or spacing of elements in the environment. By examining tangible "what" and "how much" options about space uses, networks, and technics at the scale of a site, clusters of dwellings, individual dwellings and their components, a group begins to seek alternatives that match their needs and aspirations. They would include not only comfort, amenity, and symbol but also the quantitative limits of space, financial resources, and time.

An example that examines the question of "how much" for space use at the scale of dwelling clusters is illustrated in figure 10–10. It combines the variables of dwelling height and the arrangement of house parcels for two fixed numbers (forty-eight and seventy-two) of families. The chart demonstrates the consequences on the fixed amounts of open spaces that result from alternative building heights and the subdivision and arrangement of land parcels in a variety of ways. At one extreme of the spectrum, all the land is individually controlled by each household; at the other, one half is collectively shared in a single concentrated area. The intermediate points offer two clustering options.

As dwellings increase in height, the amount of open space also increases. Thus, if a group wanted to maximize open space, they would choose the tallest dwelling in the third column. Further, if they wanted most of the open space in the public domain they would consider an alternative in the fourth row. It can be seen in the chart that answers to the question "how much" of space use begins to affect "how much" of networks. The alternatives that distribute all the land to individual families require more roads. In all the alternatives, an environmental requirement was respected. In order to insure light and ventilation, a dwelling never occupied its full lot except in those options where all outdoor space is in the public domain.

A second example that investigates the question of "how much" for networks at the scale of a residential site is illustrated in figure 10-11. This combinatorial chart maps a range of six alternatives each for pedestrian and vehicular circulation through an inner city site. The vehicular options begin by accepting the existing pattern that cuts the site in both directions through a central square. Successively, existing streets are eliminated, and the range ends with an option that proposes maintaining the square and introducing two new loop roads. Similarly the pedestrian network ranges from maintaining the existing streets to an expanded finer grid over the entire site.

Thus thirty-six options for a network are generated that convey how much land could be preempted by roads and paths as well as the number and general location of access points in the perimeter of the site. If a group wished to restrict vehicular access through the site while maximizing service to most areas within the site it would gravitate toward the sixth column. If, in addition, they selected to minimize land allocated to networks they would choose an alternative on the first row where the right of ways are not duplicated.

The third example to demonstrate answers to "what" and "how much" addresses the issue of technics or methods for construction at the scale of dwelling in figure 10-12. This investigation of the amount of self-help a dweller might contribute displays on the vertical axis two possible arrangements for one, two, and three family buildings that are three floors high. On the horizontal axis are displayed options for the extent of structure, exterior shell, and interior finishes that might be provided initially in a dwelling. The structure and shell are varied by the inclusion or exclusion of floor slabs, beams and columns, and exterior walls on the three levels; the extent of interior finishes, interior walls, lofts, and fixtures are then

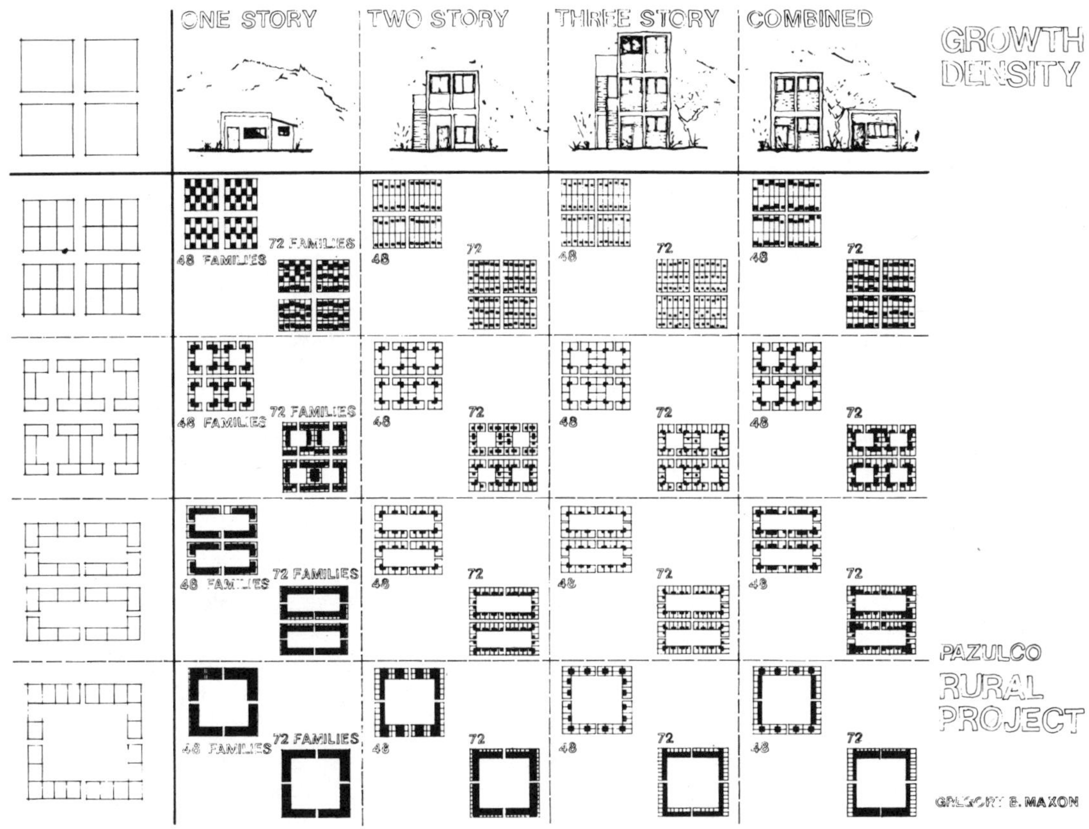

Figure 10-10
Chart representing variables that will influence space uses.

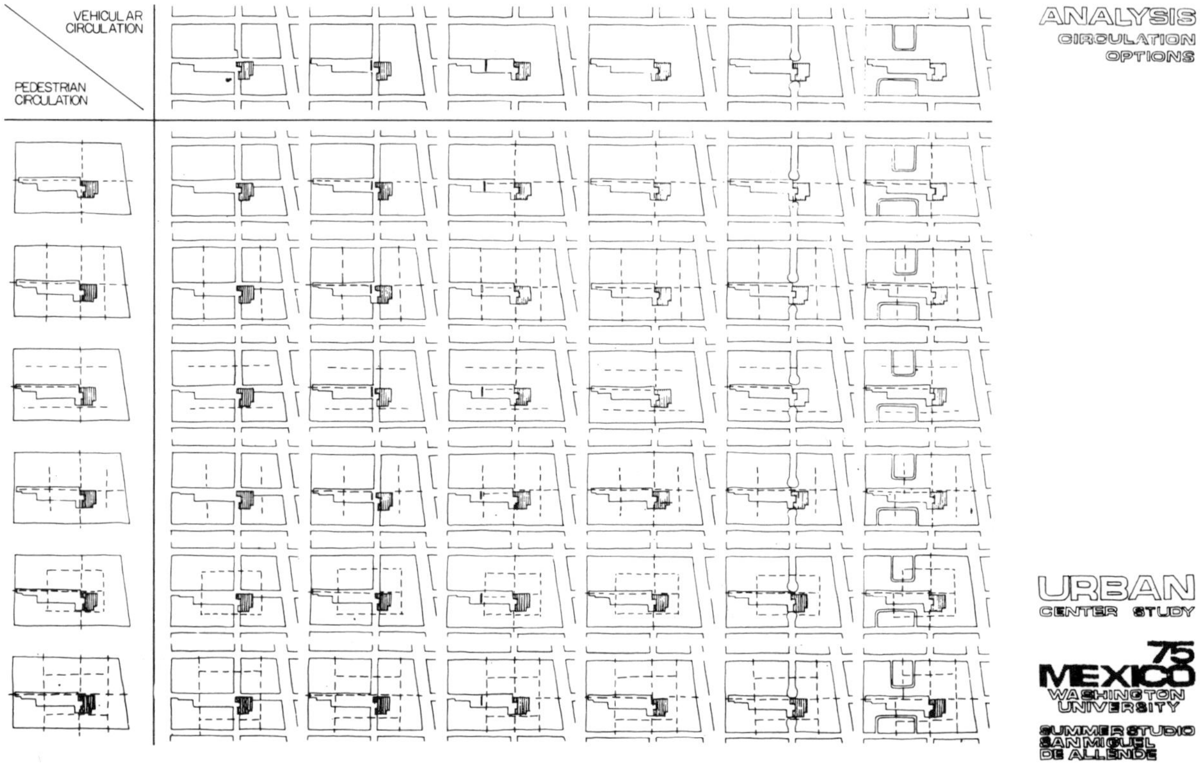

Figure 10–11
Chart representing alternative networks for pedestrian and vehicular circulation.

illustrated in six alternatives selected and noted by a dashed outline in the shell and structure options.

How much building and finishes are provided initially not only affects the amount of space but also the cost borne by a dweller. A fixed amount of resources could equally be expended in more room in an unfinished shell as in an incomplete, but expandable, finished dwelling.

The options illustrated are by far not all the possible ones. Generating alternative answers at times, as in this technics example, cannot be inclusive without creating an extensive array that is difficult to manage and could cause confusion. The expediency of only expanding one case of the structure and shell options to illustrate interior finish possibilities does not hinder discussion. The underlying logic of the argument is conveyed and readily transferable to other cases.

As these examples display, partial solutions for "what" and "how much" issues of a residential development, they demonstrate that although it is possible to focus separately on space uses, networks, and technics at several scales, these concerns often tend to overlap. Space uses influence networks and can involve technics; similarly networks and construction methods affect space uses; this reciprocity, however, should be maintained while recognizing that the categories and divisions are only a convenience to enable a group to deliberate and assess issues in a conscious way.

"What and in what relation" options

Spatial "relations" within new proposals and between new proposals and an existing context, examine relative location, adjacency, superposition, connection, and sequence. Relationships of space use, networks, and technics need not be encumbered by actual sizes or areas, but should maintain a relative size to insure that portions of the environment can be assembled in reality without losing their desired position. "What and in what relation" options at the scale of a site, clusters of dwellings, dwellings and their components engage a group to examine wide choices of functional relations that not only may match cultural needs and aspirations but also may infer social and symbolic meanings.

An example that examines the relationship of space use at the scale of dwelling clusters, illustrates in figure 10–13 house grouping options as a function of the slope of a site and whether or not a road is provided. The diagrams demonstrate linear and cluster options on the vertical axis varying the relationship to pathways. The horizontal axis at the top presents a range of alternatives that relate dwelling units, pathways, and slope conditions. Once they are combined in the chart they demonstrate different ways of entering houses parallel and perpendicular to the slope, various stacking and attaching arrangements, paths that are single or double loaded, and a general indication of retaining wall conditions.

The chart considers some environmental requirements: comfort as affected by the amount of stair

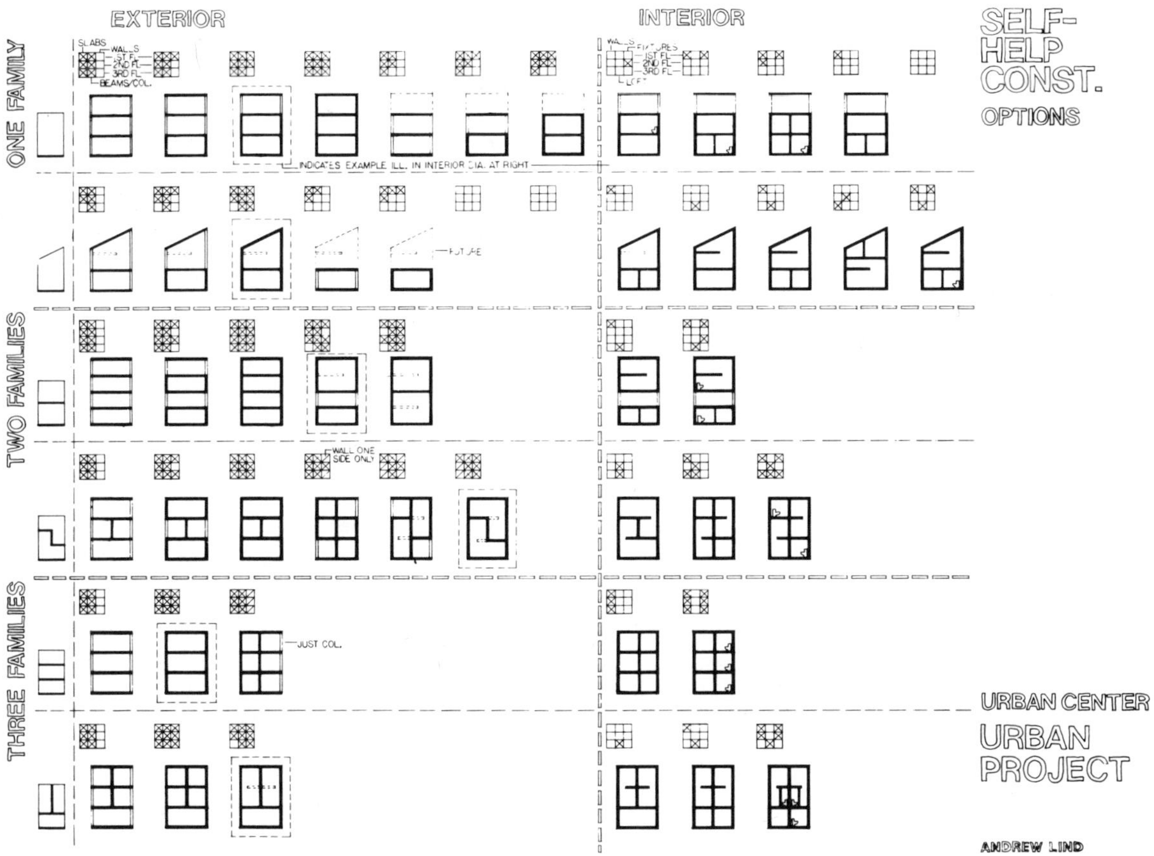

Figure 10-12
Chart representing various options for techniques or methods of construction.

climbing a solution will impose; and view orientation to sunlight and cross ventilation as affected by stacking. In a relational chart, accurate dimensions are not crucial so the dwelling units are abstracted into spherical volumes.

If a group were interested in a scheme that maximized road access to most of the dwellings in a cluster, with maximum view and minimal retaining walls, its choice might gravitate to the first two columns of the third row.

A second example of "what and in what relation" addressing the issues is illustrated in figure 10-14. It focuses on how vehicular roads, parking, and pedestrian paths affect the location and organization of dwellings and shared open spaces. On the vertical axis the options are generated by identifying the relationship between house and access either as linear or nodal, single or double loaded. These attributes are then combined with the extent of mix or separation of pedestrian paths and roads and the degree of integration and grouping of parking.

The option in the bottom right corner illustrates an arrangement that combines a central open space that is surrounded by two rows of houses. They are separated by a road exclusively for vehicles serving clusters of parking spaces between the houses. Pedestrian movement is in the central space and an outside path.

A third example, in figure 10-15, investigates methods of construction and alternative materials at the scale of house components, that is, roofs and walls. The relationship and compatibility between these two components is generated by combining a wide range of material options for each component. The spectrum includes natural as well as manufactured products, the resulting structural demands, and the required technical skills. The chart would demonstrate to a group that less permanent walls limit the number of roof options. It also shows that over time, roofs could change from less to more permanent construction. Finally, it indicates the assembly requirements inherent in each combination.

The three previous examples are useful to a group by relating spatially and graphically the issues of space use, networks, and technics without additional quantitative concerns. This partial combination contributes to a gradual development of awareness until that point where issues are familiar enough and options are reduced to permit the understanding of more complete syntheses.

"What, how much, and in what relation" options

A display that would contain the most complete descriptions of possible options would answer simul-

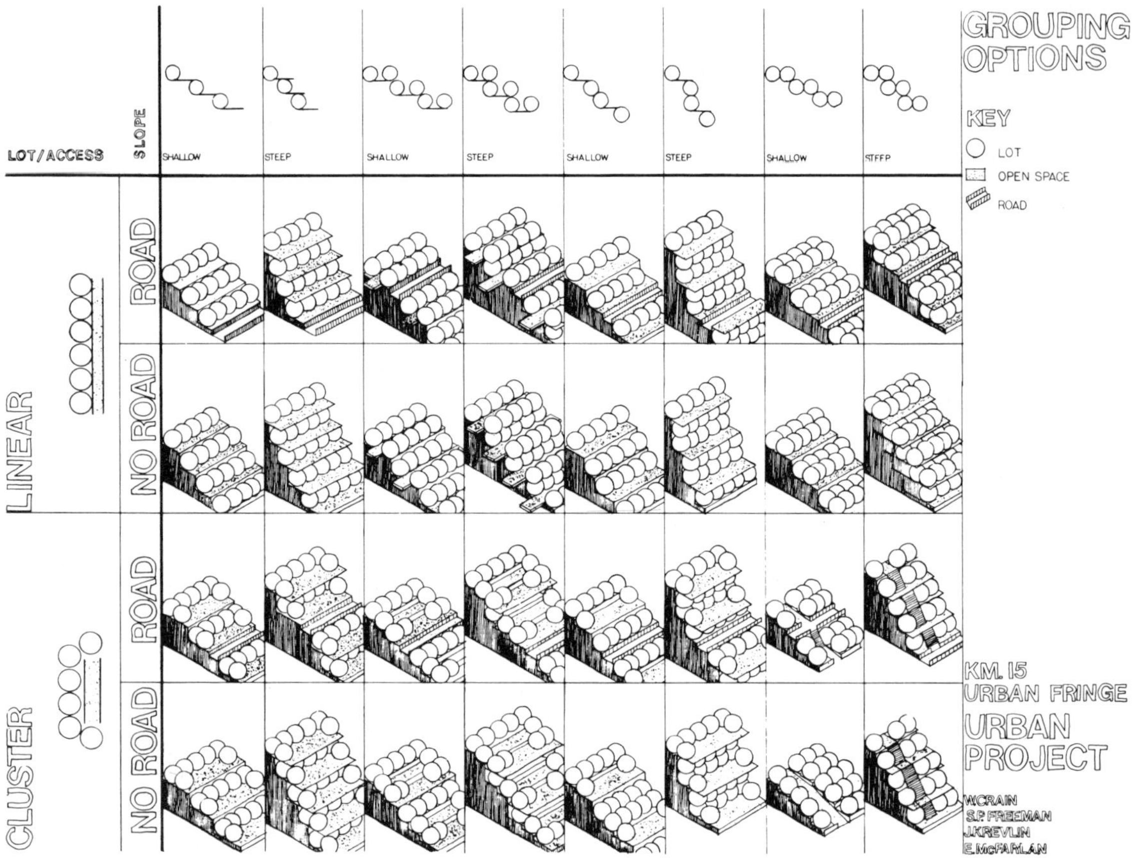

Figure 10-13
Chart representing options for space use at the scale of dwelling clusters.

taneously all the questions—"what, how much, and in what relation." A group that has the time to study thoroughly their problems would arrive at the point of asking all three questions together after having examined them separately. By saving the summary question for the end, they would generate a more manageable number of optional combined answers. Many optional answers to the separate questions will have been discarded through previous discussion and will no longer be under consideration by this time.

An example of optional answers generated for this final question is demonstrated in figure 10-16 for the topic of space use at the scales of a site and cluster in the inner portion of an urban block to house sixty families. Three different amounts of public open space (ABC) and three different ways it can be distributed on the site are arrayed on the vertical axis. These are being combined with three different amounts of space provided for private family dwellings (ABC) that can then be stacked in three different ways as demonstrated on the horizontal axis. When combined, they produce patterns of dwelling clusters to house the sixty families on the given site of 1,800 square meters. For example, the option in the center square of chart C displays what might happen if the site were organized into four clusters of sixteen families. Each is located around a public space of 180 square meters, where each dwelling contains a maximum of 30 square meters arranged so that one family is stacked above another. With such small dwellings, not all sixteen would need to stack in order to fit comfortably on one quarter of the available 1,800 meter site. It can be seen how some very important environmental requirements and some considerations of technics influenced the formation of options. Seven major requirements were considered: 1) families should have their own dwelling space; 2) a minimum room size should be 10 meters square with a minimum room width of 3 meters; 3) every room should have at least one exterior wall to receive windows for light and ventilation; 4) every family should have private outdoor space open to the sky of 10 meters square minimum either on the ground or roof; 5) no building can exceed three floors, or about 7.2 meters; 6) no buildings can be closer than 5 meters; and 7) the minimum width of public circulation should be 2 meters.

The lower right hand chart examines options for pedestrian circulation in a cluster of dwellings. Two locations for entering and passing through a cluster are combined with the placement of stairs and the three stacking options for dwellings. Charts ABC can be transformed to accommodate these circulation alternatives. It should be noted that the site is so constrained that vehicular access and parking were excluded. In fact when public space is reduced to the minimum necessary for circulation and private dwellings are maximized to 50 meters square, as in chart A, there are no possible options to cluster dwellings around public open space.

Another example of the consequences of asking all three questions at once is exemplified in figure 10–17. "What, how much and in what relation" has been asked by a self-help construction cooperative of forty-five families about the basic topic of space use at the scales of dwelling and cluster for a steep hillside site on the edge of a major metropolitan region. Once again, however, when generating optional answers to all three questions at the same time, all the topics of space use, networks, and technics inevitably are addressed. Across the top of this chart are arrayed five possible arrangements for distributing a dwelling on the given slope. The first two diagrams depict dwellings that may ultimately reach 80m^2 by the owner gradually building additions over time. The other three indicate larger lots that permit two families to occupy a lot in the future since the initial owners can subdivide their lots. The five possibilities are composed of basic modular volumes determined by the slope of the hillside and the approximate spanning limitations of the materials most likely to be used in self-help construction. This is noted in the upper left hand corner. The two possibilities arrayed on the vertical axis indicate in section the consequences of entering a dwelling from above or below as it sits on the hillside. The chart combines these two sets of options for distribution and access, by also considering five additional environmental requirements: 1) the maximum height any family can and may build is two floors; 2) dwellings built on the downhill side of any public path running parallel to the hill cannot be built more than one floor high above the path to permit sunlight to enter the path since the hill faces south; 3) dwellings built on the uphill side of a path may not exceed the highest grade elevations of their lots to prevent them from blocking the view and sunlight of houses behind and above them; 4) each dwelling must include at least one outdoor patio on the ground and one rooftop terrace, each equivalent in area to one basic module, to insure proper light, ventilation, and useful outdoor space; and 5) all rooms except one must have at least one wall open to the outside to accommodate windows. The resulting alternatives display some of the consequences on the organization of dwelling and cluster when responding to the site, respecting these five environmental requirements and two additional concerns. On the vertical axis is noted a legend labelling the possible phases of construction suggested in the alternatives. Also noted are initial core house plans that must be built adjacent to the paths to minimize utility lines. The numbers in each diagram represent the amount of modular volumes available for enclosed habitable space. The bottom single line on the chart displays a variety of ways by which public circulation may pentrate different dwelling clusters.

A final example of an array of options that demonstrates the nature of alternatives generated by the last

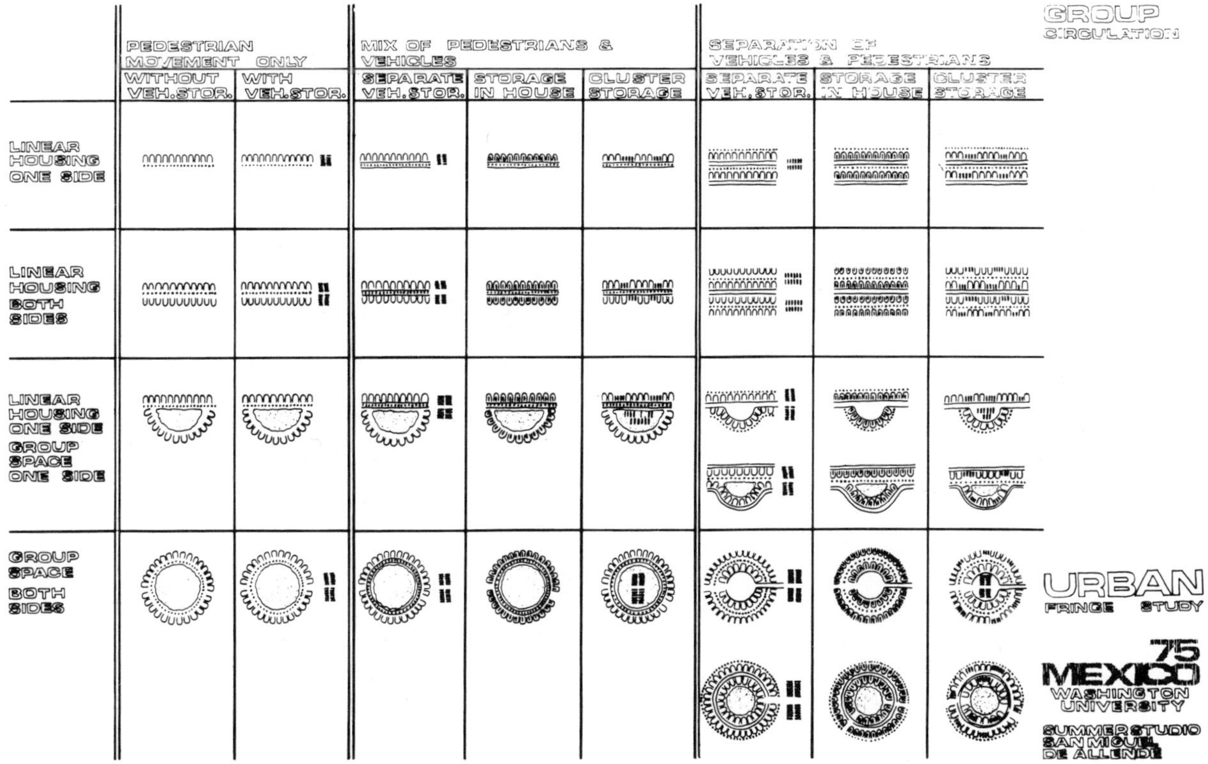

Figure 10–14
Chart representing options for networks at the scale of dwelling clusters.

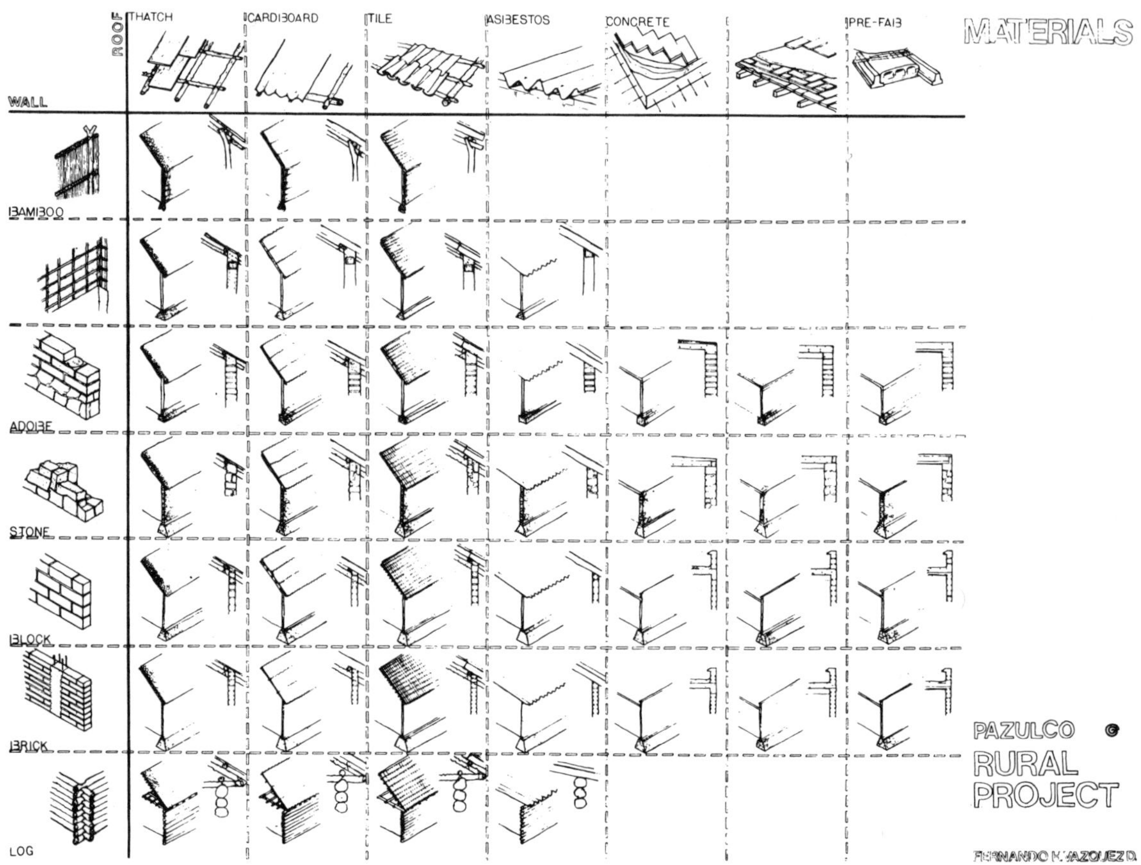

Figure 10–15
Chart representing alternative methods of construction and materials at the scale of house components.

composite question is shown in figure 10–18. It addresses the topic of space use at the scale of dwelling components for pre-established modular building volumes two floors in height that accommodate two family dwellings, partially overlapping. Six possible distributions of the living and sleeping activities are displayed for two families, one on each axis. The dwellings must respect the dimensions of a volume established by the construction technique to be used and a previously selected cluster pattern. Some environmental and technical requirements also respected were: 1) all rooms must have at least one wall open to the exterior to receive an adequate area of windows; and 2) all families must have access to the ground and to their roof. The thirty-six resulting alternatives not only display various locations for activities and the amount of space devoted to them in a dwelling, but also the degree of adjacency for these activities between dwellings.

To reiterate, this process is based upon the premise that the gradual layering of complexity facilitates the involvement of people in preparing residential development proposals. By examining a variety of alternatives at each stage of complexity a group not only broadens its awareness of what could be, but also prepares itself for more informed judgments about what should be. Therefore, the process enables participants to reduce the field of options in a conscious and arguable manner and at the same time address those remaining in more detail. The continuous presentation of multiple opportunities confronts the participants with the task of always re-examining their existing assumptions and conventions about how they live, how they relate to each other, and how they affect and are affected by their institutional contexts. To obtain full benefit from such a process, a group must then evaluate and select their courses of action from among the options, with the same rigor and inclusiveness that they used to generate them.

Evaluating Physical Options

When discussing options before selection, a group would need to be certain that it kept in mind its initial tentative intentions about the *internal* relations it preferred among its membership and the *external* connections it wished to encourage with its institutional framework. Options selected from the many responses to "what, how much and in what relation," no matter how small in scale, when implemented may alter the desired relations among the membership of the group as well as the group's external connections to existing institutions. Even during the whole process of preparing the residential development project, the initial attitudes about internal and external relations may

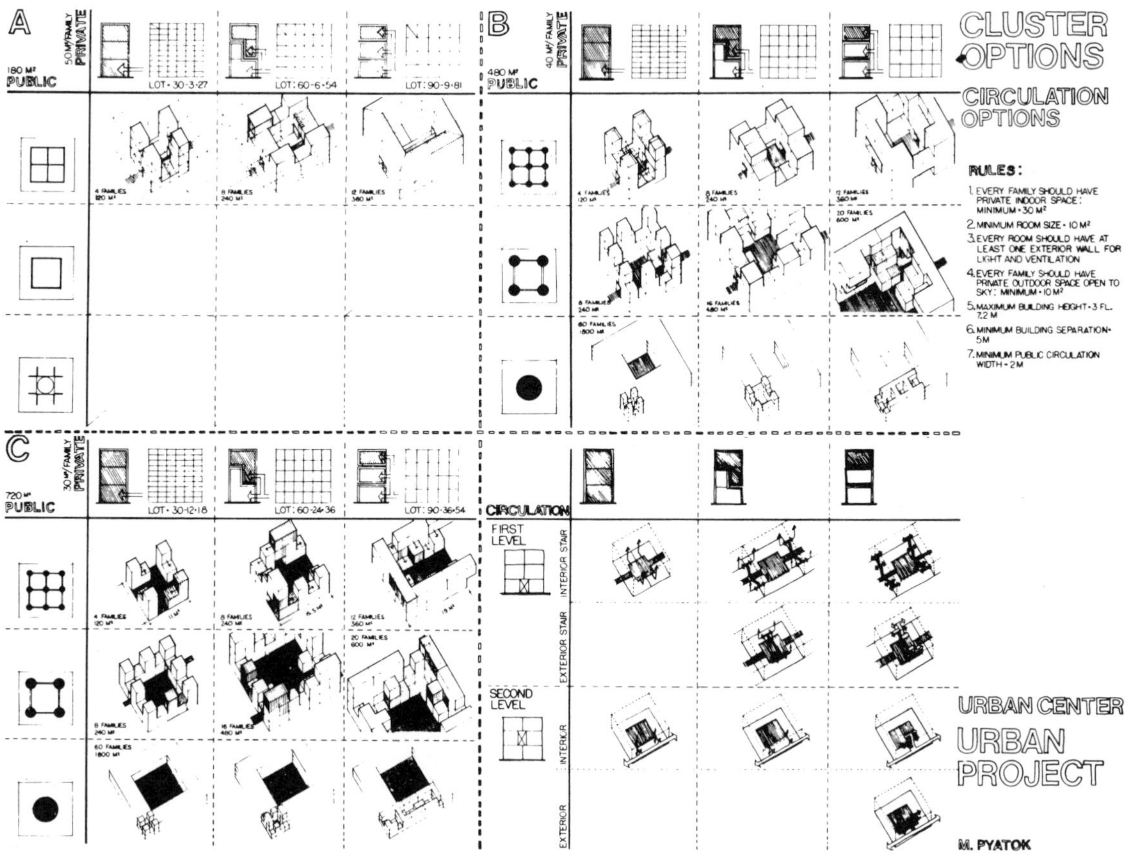

Figure 10–16
Chart representing options for space use at the scales of a site and cluster in the inner portion of an urban block.

change. So the awareness of these evolving attitudes is important because it provides the foundation for deciding upon an option, whether it be about space use and networks at the scale of the site, or a technical question about the components of a dwelling.

To test options on how well they may reinforce the desired internal or external relations, a group should focus its attention on the impact it will incur on: 1) political and economic arrangements; 2) social and cultural customs; 3) standards of environmental comfort; and 4) the acquisition and consumption of resources. When discussing any option generated to answer questions of "what, how much and in what relation," of any topic at any scale, these four issues may serve as an outline to insure a more inclusive review.

Each of the four main issues subsumes innumerable questions to use for evaluation. The following list identifies a small sample of questions that should be considered to understand and assess the impact of an option upon the internal and external relations of a group.

Internal Relations

A. Political and Economic Arrangements
   1. Whose labor is necessary to manage and maintain the resulting building and land use patterns?
   2. How will the necessary property regulations be enforced and by whom?
   3. How will the future growth and structure of the organization be affected?
B. Social and Cultural Customs
   1. Will new relations and obligations be required with neighbors and relatives, and are they too foreign to existing patterns?
   2. Will the continuation of existing desired customs and rituals be impeded?
   3. Will the appearance and arrangement of the project signify the group's identity and aspirations?
C. Environmental Standards
   1. Will sufficient light, ventilation, and heat be provided?
   2. Will enough open space for family or group outdoor needs be available?
   3. Will existing natural features be preserved or ignored?
D. Resources
   1. How much energy will be conserved?
   2. How much time and labor will be needed for building and maintaining the project?
   3. Will the available labor skills be sufficient or will special imported help and machinery be required?

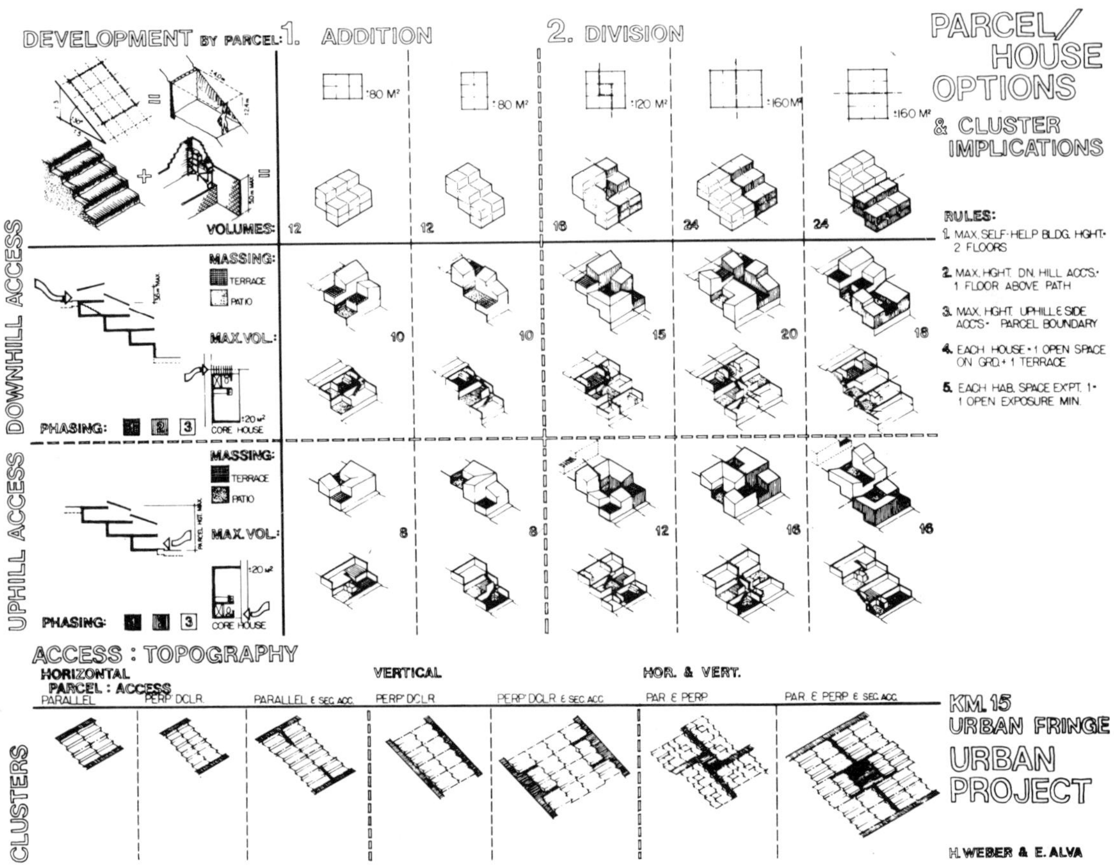

Figure 10–17
Chart representing problems of asking questions "what, how much, and in what relation" at the same time.

These same four general issues can generate a variety of evaluation questions about an option regarding its effect upon the group's relation to its institutional context.

External Relations

A. Political and Economic Arrangements

1. What obligations to banks and insurance institutions will be demanded?
2. What will be acceptable to government regulatory agencies—building codes, zoning ordinances, police protection?
3. What will be the effect on land values, taxation rates, existing services, and employment opportunities?

B. Social and Cultural Customs
   1. Will the desired integration or isolation of the group from its neighboring areas be achieved?
   2. Will a compatible appearance with neighboring areas be created?
   3. Will existing patterns of life important to long-term residents of neighboring areas be disrupted?

C. Environmental Standards
   1. Will any air, water, or noise pollution be imposed on surrounding areas?
   2. Will existing traffic and transportation systems be overloaded?
   3. Will needed open space for existing neighboring areas be provided?

D. Resources
   1. How is the existing labor market of the region affected?
   2. How is the existing material market affected?
   3. Will existing resources natural to the region be used?

All four categories of evaluation questions are significant to consider when discussing options generated at any scale, as will be demonstrated for the topics considered in figure 10–7 and figure 10–12.

A citizen group reviewing the general circulation options for a rural village in figure 10–7 might consider the following sample of questions.

Internal Relations

A. Political and Economic Arrangements—which options:
   1. will require the least amount of administrative difficulties, having the shortest roads, fewest intersections, least potential accidents between users of varying speeds?
   2. have the potential of providing equal access to all existing properties?

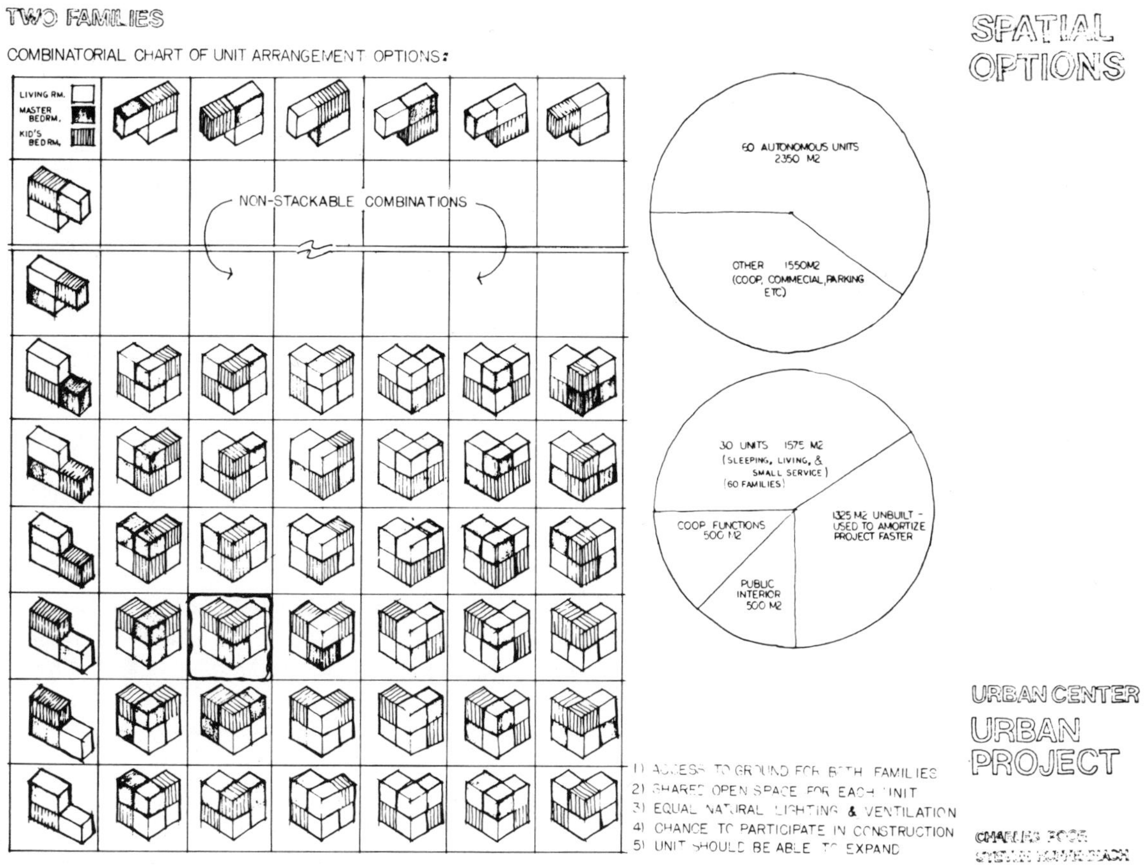

Figure 10–18
Chart representing options for space use for pre-established modular building volumes.

3. best define the maintenance obligations for different portions of the network—paths, lanes, streets, highway?

B. Social and Cultural Customs—which options:
1. will villagers accept if they need to mix their animals in outlying communal corrals in order to reduce the size of the village and its network of roads?
2. will the villagers comply with if they need to separate paths for animals and vehicles in order to reduce health and safety hazards?
3. may eliminate potential visitors from the village center and affect local shopkeepers?

C. Environmental Standards—which options:
1. achieve a balance between preferences for health and safety on the roads and convenience of movement?
2. reduce the inconveniences of intersecting and mixing users of different speeds?
3. create the least difficulty in accommodating topographical conditions?

D. Resources—which options:
1. require the least amount of labor and materials to build and maintain?
2. would facilitate the installation of other utilities such as water supply and waste disposal?
3. can be constructed most easily in phases without disrupting the life of the village?

External Relations

A. Political and Economic Arrangements—which options:
1. may require more or less negotiation with government agencies to determine final decisions?
2. may require more dependence on governmental agencies for upkeep?
3. facilitate the use of the town's commercial enterprises by outsiders?

B. Social and Cultural Customs—which options:
1. provide easy opportunity for the people of the region to use the village?
2. best maintain exclusion of passersby from areas of the village requiring more privacy?
3. least inconvenience vehicles passing through the town to other destinations but also display the full life of the village?

C. Environmental Standards—which options:
1. reduce the impact of noise and air pollution of passing busses and trucks on the dwellings neighboring the main roads?
2. reduce the potential hazards of village vehicles mixing and intersecting with high speed through traffic?

D. Resources—which options:
   1. may require the importation of materials and labor to construct and maintain the network?
   2. may cause drainage problems affecting the river and areas downstream from the village?

Another example of assessing options by considering these four major issues as they may affect the internal and external relations of a group will be demonstrated for the dwelling construction possibilities of figure 10-12. This chart arrays options to demonstrate which portions of a dwelling could be built initially and which could be built later by individual families.

Internal Relations

A. Political and Economic Arrangements—which options:
   1. achieve an acceptable balance between the group's economic resources, which would finance major construction, and the individual family's income, which would finance small scale incremental additions?
   2. for partial construction would create less disruption and argumentation between neighboring families undertaking future additions and renovations?
   3. equally distribute among the families the inconvenience of future construction?

B. Social and Cultural Customs—which options:
   1. provide families with ample opportunities to complete their dwellings, both inside and outside, in ways that have a significant effect on their way of life?
   2. minimize difficult construction responsibilities of each family, eliminating unnecessary stress and frustration?
   3. might generate workable collaboration and sharing of responsibility between families during future construction additions?

C. Environmental Standards—which options:
   1. may cause excessive noise and dust in a dwelling during future expansions or alterations?
   2. for initial construction require too low a level of comfort for a family while it saves to upgrade the condition of its dwelling in the future?
   3. create fewest hazards for children even though only partially constructed when first occupied?

D. Resources—which options:
   1. rely upon the skills already possessed by the families?
   2. allow families to use temporary, but easily available materials, until they can purchase the material and/or labor for more stable, permanent construction?
   3. require too much initial permanent construction for dwellings and may exhaust a family's economic resources necessary for other needs?

External Relations

A. Political and Economic Arrangements—which options:
   1. may have difficulties gaining approval by insurance institutions and being financed by banks?
   2. will meet resistance from local government authorities enforcing building codes?
   3. rely too heavily on family self-help labor, reducing employment opportunities for local tradesmen?

B. Social and Cultural Customs—which options:
   1. may cause unwanted stress on neighboring areas due to the inconveniences introduced by long-term incremental construction?
   2. may permit families to finish their dwellings in a way that satisfies their need to fit the character of the surrounding neighborhood?
   3. may stigmatize the project too long with the image of incompleteness, chaos, and instability, instilling a negative reaction and avoidance among the long-term residents of neighboring areas?

C. Environmental Standards—which options:
   1. may create too much noise and air pollution in surrounding neighborhoods during the long-term construction process?
   2. best control erosion and drainage during construction to minimize the soiling of surrounding areas?

D. Resources—which options:
   1. take advantage of the skills in the local labor market?
   2. permit families to use materials easily available in local markets?

When evaluating options generated for "what" and "in what relation" questions, qualitative or approximated evaluation responses would be most frequent. When "what, how much and in what relation" options are evaluated, there is more opportunity to make quantitative comparative judgments. However, what is most important to understand about this proposal for evaluating options is where it focuses the attention of those making decisions. Contrary to the beliefs of architects, the final appearance and configuration of the physical *product*—the form itself—is never judged in this evaluation proposal as the prime causal agent that affects human relations within a group. Certainly this process of continual creation will address issues of appearance and image along with others of social and cultural significance. What is primary is the very *process* of creating, implementing, and transforming a residential project and how these stimulating activities of life will affect the external and internal relations of a group who undertake them.

The architectural product is important as an influence on individual and social behavior insofar as it instigates a complex network of inventive human interaction to bring it into being, to maintain it, to use it, and to transform it over time. In that complex exchange, the physical environment is sometimes an unnoticed backdrop, sometimes a helpful participant, and sometimes an annoying but necessary presence. But most important of all, it becomes to those who use it a reminder of these many creative activities of their lives that have enveloped it, rather than becoming a determinant that envelops human activity. In this sense, the environment becomes a symbol to its users.

Hence, evaluating an array of proposed physical products on the basis of how well they may symbolize the future life of a people is asking for mystical powers of cultural foresight. Symbolism is a label that does not describe a property inherent to architectural form itself, but is a label of the process of exchange between form and maker/user, a relationship that evolves over time between the two because they also are evolving. Therefore, a dwelling cannot be "designed" to be symbolic before it evolves in the integrated process of making/using, nor can its *full* symbolic meaning be sensed through historical retrospection after this process ends. Neither designer nor historian can sense the full meaning of a dwelling unless they were participants in the process of creating, maintaining, using and transforming it. At best, they engage in interesting exercises that interpret meaning in their terms. For this reason, symbolism need not be an issue for discussion while a group is evaluating physical options. For the very act of participation by a group in generating, evaluating, and selecting physical options for their dwelling environment is the necessary exchange between users and the emergent physical form to establish already a symbolic union between the two. With dweller participation in design, the process of symbolization commences.

Selecting Design Alternatives

When the design of a residential development is pursued as a social process and a collective creative endeavor, a group, depending on the time it has available and the priorities it has established, may vary its degree of involvement. The selection of physical proposals that assemble the various options for pieces of the project at the various scales, therefore, can be either an evolutionary process that insists on full involvement of a group or one that is more delegated to the professionals working with a group.

In an evolutionary process, the intent of a group may be to develop full understanding by engaging in the design activity as a continuous debate over an extended period of time that requires a regular exchange of information. Judgments and definitions change as a function of criticism and new awareness that alters previous conceptions and at the same time progressively reduces options. This process comprises the gradual elimination of alternatives and the gradual assembling of those considered appropriate into a final developed project. Throughout the process the professionals working with the group contribute to the expansion of alternatives, the generation of issues and their representation, and the structuring of a provocative discourse that assists a group in transcending existing perceptions of "what ought to be."

In a delegated process, the objectives of a group may be to achieve the more immediate need of a physical proposal while devoting their time and energy to other endeavors as well. Such a process, to some extent, forgoes the regular deliberation of issues and the full involvement of a group. The design professionals, therefore, are delegated and assume the responsibility of generating and selecting partial solutions that are then assembled in a range of alternative completed proposals. The range for these proposals must encompass enough variety to include opposite extremes that would insure full debate before selection by a group. A group, however, not only must have the opportunity to understand and review the full sequence of preliminary choices made by the design professionals but also the responsibility to judge the assembled alternative residential proposals.

An evolutionary process

The design of a residential development by continuous deliberation with the full involvement of a group could be structured by following quite literally the sequence of topics and issues presented previously. What follows is a sketch scenario of such a process, which if pursued seriously, may consume weeks of deliberation.

A group might begin by defining itself. It would examine alternative internal organizations and relations to the surrounding institutional context. In order for the group to meet its objectives for the provision of shelter, food, energy, transportation, some employment, and some commodities, it resolves to form a cooperative. The cooperative intends to pool its resources, money, skill, and time, and thereby reduce its dependence on its institutional context. At the same time, the group develops a collective social alternative that retains the autonomy and identity of the nuclear family. Thus, in its "economic design," the cooperative will not only manage funds for the financing of loans, the collective buying of building materials, food, fuels, vehicles and some commodities, but also coordinate building contracts, self-help labor, the allocation and distribution of house parcels, and even administer communal child care and recreational programs and facilities.

When the group begins to define the physical issues for a residential development it might choose to investigate its design by addressing the questions of "what, how much and in what relation" for space use, networks, and technics together for each scale rather than separately. They then would establish a sequence that would examine the selection of a site, the links to the site context, the organization of the site, the dwelling clusters, and lastly, dwellings and other facilities. Their intent is to make sequentially related decisions that would determine courses of action and reduce options as they investigated successively smaller scales of issues.

When selecting a site, a group might assess the impact of various locations on their preferred internal and external relations. The filter of political and economic arrangements, social and cultural customs, environmental standards, and consumption of resources would be used to review and evaluate the options.

From a choice of small and expensive inner city parcels and larger urban fringe lots the group chooses an inexpensive, very steep site of about one hectare or 2.5 acres, illustrated in figure 10-19. Their selection is influenced by the desire to accommodate 40 families on house lots not less than 1,000 sq. ft. or $100m^2$ and allocate less resources to land purchase, even if it means reducing their accessibility to services and increasing their transportation costs. The removed location and difficult topography are seen as incentives to stimulate cooperation. Even the required extensive excavation work to terrace the site is seen as an opportunity to develop a local self-help, building materials industry.

In relating to the surrounding physical context the group rejects making an isolated enclave and instead chooses to consciously develop the site as one segment of an extended settlement that will be connected and become accessible to adjacent areas. Thus in organizing the site, the group selects a road network that offers vehicular access both at the top and bottom of the slope and concentrates collective facilities, not in the center but along those two edges where they will be shared by others. The center of the site is assigned primarily to residential parcels reached only by a grid of pedestrian paths and stairs and a vehicular service road that connects the perimeter streets at one edge of the site. This organization recognizes the intention to de-emphasize the need of private transportation, the desire to maximize house lots by minimizing a road network, and the reality of a very steep topography that would impose severe construction and cost constraints on road building.

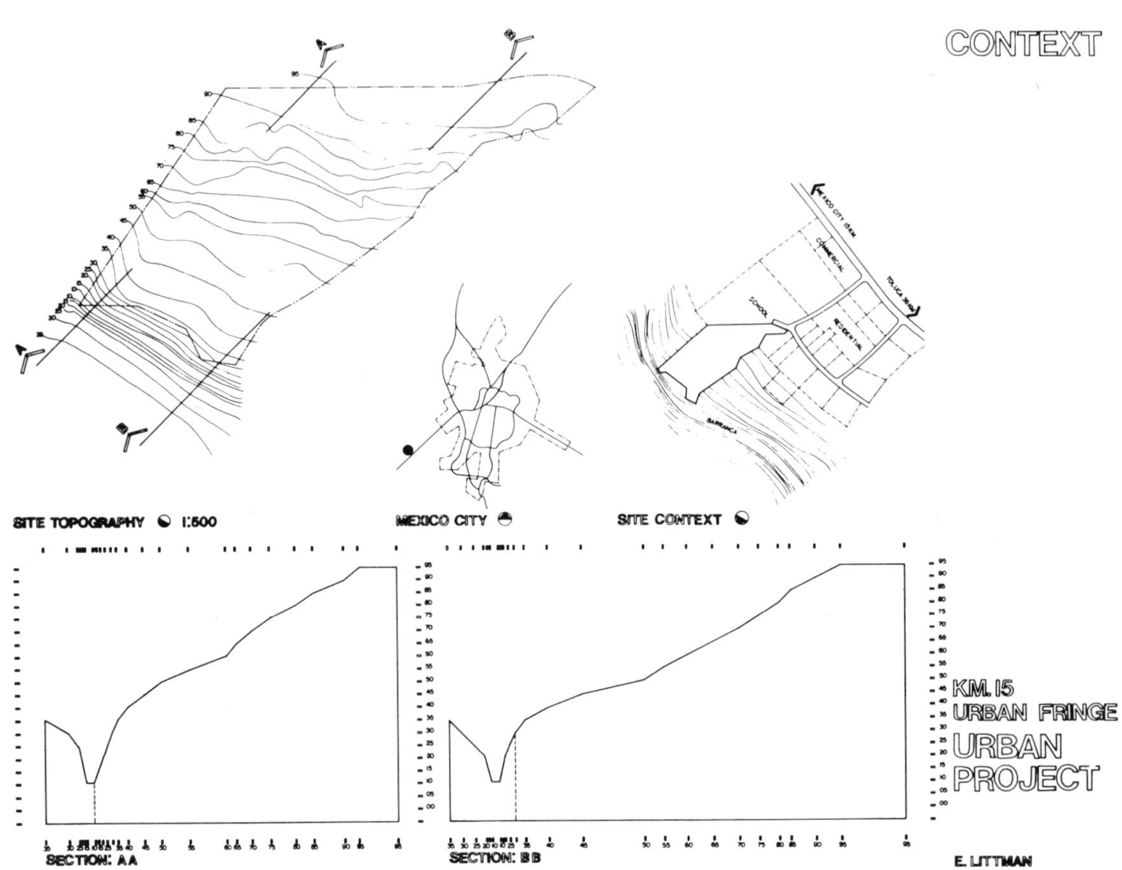

Figure 10-19
Chart of site chosen from a representation of evolutionary process.

Quantitative decisions about the site focus on the number and size of parcels. Given the previous decisions on the road and path network and the land assigned to shared space and facilities, about half of the land remains to be distributed for house lots. The choices range from sixty-five minimum lots of $80m^2$

to just forty lots of $130m^2$. Development options suggest a partial use of the site with smaller parcels to accommodate future growth in the cooperative or growth by making fewer parcels large enough to permit their use by two households. The group selects the latter option that creates the potential to maximize the total number of houses to eighty on an initial forty lots, each $130m^2$.

With the site organized, decisions at the scale of a cluster of dwellings consider how to group houses, how to provide additional open space, and how to relate houses to paths and a utility network. The group rejects any additional open space and intricate proposals to reduce the paths because they interfere with direct surface drainage systems on the steep site. The selected alternative is a regular grid of stairs and horizontal paths with open gutters that group houses in small blocks adapted to the topography. Subsequent discussions about the shape of lots and form of houses causes the group to select square parcels grouped in fours that can have access to both a stair and a path and lend themselves better to be shared by two houses around a common interior court. When the dwelling scale was examined, this solution was selected over more conventional wide front, shallow parcels that created too many paths, or narrow front, deep parcels that made the second house in the backs of lots inaccessible. The interlocking of two L-shaped houses around the court could give path frontage to each house and would maximize the individual house area on a smaller parcel by sharing with one other house the circulation and other services if so desired. The square parcel also was considered to be more adaptable to incremental construction over time, allowing rooms to be added as needed beginning with a core house and enabling the building of retaining walls and terraces to be phased in three stages.

Upon further investigation of house arrangement and building construction methods, the group chose not to establish a uniform or fixed design to be imposed on each dwelling. Instead it established a set of guidelines that would permit member families to pursue their initiative while assuring that no single individual action would be detrimental to others and would not compete with other construction work requiring the collective effort of the group. Thus, a zoning envelope was selected that insured no house would block the view, sunlight, and ventilation of another, and would always locate plumbing facilities closest to the path and utility network. The walls and floors of permanent construction would rely on concrete masonry units, small beams, and slabs fabricated from the aggregate excavated on the site. All other construction choices including non-bearing roofs and walls, windows, and other details would be made by each individual family.

The resulting design proposal, which evolved from the previous hypothetical set of decisions, is illustrated in figure 10–20. It represents in tangible form the synthesis of physical arrangements and relationships to match particular political and economic arrangements, social and cultural customs, and comfort needs, all within available resources. It reflects as a physical design proposal the group's internal organization as a social structure and its external linkages to the institutional context. Within each partial decision are contained carefully made and informed trade-offs and compromises that have considered shelter as only one endeavor for gaining control over all facets of life. The design proposal, therefore, is not an ideal universal solution; it is only the most appropriate resolution reached within the constraints of resources, time, and patience.

A delegated process

The design of a residential development can be delegated by a group to a professional without their continuous involvement. The process of defining issues, generating options, and selecting solutions need not differ in principle from the previously described evolutionary endeavor. Who undertakes the sequence of decision making and his/her synthesis or assembly phase would differ. Professionals would systematically and rigorously generate and investigate "economic design" alternatives, define physical issues, and generate partial solutions in an explicit and communicable manner. Not having the benefit of a continuous dialogue and feedback that progressively reduces options, professionals would have to synthesize a range of completed proposals. These would be generated by combining extreme opposite possibilities for significant variables. Since the combinatorial possibilities are innumerable, the selection of options made by design professionals must attempt to be as inclusive and explicit as possible in order to insure that a client group can engage the evaluation of design proposals with information, choice, and deliberation.

An example of a delegated process illustrates the responsibilities that would be assumed by design professionals working with a group to develop a site. The example uses the same site previously described, which the group had acquired on their own. In this case, because most members of the group are too busy with employment and other esponsibilities, they are unable to devote any time to extensive and regular involvement in the design process and prefer to delegate the task to professionals. The professionals should undertake a thorough generation of alternatives for the economic and physical designs very much like that presented previously. Then, they would have to identify those critical variables that will help to structure significantly different design proposals.

In this example the important determinant for the professional to consider is the influence of land and house ownership on the physical form and management of the project. This first variable concerning ownership suggests a range from private property

Figure 10-20
Graphic representation of final design proposal—alternative number one.

options at one end, to a regional cooperative at the other, with a more local cooperative in the middle. A group seeking private property for each family may pool capital for obtaining loans for management and construction and would focus its efforts to maintain the identity and autonomy of each house parcel. A more cooperative venture might seek to expand the shared responsibilities of a group not only to other aspects of life but also to join other cooperatives to form a regional network. The residential development would be owned and managed collectively and might include more extensive nonresidential functions such as cooperative employment and service opportunities. The professionals, therefore, would postulate three sets of schemes: those advocating private property; those that could serve the needs of a local cooperative; and those that would integrate into a regional cooperative.

How much of the site is developed, the size of parcels, and the extent of a road network offer a second layer of variables to structure several alternative design proposals. The professionals could develop options that allocate larger parcels to all members of the group using all the land. The parcels could then be subdivided to accommodate new families at a later date. Other options could develop only part of the land initially with smaller parcels and develop the remainder in a second phase. Within these options for initial land use, another variable is the extent and location of vehicular access. Cars could be excluded or cars could service all dwellings.

The two land-use alternatives, when combined with the three ownership possibilities, produce six basic schemes. These six schemes, C2 and C1, B2 and B1, and A2 and A1 are illustrated in figures 10-21, 10-22, 10-23, 10-24, and 10-25. The first scheme in each pair represents phased development with small lots; the second, full initial development of the site with larger parcels that can be subdivided in the future. The variable concerning the road network was introduced as follows: schemes A2 and B1 limit car access to the site; B2 provides partial vehicular access and A1 maximizes it; both schemes C1 and C2 respond to their requirements for regional integration by linking the site to its surroundings. The alternative road networks, in turn, offer several options for the location and amount of nonresidential facilities. Some are concentrated at the entrance to the site as in A1, A2, and B1; others are extended along the service road as in B2 and C1; and in the case of C2, commercial facilities are located on the road at the top of the site and institutional services occur along the bottom road.

The six basic schemes then serve to illustrate additional variables at the scales of a dwelling cluster and a dwelling. Some schemes, A1, B1, and C2 concentrate the public open space in one area and minimize shared space immediately adjacent to dwellings. Scheme B1 concentrates open space in a ramp system that extends the length of the site to facilitate service access; and scheme A1 provides a central place along the road. Schemes A2, B2, and C1 cluster the open space and group dwellings around smaller public courts distributed throughout the site; scheme A2 encourages the use of small courts by providing a collective water supply in each court.

At the scale of a dwelling, the number of terraces on a house parcel that result from the shape of the lots on the slope is used to develop additional variation between proposals. Schemes A2 and B2 use rectangular parcels that can be subdivided into two square house lots of two terraces each. Schemes A1, B1, and C1 offer single house lots on two terraces. Scheme C2 provides a special case of two smaller house parcels sharing three terraces with a common courtyard in the center of the middle terrace. All schemes also suggest that families could build their own houses beginning with an initial core house adjacent to a path and utility network. Only one scheme, A2, proposes both a uniform and regulated construction system for each family.

In this manner of combining a few variables, the six schemes cover a wide range of possibilities. The ranges of some of the variables were fully represented while others were introduced more casually among the suggested schemes. The variables that were selected and represented were chosen by professionals after they fully investigated the options for "what, how much and in what relation" for the various scales of site, cluster, dwelling, and component. Thus, when the schemes are presented to the client group, it is essential that the professionals include not only the range of illustrative solutions, but also the preliminary investigation and the reasons for stressing the few selected variables. This information will help a group to assess the schemes and to alter them to suit more closely their interpretation of how the project should be developed to meet their desired internal relations as a group and their relations to the external institutional context. As in the more evolutionary process, they would evaluate the schemes or parts of any scheme considering their influence on political and economic arrangements, their effect on social and cultural patterns, their fulfillment of environmental comfort standards, and their consumption of resources.

The few sample design schemes should not be seen as unchangeable proposals or as the only alternatives. They are only illustrative examples of combinatorial possibilities that demonstrate that parts of schemes can be mixed and matched and entire schemes altered. Conventional design practice would also offer alternative proposals. What distinguishes this process from conventional practice is that the factors that can be varied to generate proposals are surfaced and their full range is made explicit. This exposure puts in the hands of dwellers the opportunity to dismember, analyze, and evaluate proposals made by professionals and subsequently to modify or reassemble their own proposals.

(Text continues on page 203.)

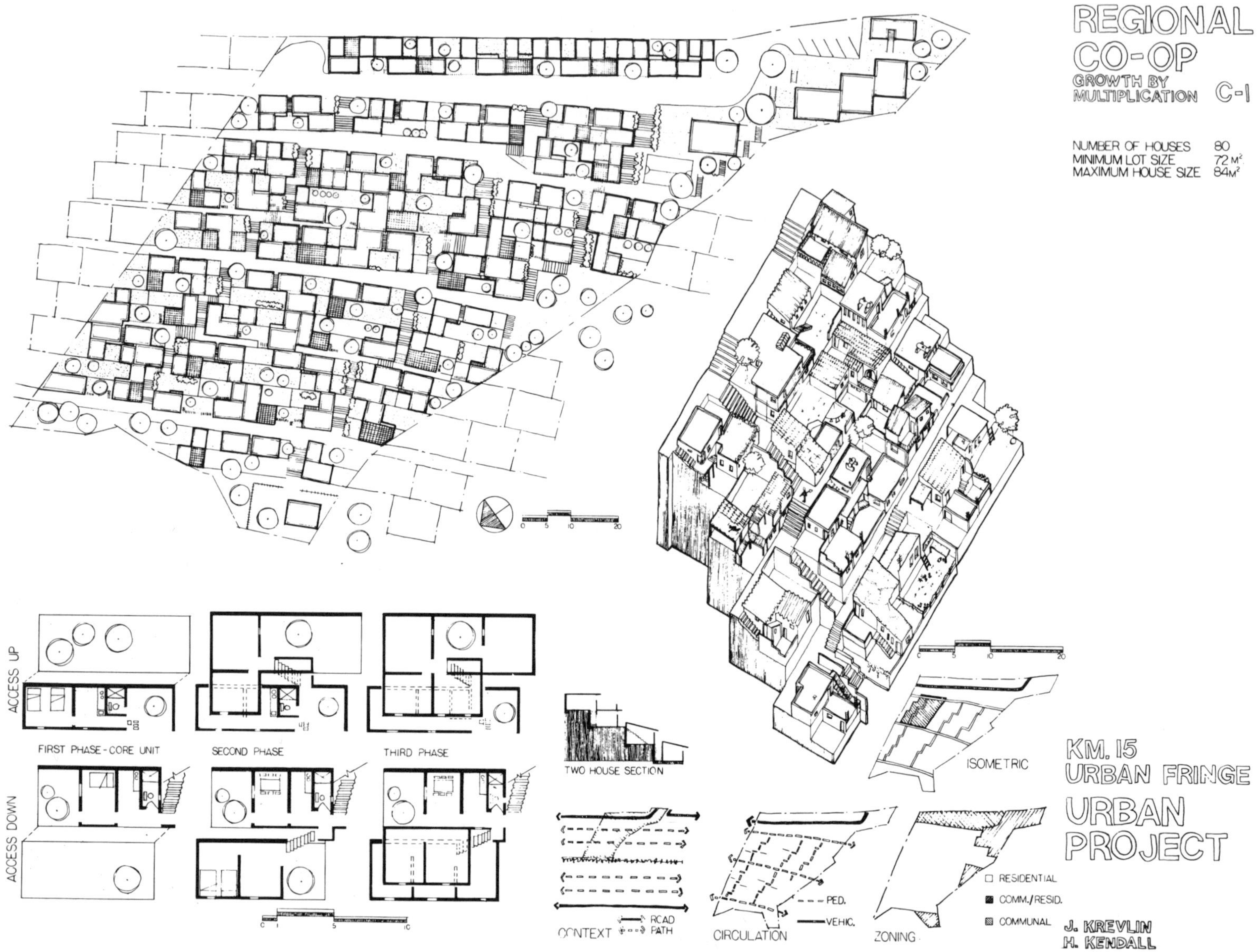

Figure 10–21
Graphic representation of design alternative number two.

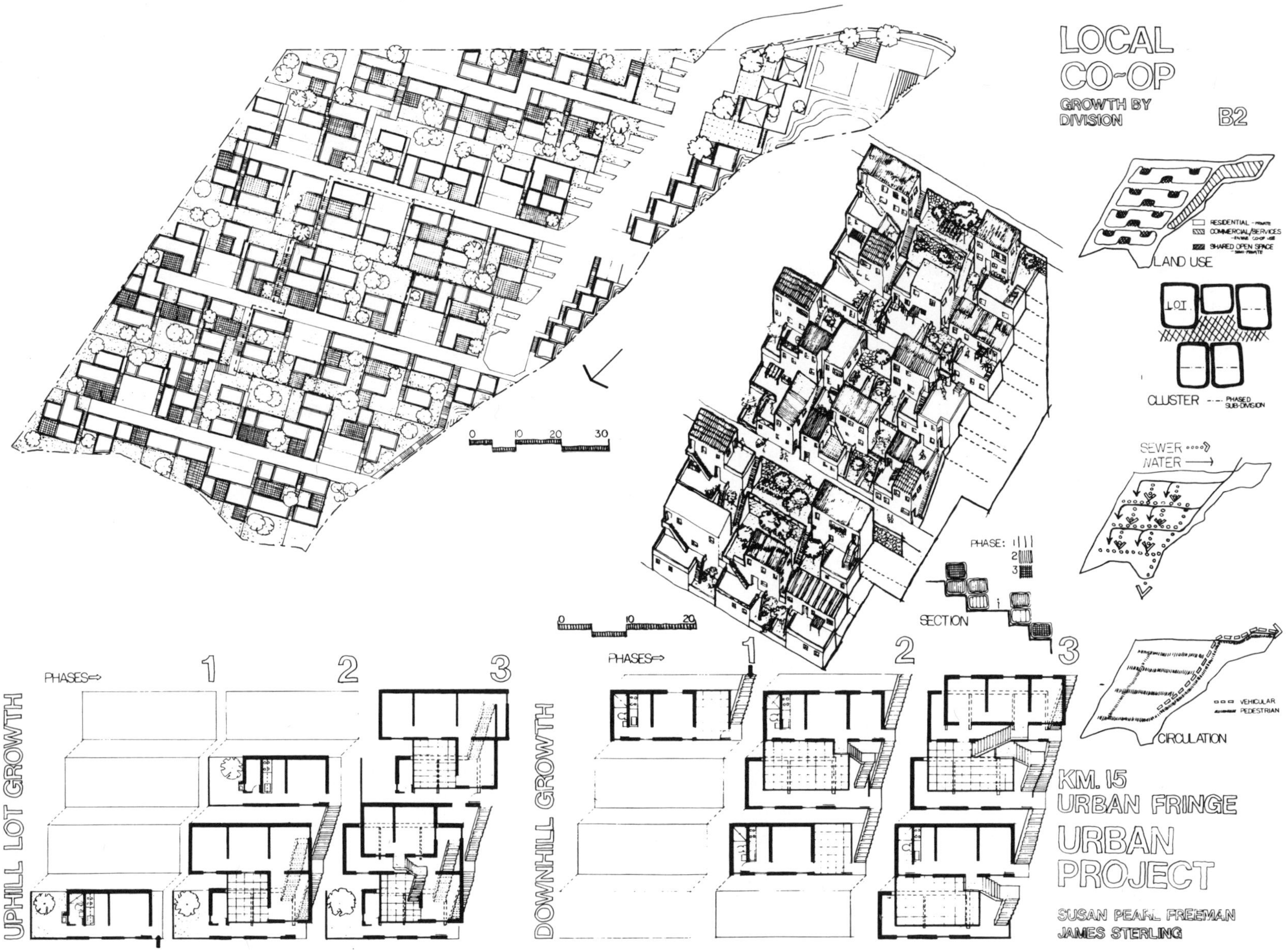

Figure 10–22
Graphic representation of design alternative number three.

Figure 10–23
Graphic representation of design alternative number four.

Figure 10-24
Graphic representation of design alternative number five.

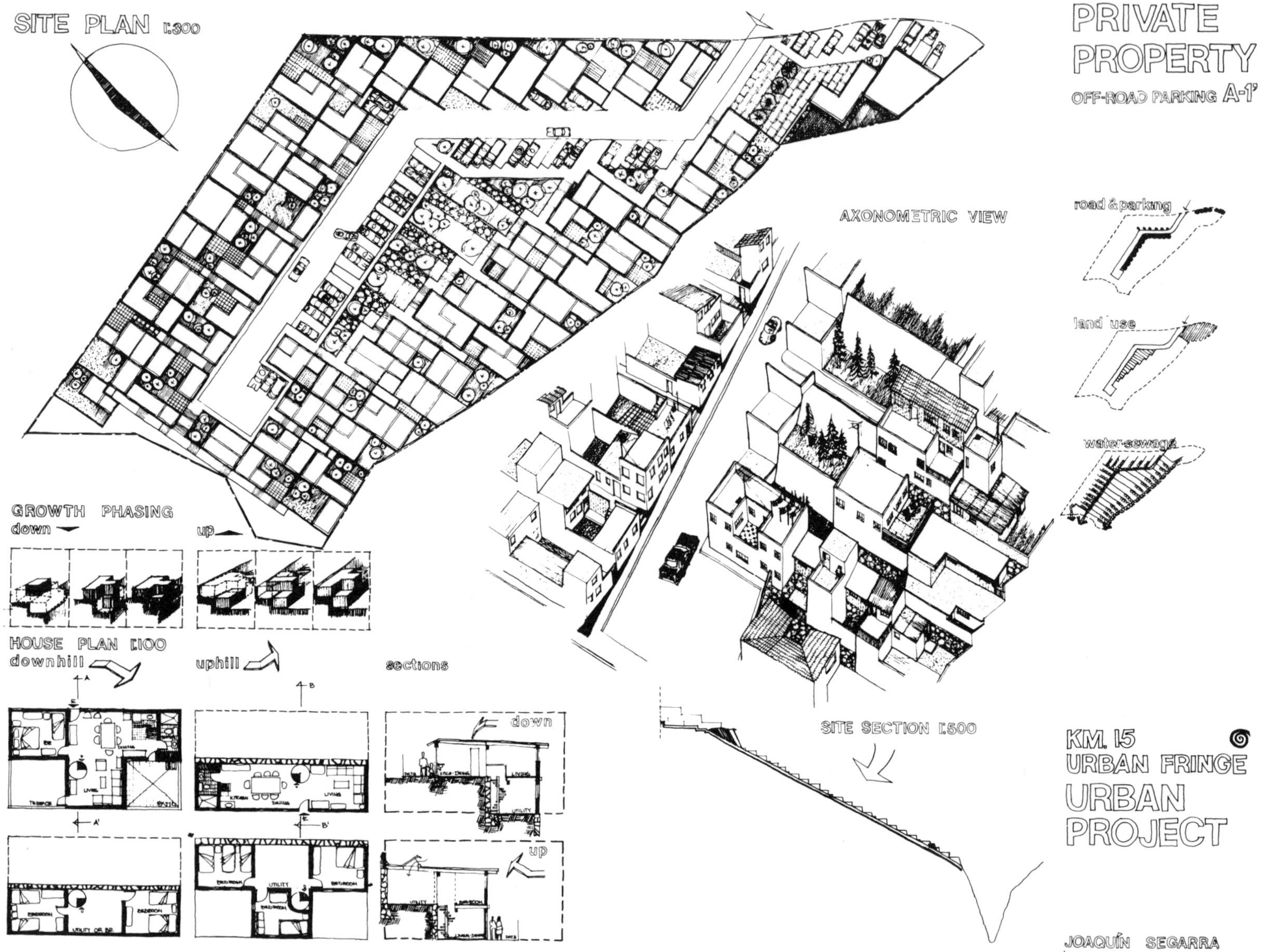

Figure 10-25
Graphic representation of design alternative number six.

CONCLUSION

When shelter is reduced to a market commodity that is designed, produced, and distributed only by professionals and those who control capital, people become alienated from their own dwelling environments even when they own and manage them. The loss of opportunity to participate in all the activities that bring dwellings into being impairs their abilities not only to contribute creatively to their own lives but even to perform the simple tasks of ordinary maintenance. In addition, such exclusion prevents people from understanding how the production system—that complex web of banks, insurance companies, real estate speculators, building and zoning codes, technology, and the compliant cadre of professionals and building trades—brings about shelter. The lack of understanding fosters a consumer mentality that embraces without question the production of shelter as a commodity. It also frustrates the creative potential of those who are aware that they have lost control over their lives and culture. Even the efforts to involve people in specifying the design of the product only continues to define them as merely "users" and consumers. They remain crippled by an inability to conceptualize an alternative process for the production of shelter. What is remarkable, however, is that even in the face of forces that militate against the contribution of dwellers, there are isolated cases of individuals who assert their presence through the creation and building of their own idiosyncratic dwellings. Such dramatic displays of individual initiative and inventiveness demonstrate that present in people are, and have historically always been, the desire and skills to shape their dwelling environments. Separate and individual efforts, however, are not adequate enough to alter the continued drive to centralize the production of dwellings as packaged commodities and the development of land as speculative ventures. Those who are aware of their loss of control need to join efforts and cooperate to gain sufficient leverage to alter their condition. (See figures 10-26 and 10-27).

Figure 10-26
Netzahualcoytl: Self-help dwelling in Mexico City (Photo by M. Pyatok)

Figure 10-27
San Raphael Chamalpa: Self-help settlement in Mexico City (Photo by M. Pyatok)

Cooperation inevitably requires a shared language to analyze and criticize "what is" and to conceptualize "what could be." To insure that the making of the physical environment achieves its full cultural and political potential, the language for dialogue about "what could be" must simulate the same physical and visual images that will exist in reality. In order for these images to have meaning and significance, they must emerge from those who will eventually use them when they become physical realities. Just as the abstract symbols of verbal language impart no meaning without frequent use and a collectively agreed upon semantic content and syntactic structure, so too the physical environment will impart no meaning unless people are often interactively engaged in the act of shaping their environment, using a shared language from which the environment will eventually be made.

To reiterate, most people's contact with their environment has been reduced to the experience of a user of a commodity. The existing social relations do not permit people to become makers of the environment, involved in the critical processes to bring it into being continuously. These processes nourish rich symbolic attachment and necessitate a shared language. Thus, to evolve a language capable of constructing a new reality, its makers must be critically aware of how the existing language for describing and discussing the environment has imbedded within it both the content of existing reality and its justification, making it difficult to criticize itself. For this reason the design and development of residential settlements require that their makers/users re-examine their own internal relations as well as their links to the existing context. They need to identify how these relations constrain and handicap their very ability to conceptualize an alternative. The creation of an alternative shelter process and product, the re-examination of social and political relationships, and the formation of a language to accomplish this must all organically evolve together.

In the current production of shelter, professionals employ a language for conceptualizing that is detached from the reality of dwellers but attached to the reality of those who presently own and produce shelter as a commodity. The alternative presented here attempts to reunite professionals and laymen in creating an appropriate shared language for the mutual task of re-examining existing relations and conceptualizing an alternative shelter process. This proposed language is, of course, not yet genuinely internalized and shared to produce spontaneous discourse. It is still a mechanical surrogate that reflects the separation by class and training that the ongoing context has created between its authors and the laity. Its intent is merely to begin.

(The work demonstrated in this proposal was undertaken by students in the Mexico Summer Design Program sponsored by Washington University in 1975 and 1976. The authors wish to acknowledge the contributions of:
Summer 1975—Marilyn Cote, Glen Garrett, Domingo Gonzales, Karl Grice, Thomas Hotaling, George Kimmerle, Andrew Metter, Bruce Osen, Lisa Roberts, David Schraer, Richard Springwater, Shinichiro Watari, Dewey Webster, Richard Wise, Elizabeth Yost.
Summer 1976—Joyce Bond, Terry Cohn, Warren Crain, Fernando Diaz, Joan Duncan, Tobias Flatow, Susan Freeman, Alberto Gomez, Heidi Helf, Daniel Jay, Harry Kendall, Joan Kirshheimer, Joan Krevlin, Ken Levien, Andrew Lind, Gregory Maxon, Edward McFarland, John Ostrum, Charles Poor, Sergio Queda, Margarita Romirez, Steven Ruffenach, Joaquin Segarra, James Stirling, Zaida Tarazona.)

11

Yerba Buena Planning Ballot

Charles Tuner, Jr.
Director

Lawrence Adams
Andrew Beckerman

Community Design Center
San Francisco, California

The Yerba Buena Center Urban Renewal Project has been the focus for more than twenty years of planning, controversy, litigation, and dissension. In the past year alone more than five different alternative plans for developing YBC have been proposed. In turn, these plans have been the subject of more controversy, numerous meetings, committees, speeches, press releases, and debates, all of which tended to raise the level of public information and interest without providing any effective way for people to make decisions, register opinions, or select from the potpourri of plans and proposals—a situation that has the effect of giving more importance to the views of a few protagonists than those of ordinary citizens.

The Yerba Buena Planning Ballot was conceived as a way to overcome this situation, to broaden active citizen participation by providing others—people who do not or cannot attend or speak out at public meetings—with a medium to express their feelings and ideas. The ballot was designed, in this instance, to provide residents of the south of Market area (the area most directly affected by the project, and also an area with a large population of low-income minority families and elderly residents whose needs and aspirations received the least consideration and whose views are not often solicited or otherwise obtained) with a means of expressing their preferences on what should or should not be built in the Yerba Buena Urban Renewal Project.

The Yerba Buena Planning Ballot is made up of three parts: three propositions outlining the major alternative plan concepts; thirteen policy statements dealing with how the project should be planned, financed, and administered; and a section containing a map of the project area and a selection of land uses from which people can devise their own plan for YBC. The detachable part of the ballot can be mailed or placed in one of the YBC ballot boxes located in the south of Market area. Together, the three parts of the ballot provided people with a way to organize and express their preferences on the most important facets of a large complex urban renewal project.

The Yerba Buena Planning Ballot enables people to design their own plan for YBC. Citizens are able to express their preferences at several levels: for overall concepts; for policies; for individual land uses; and for activities. In allowing for the interrelatedness of design, policy, and administrative considerations in decision making, the ballot provides a realistic context for citizens to order their preferences and serves as a tool for planners, designers, and policy workers to measure public reactions and preferences on a variety of planning issues.

COMMUNITY DESIGN CENTER
215 HAIGHT STREET
SAN FRANCISCO, CALIFORNIA 94102

PLACE POSTAGE HERE

(FOLD HERE)

(select ONE from each group)

CONVENTION CENTER

a. ____ b. ____ c. ____ d. ____

PARK AND OPEN SPACE

a. ____ b. ____ c. ____ d. ____ e. ____

HOUSING

a. ____ b. ____ c. ____ d. ____ e. ____

OFFICE BUILDING

a. ____ b. ____ c. ____ d. ____

INDUSTRY

a. ____ b. ____ c. ____

APPAREL MART

a. ____ b. ____

SPORTS FACILITY

a. ____ b. ____

OTHER LAND USES (select those which you want)

a. ____ b. ____ c. ____ d. ____
e. ____ f. ____ g. ____ h. ____
i. ____ j. ____ k. ____ l. ____
m. ____ n. ____ o. ____

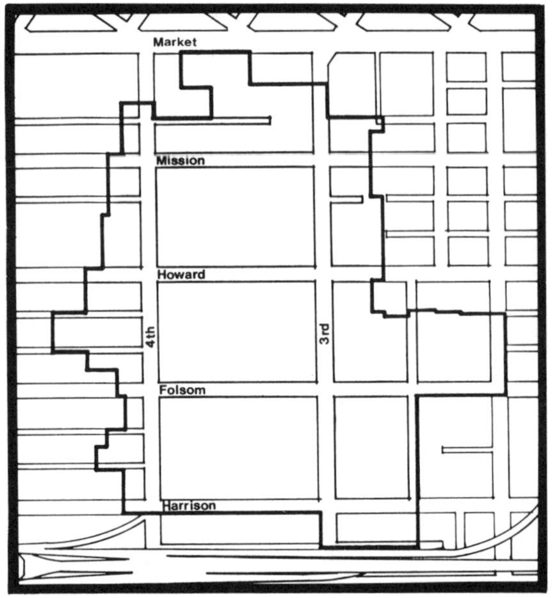

If you would like, use the map above to sketch your own plan for YBC.
Please indicate on the attached ballot your choices of the items listed below:

MAJOR LAND USES FOR CENTRAL BLOCKS (select ONE from each group)

CONVENTION CENTER

a. ABOVE GROUND CONVENTION CENTER
b. UNDERGROUND CONVENTION CENTER
c. UNDERGROUND CONVENTION CENTER WITH PARKING GARAGE
d. NO CONVENTION CENTER AT ALL

PARK AND OPEN SPACE

a. TIVOLI TYPE PARK WITH ADMISSION FEES TO COVER COST OF DEVELOPMENT AND OPERATIONS
b. TIVOLI TYPE PARK FINANCED FROM RENTS AND PROPERTY TAXES
c. LARGE PUBLIC PARK
d. SMALLER NEIGHBORHOOD TYPE PARKS AND OPEN SPACE
e. NO PARK FACILITY AT ALL

HOUSING

a. 2,000 UNITS OF LOW & MODERATE RENT HOUSING
b. 2,000 UNITS OF MARKET RATE HOUSING (RENTING FOR $600.00 OR MORE PER MONTH)
c. A COMBINATION OF LOW & MODERATE INCOME HOUSING AND MARKET RATE HOUSING
d. 100 UNITS OF LUXURY HOUSING
e. NO NEW HOUSING AT ALL

OFFICE BUILDING

a. 4 MILLION SQ.FT. OFFICE SPACE
b. 3 MILLION SQ.FT. OFFICE SPACE
c. LESS THAN 2 MILLION SQ.FT. OFFICE SPACE
d. NO OFFICE SPACE AT ALL

INDUSTRY

a. 200,000 SQ.FT. LIGHT INDUSTRY
b. 100,000 SQ.FT. LIGHT INDUSTRY
c. NO LIGHT INDUSTRY AT ALL

APPAREL MART

a. 1 MILLION SQ.FT. APPAREL MART
b. NO APPAREL MART AT ALL

SPORTS FACILITY

a. LARGE SPORTS ARENA
b. NO SPORTS ARENA AT ALL

OTHER LAND USES FOR THE YBC PROJECT (select those which you want)

a. 45,000 SQ.FT. ENTERTAINMENT CENTER
b. 20,000 SQ.FT. COMMUNITY-ORIENTED SHOPS AND STORES
c. 1,000 ROOM HOTEL
d. PUBLIC PARKING GARAGE FOR 1,800 CARS
e. SEVERAL SMALL URBAN PLAZAS
f. MEDICAL CENTER
g. RESTAURANTS
h. NIGHT CLUBS & BARS
i. 2,200 SEAT THEATRE/CONCERT HALL
j. 400 UNITS TOOR ELDERLY HOUSING
k. DOWNTOWN HIGH SCHOOL
l. PEDESTRIAN MALL
m. TENNIS COURTS
n. COMMUNITY CENTER
o. CHILD CARE CENTER

YERBA BUENA PLANNING BALLOT

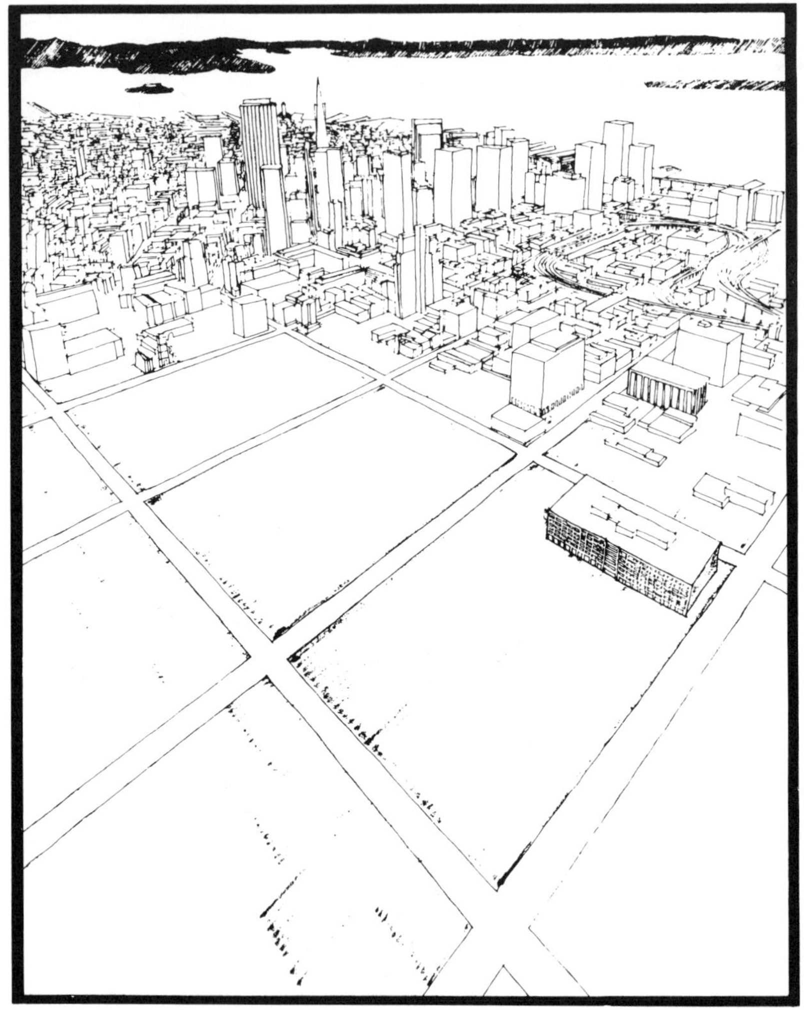

PREPARED BY THE COMMUNITY DESIGN CENTER

INTRODUCTION

ISN'T IT ABOUT TIME SOMEBODY ASKED YOU?

What do you think should be built in the Yerba Buena Center urban renewal area? What kind of place should YBC be? What would be best for all of San Francisco?

When everyone joins in planning Yerba Buena Center, we'll get the kind of YBC everyone wants. Help decide. Cast your ballot now.

Planning YBC has been going on for more than 20 years. During this time no one has ever bothered to ask the people who live in San Francisco to vote on what they wanted to see built in YBC; to vote on what they thought San Francisco needed most.

In fact, a great effort has been made to keep people from voting on which things in YBC should be built and paid for with public money. This is one of the reasons why YBC has come to so little in 20 years and why most of the land is now vacant and there is so little agreement on what happens in YBC.

What do you think should be built in the YBC urban renewal area?

At the present time, the Yerba Buena Project Area is just a few buildings surrounded by a lot of vacant land.

Vacant land is valuable in San Francisco, too valuable to be used only as a vast parking lot. And too valuable also to let a few fat cats decide how it's going to be used to make themselves some money, and let you pay for it.

San Francisco has too many problems and needs for that! The people of San Francisco should decide what's going to be built in YBC.

We need to decide what's important. YBC cannot at once be all things to all people. However, it can be something for everyone. We need to set our priorities; decide what comes first: residents or tourists, housing or convention centers, open space or parking lots, offices or hotels, stores or theatres. Then we can go ahead and have it built.

We think your answers can help everyone interested in YBC to develop plans that work, plans that people will support, plans that represent the way people feel about development in the South of Market area.

INSTRUCTIONS

The Yerba Buena Planning Ballot has some of the important questions we need to answer. You don't have to be an expert to answer them. All you have to do is select the answers that come closest to what you'd like to see built in the Yerba Buena Project.

This ballot is intended to provide you with an opportunity to express your opinion and to put together your own alternative plan for the facilities, services and activities in the Yerba Buena Project.

The YBC Planning Ballot is made up of a combination of *3* PROPOSITIONS, *13* STATEMENTS, and a SELECTION on the back page for you to make your own plan. Be sure to check each Statement (agree or disagree) and select ONE (1) Proposition on the ballot, as well as make the choices for YOUR PLAN.

When you have marked your ballot, then tear it along the perforated line and deposit it in one of the ballot boxes at the following locations:

CANON KIP
705 Natoma St.

NEW ARRIVAL CENTER
1173 Mission St.

SALVATION ARMY SENIOR ACTIVITY CENTER
360 4th St.

SOUTH OF MARKET HEALTH CENTER
551 Minna St.

SOUTH PARK COMMUNITY CENTER
64 South Park

CLEMENTINA TOWERS
320 Clementina St.

ALEXIS APARTMENTS
390 Clementina St.

SILVER CREST
133 Shipley St.

OR MAIL IT directly to:

COMMUNITY DESIGN CENTER
215 Haight St.
San Francisco, Ca. 94102

ALL BALLOTS SHOULD BE MAILED OR DEPOSITED IN A BALLOT BOX BY AUGUST 12, 1976.

PROPOSITION A

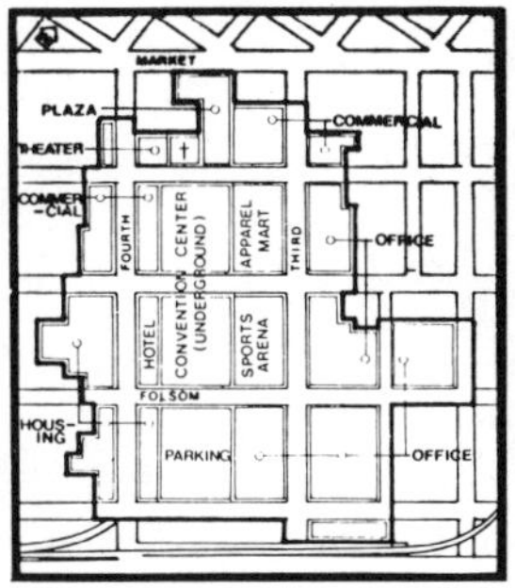

YBC should be built as a convention, sports, commercial complex with the following features:

- 330,000 sq.ft. UNDERGROUND CONVENTION FACILITY
- 16,000 seat SPORTS ARENA
- 2,200 seat THEATRE
- 1,000 room HOTEL
- PEDESTRIAN MALL
- 450,000 sq.ft. COMMERCIAL SPACE
- 1 million sq.ft. APPAREL MART
- PUBLIC PARKING GARAGE for 1,800 cars
- 4 million sq.ft. HIGH-RISE OFFICE SPACE
- 400 units TOOR ELDERLY HOUSING

PRO'S

- The convention center complex would provide San Francisco with new and competitive facilities to attract and accomodate additional conventions each year.
- Increasing tourism and conventions will benefit 60,000 people in San Francisco directly or indirectly.
- The sports arena, in addition to supplementing the convention facilities, could also bring major league hockey and basketball to San Francisco.
- The convention/sports/commercial complex will provide 3,500 new jobs.
- This plan would enable San Francisco to attract the largest conventions which cannot now be accomodated in existing facilities as well as accomodate several conventions simultaneously.

CON'S

- The convention complex will not pay its own way. It is highly unlikely revenue from leases, tax increments and other revenue measures will be enough to support the repayment of the cost of construction and current operating cost of the convention com-
- A convention center financed according to this plan would increase the City's bonded indebtedness without public approval.
- Money from the City's property taxes (General Fund) would be needed to pay for YBC public facilities such as the convention center and sports arena.
- San Francisco taxpayers would end up paying for a convention facility conceived for the direct financial benefit of a favored group of hotel owners, restauranteurs, and other business interests.
- This plan would require repayment of tax-exempted bonds with revenues from the Center and the City's property taxes.

PROPOSITION B

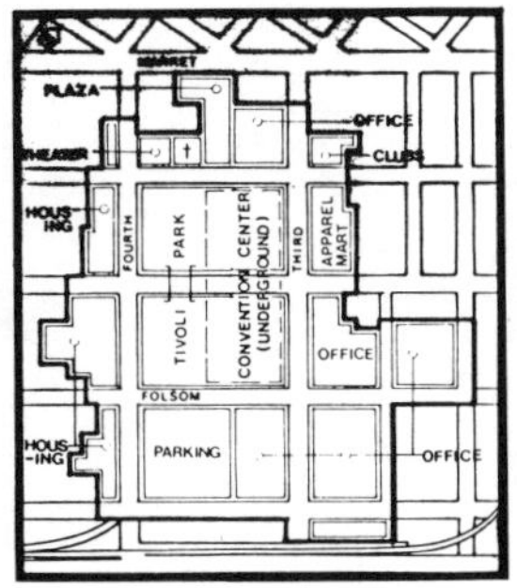

YBC should be developed as a "Tivoli"-type garden, amusement park and restaurant complex with the following features:

- 18 acre "TIVOLI" PARK COMPLEX
- 400 units TOOR ELDERLY HOUSING
- 3.5 million sq.ft. HIGH-RISE OFFICE SPACE
- PUBLIC PARKING GARAGE for 1,800 cars
- RESTAURANTS, BARS & CLUBS
- THEATRE
- 1 million sq.ft. APPAREL MART
- 330,000 sq.ft. UNDERGROUND CONVENTION FACILITY

PRO'S

- The park complex and other development would provide approximately 2,800 new jobs.
- The park complex would provide a variety of recreational and cultural opportunities for residents and workers in the immediate vicinity, as well as other San Franciscans and visitors to San Francisco.
- The park complex would pay for itself and generate additional income for the City. The additional income could cover part of the costs of the convention facility and pay for community services in the area.
- The park complex will bring major open space into the downtown area.
- Development of the park complex would still allow development of other facilities such as a convention center.
- Revenues from the "Tivoli" part of the plan could be used to assist development of rehabilitated housing and community facilities in other areas in the South of Market.

CON'S

- The park complex would prevent development of an above ground convention facility. Putting this facility underground would be more expensive.
- A "Tivoli"-type park complex would use prime commercial land for open space and limit expansion of downtown activity.
- A "Tivoli"-type park complex would not satisfy all the recreation needs of South of Market residents.
- This plan does not provide the entire solution to the financing of the convention facility.
- It may be difficult to find a public or private organization to develop and operate the park complex.
- The development of the "Tivoli"-type plan would require moving the proposed apparel mart to a less desirable site in YBC which would jeopardize the development of such a mart.

PROPOSITION C

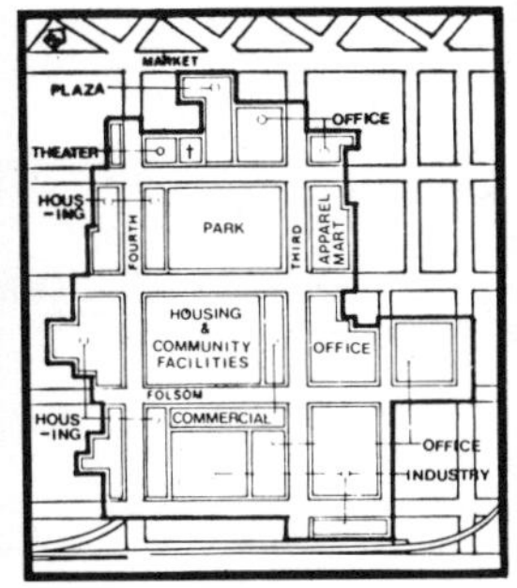

YBC should be developed as an urban residential and working community with the following features:

- 400 units TOOR ELDERLY HOUSING
- 2000 units LOW AND MODERATE INCOME HOUSING
- 100 units LUXURY HOUSING
- 20,000 sq.ft. community-oriented COMMERCIAL SPACE (Shops & Stores)
- 9 acre PUBLIC PARK for residents and downtown workers
- 1 million sq.ft. APPAREL MART
- 200,000 sq.ft. LIGHT INDUSTRY
- CLINICS, SCHOOLS, CHILD CARE CENTERS, LIBRARY AND OTHER COMMUNITY SERVICES
- 3 million sq.ft. HIGH-RISE OFFICE SPACE

PRO'S

- An urban residential-working community would provide a balanced neighborhood with housing, working and recreational opportunities for all San Franciscans.
- Much needed housing would be provided for all income levels.
- The combination of living-working would ensure continuous 24 hour use of the YBC area and provide more pedestrian amenities.
- Community facilities would serve the larger South of Market area as well as residents of YBC.
- New employment opportunities for 1,500 commercial, industrial and service workers in San Francisco would be provided.
- Approximately 55 acres of land would be returned to the City's tax rolls and generate income for the City's General Fund.
- This plan does not require bonded indebtedness of the City for any facilities except the public park.
- The 9 acre public park would be the only major open space in the South of Market area and would serve both residents and downtown office workers.

CON'S

- Given current construction and financing costs, it would be difficult and would take a long time to develop housing in the area.
- There is not enough demand for existing industrial space in San Francisco to warrant using land in YBC for industry.
- It would be too difficult to develop industrial uses in YBC.
- An urban residential-working community would require major land use changes and substantial replanning of YBC.
- This plan ignores San Francisco's need to attract visitors and conventions.

STATEMENTS ?

1. In order to increase San Francisco's share of the convention market, a larger modern convention facility/exhibition hall should be built in YBC to accomodate those conventions that cannot be adequately handled elsewhere in the City.
2. The use of any public funds to pay for the construction of facilities such as the convention center, exhibit hall, malls, parking garages and sports arena must be approved by San Francisco voters.
3. The YBC convention complex should be financed with private funds and an increased hotel tax, thus placing the cost on those who will most directly profit from the convention business.
4. A new 16,000 seat sports arena should be built at YBC to accomodate major league sports such as basketball, ice hockey and boxing matches.
5. At least 18 (about 20%) of the 87 acres in YBC should be devoted to parks and open space to serve South of Market residents and those employed in the area.
6. Land on the western side of YBC should be reserved for the development of at least 1,600 additional units of low and moderate income housing.
7. Space for light industry such as printing, apparel manufacturers, motion picture production, bakeries, candy factories and other downtown support industries should be reserved in the southern part of YBC.
8. YBC should include neighborhood services and facilities such as community centers, health clinics, shopping centers, parks and educational facilities for use by those living in the areas near YBC.
9. Existing residential communities adjacent to YBC should be protected from the impact of expensive development which would spur demolition and displacement of low rent housing and commercial space.
10. High rise office buildings should be located along the mass transit corridors on Market and Third Streets and should conform to the height and bulk limitations in the Master Plan.
11. In order to comply with the City's "public transit" first policy, new off-street parking garages should be minimized in the YBC area.
12. Priority should be given to YBC [illegible] that demonstrate an existing affirmative [illegible] plan with a satisfactory [illegible] a job training program.
13. A majority of both [illegible] jobs created by YBC [illegible] cisco residents.

PROPOSITION:
A B C

Select the one that comes closest to what you would like to see built in YBC.

BALLOT

1. AGREE DISAGREE
Comments

2. AGREE DISAGREE
Comments

3. AGREE DISAGREE
Comments

4. AGREE DISAGREE
Comments

5. AGREE DISAGREE
Comments

6. AGREE DISAGREE
Comments

7. AGREE DISAGREE
Comments

8. AGREE DISAGREE
Comments

9. AGREE DISAGREE
Comments

10. AGREE DISAGREE

11. AGREE DISAGREE

12. AGREE DISAGREE

13. AGREE DISAGREE

A B C

SAMPLE

12

Capital City Readout, AAAS Popular Exposition in Washington, D.C.

Thomas V. Vonier
American Institute of Architects Research Corporation

Richard A. Scribner
American Association for the Advancement of Science

Excerpts reprinted from *Community Information Expositions,* Thomas Voiers and Richard Scribner, Washington, D.C.: American Association for the Advancement of Science, 1973.

During its December, 1972, annual meeting in Washington, D.C., the American Association for the Advancement of Science (AAAS) offered to citizens of the region an exposition on the role of science and technology in addressing local social problems. Using a variety of communication techniques in exhibits, forty public and private organizations portrayed problems of the region and efforts being made to solve them. Organizations and their exhibits were grouped into three theme areas:

- Planning for social, physical and technical resources.
- Community health and well-being.
- Use, operation, and management of natural resources.

To involve visitors further in considering policy-oriented exhibits, organizers of the exposition prepared a ballot on twenty-two issues, which allowed visitors to register their responses to topics raised in the exhibits.

Over 5000 people attended the exposition, "Capital City Readout," which remained open to the public for five days and four nights. This chapter describes the objectives, development, organization, and physical design of the exposition as well as the participants, their contributions, and activities.

OBJECTIVES AND GOALS

Our ultimate objectives were to:

- Increase citizens' understanding of the nature and scope of community problems, resources, and opportunities.
- Increase citizen participation in identifying and solving problems and developing opportunities.
- Increase people's understanding of both the beneficial and detrimental aspects of science and technology in society, and thereby

- Increase public appreciation of the role of science and technology in aiding community efforts to identify and solve problems, provide options, and develop opportunities.

The goals of Capital City Readout were to:

1. Establish a new mode of communication on current and specific social issues confronting a particular metropolitan region and demonstrate the viability of such expositions.
2. Present a balanced approach to a number of important problems of the region, and do this from the viewpoint of organizations working to identify and solve problems in their social context.
3. Provide an opportunity for person-to-person communication between members of the public and representatives of organizations about displayed information.
4. Enable citizens to register their views on issues raised by the exhibits and to report these views to exhibiting organizations.
5. Develop organizational procedures and communication techniques for use in other regional information centers.
6. Stimulate subsequent similar efforts in other communities by exposing scientists, engineers, educators, government officials, media representatives, and citizens to an example of problem-oriented communication.

The strategies employed to meet these goals included efforts by the exposition coordinators to:

- Seek out organizations with opposing views and encourage them to participate. (goal 2)
- Make organizational representatives available at each exhibit. (goal 3)
- Develop a ballot on current issues of concern to local residents and institutions. (goal 4)
- Produce a report on the event and disseminate it to people interests in the results. (goal 5)

EXHIBIT CRITERIA

On the basis of the project's objectives, the exposition organizers established guidelines for development and sponsorship of exhibits. To be approved for use in the exposition, each exhibit had to be:

- Sponsored by an organization with local concerns and programs;
- Attended at all times by at least one representative of the sponsoring organization; and
- Focused on a specific problem, and on efforts being made to solve it, in one of three theme areas (see the introduction to this chapter for the list of theme areas).

Many sponsoring organizations had concerns and programs that were broader and more numerous than those they presented at Capital City Readout; however, project organizers felt that they would obtain better exhibits if organizations limited the scope of their exhibit coverage.

DEVELOPMENT

A working team composed of five members, each with separate but related responsibilities, carried out the project. A three member editorial committee[1] was established to develop and refine project communication objectives, determine and interpret policy, assist in preparation and review of public statements concerning the exposition, and review and evaluate the applications submitted by organizations interested in participating.

In addition, two full-time project coordinators[2] shared responsibilities to develop interest among a range of local organizations, follow up with most promising potential participants, assist participants with display techniques, plan and design layout of the exposition site, plan and design the citizen feedback device (issue ballot book), and function as executive staff to the editorial committee.

Preparation for the exposition required eight months, although it could be done in less time . . . and involved five major steps:

- On the basis of objectives for the exposition, the organizers formulated an initial list of 200 potential participants, and mailed letters that explained the exposition concept and invited organizations to participate.
- The coordinators began discussions with interested respondents to the initial mailing. Suggestions made by several of these contacts led to approximately fifty more organizations (usually volunteer citizens' groups) not included on the initial mailing list.
- Interested organizations received application forms requesting that they (1) state the nature and objectives of their organization, (2) identify the problem, and efforts being made to solve it, they wished to examine, (3) explain the proposed intent and design of their exhibit.
- The coordinators and the committee met almost weekly for three months (beginning approximately five months before the exposition) to act on applications. No applications were accepted one month prior to the exposition, which was somewhat later than originally anticipated.
- Groups whose applications gained approval received notices of acceptance. The editorial committee sent suggestions for improvement and invitations to resubmit to groups whose applications failed to meet exhibit criteria. The AAAS received approximately sixty applications; forty gained final approval.

Project coordinators carried on intensive discussions with participants until the time of the exposition, and provided exhibit design assistance to organizations requiring it. Planning of the exposition site, design of the ballot book, and preparation of local promotional campaigns occurred during the same eight-month period.

EXHIBITS AND SPONSORS

Forty organizations participated in Capital City Readout. Exhibits were located in the exposition hall according to their relevance to one of three theme areas. Exhibit sponsors and the location of their exhibits are shown in the diagram. . . .

● = EXHIBITS PREPARED ESPECIALLY FOR USE IN CAPITAL CITY READOUT
○ = RECIPIENTS OF FINANCIAL AID FOR PREPARATION OF EXHIBIT

PLANNING FOR SOCIAL, PHYSICAL, AND TECHNICAL RESOURCES

○● 1. COALITION ON OPTIMUM GROWTH (CO-OPT) problems of urban growth, development, and congestion
○● 2. CITIZENS TRANSPORTATION COALITION citizen participation in regional transportation planning
3. FEDERAL CITY COLLEGE SCIENCE VAN inner city science, education, and environmental monitoring
● 4. DISTRICT OF COLUMBIA REDEVELOPMENT LAND AGENCY urban renewal and reconstruction
○● 5. GEORGE WASHINGTON UNIVERSITY DEPARTMENT OF URBAN AND REGIONAL PLANNING planning neighborhoods
○● 6. LEAGUE OF WOMEN VOTERS OF THE NATIONAL CAPITAL AREA voter education and citizen participation
○● 7. ARLINGTON COALITION ON TRANSPORTATION transportation planning and protection of natural resources
8. NATIONAL CAPITAL PLANNING COMMISSION development of the Potomac River
● 9. UNITED STATES NATIONAL PARKS SERVICE summer urban recreation programs for the District of Columbia
○●10. WORLD FUTURE SOCIETY the influence of present trends and activities on the future
11. WASHINGTON TECHNICAL INSTITUTE human resources dimension of Washington area problems

USE, OPERATION, AND MANAGEMENT OF NATURAL RESOURCES

12. UNITED STATES ATOMIC ENERGY COMMISSION nuclear reactor power plant licensing procedures
13. UNITED STATES DEPARTMENT OF INTERIOR, BUREAU OF MINES solid waste recycling potential
14. CHESAPEAKE AND POTOMAC TELEPHONE COMPANY mineral resource conservation
○●15. CENTER FOR SCIENCE IN THE PUBLIC INTEREST reordering regional scientific priorities
○●16. SIERRA CLUB, POTOMAC CHAPTER strip mining and pollution of the Chesapeake Bay
○●17. WASHINGTON GAS LIGHT COMPANY the natural gas shortage
○●18. CHESAPEAKE BAY FOUNDATION ecology in the Chesapeake Bay
●19. UNITED STATES NATIONAL OCEANIC AND ATMOSPHERIC ADMINISTRATION monitering of urban weather
●20. DISTRICT OF COLUMBIA DEPARTMENT OF ENVIRONMENTAL SERVICES solid waste analysis and treatment
○●21. CONCERN, INCORPORATED the quality of local drinking water and other urban services
22. POTOMAC ELECTRIC POWER COMPANY (PEPCO) energy policy and the consumer

COMMUNITY HEALTH AND WELL-BEING

●23. PRETERM private, stress-free abortion services
24. AMERICAN NATIONAL RED CROSS blood donation and hepatitis
○●25. PLANNED PARENTHOOD OF METROPOLITAN WASHINGTON, D.C. contraception and the control of birth rates
26. UNITED STATES NATIONAL AERONAUTICS AND SPACE ADMINISTRATION medical applications of space hardware
○●27, MITRE CORPORATION interactive cable television for urban communities
○●28. WOODROW WILSON INTERNATIONAL CENTER the needs and characteristics of local Latin Americans
29. ANACOSTIA NEIGHBORHOOD MUSEUM, SMITHSONIAN INSTITUTION contributions of blacks to science
○●30. NATIONAL CAPITAL AREA CHILD DAY CARE ASSOCIATES comprehensive child development
○●31. UNIVERSITY OF MARYLAND INSTITUTE FOR CHILD STUDY child classroom behavior and teacher training
●32. APPLIED URBANETICS investment in and results of drug abuse treatment programs
33. UNITED STATES NATIONAL BUREAU OF STANDARDS large research organizations and social concerns
●34. DISTRICT OF COLUMBIA DEPARTMENT OF HUMAN RESOURCES the provision of comprehensive social services
●35. GEORGETOWN UNIVERSITY PUBLIC SERVICES LABORATORY lead poisoning among local children
36. UNITED STATES DEPARTMENT OF HEALTH, EDUCATION AND WELFARE development of aids for the handicapped
○●37. WASHINGTON CENTER FOR METROPOLITAN STUDIES regional demographic patterns

•38. UNITED STATES BUREAU OF THE CENSUS the uses of census data
39. UNITED STATES GOVERNMENT PRINTING OFFICE reading material oriented to exhibit topics
•40. DISTRICT OF COLUMBIA PUBLIC LIBRARY VIDEOTAPE CENTER recordings of related lectures and symposia
a. CHRYSALIS CORPORATION MEDIA BALLOONS related movies and slide presentations
b. ISSUE BALLOT DEPOSITORIES

INTERACTIVE PROCEDURES

Capital City Readout had two major interactive components. First, representatives of sponsoring organizations attended all exhibits. Thus, visitors had an opportunity to seek direct personal dialogue with knowledgeable individuals, and, where necessary, to seek clarification of information presented in exhibits. Organizational representatives were sometimes present in order to assist visitors with operation of special equipment, as in the cases of the Bureau of Census and the National Oceanic and Atmospheric Administration. Attendants also advised interested citizens on how to join an organization, where to go for further information, and how to take advantage of various services and programs.

Second, each visitor received, upon entering the hall, a booklet containing twenty-two statements on topics of current local interest, many of which received attention in exhibits. By tearing a stub from the booklet and placing it in the appropriate container visitors could ostensibly make known, to exhibiting organizations, their views on local issues ranging from dispensation of methadone to heroin addicts to curtailment of freeway construction. Some 2200 visitors performed the balloting operation.

Local residents received yellow ballot booklets and visitors from out of town received green, so that differences of opinion based upon residency might be revealed in result tabulations. One thousand of the booklets (500 green and 500 yellow) were coded to correspond to personal profiles of respondents, which were obtained through registration forms provided at the hall entrance. The profiles contained information on most of the independent personal variables used in standard statistical surveys, including age, sex, marital status, level of education, occupation, and residency.

It is important to note here that unfortunately the

Charles del Vecchio, THE WASHINGTON POST

ballot consisted of rather hastily assembled and informal statements and in no sense constituted a professional attitude survey instrument such as social psychologists use. For that reason, we offer it here as primarily a technique—and a valuable one—for enhancing interaction. . . . The potential value of reliable information gained through a good survey instrument would, of course, be enormous to citizens and to institutions serving them. These and related matters are discussed in later sections.

PROMOTION

Because Capital City Readout sought to attract a large and diverse audience, the project coordinators used various promotion techniques. The context of the 1972 Annual Meeting provided an opportunity for advance notice to reach a national audience through articles and announcements placed in various regular AAAS publications:

- *Science* magazine, which has a weekly circulation of 160,000, published several notices and a statement prepared by the project coordinators, which described concerns to be addressed by the exposition.
- *AAAS Bulletin,* a newsletter mailed to 130,000 Association members, published an article about the exposition. The article included descriptions of several exhibits being prepared by local organizations.
- *AAAS Annual Meeting Program,* distributed to approximately 9,500 meeting attendees and others, contained articles about the exposition, a full list of participating organizations, and a description of topics to be treated by exhibits.

The exposition's focus on local concerns led to interest among local print and broadcast media. In addition to publicity gained through coverage of opening day activities on television newscasts and in daily Washington newspapers, the exposition received advance promotion through local radio "talk shows" in which the project coordinators participated. Articles and announcements concerning the exposition appeared in newsletters and magazines circulated by organizations participating in the event, including among them Planned Parenthood, the World Future Society, the Sierra Club, and the Center for Metropolitan Studies. . . .

The project coordinators used several other local promotion techniques for the popular exposition:

- The Smithsonian Associates and the Washington Center for Metropolitan Studies, whose combined mailing lists carry a total of over 12,000 names and addresses, mailed leaflets describing the exposition and inviting attendance.
- The Woodrow Wilson International Center distributed leaflets printed in Spanish to local Latin American residents.
- Elementary and secondary schools in the District of Columbia and surrounding Maryland and Virginia counties distributed promotional packages to students. Ten thousand packages reached 600 schools throughout the region. Each contained prepared statements for teachers to read to their students, posters and leaflets describing the exposition, and a registration form that enabled students with special interests to receive further information.

STAFFING

It is customary for high school science departments from the host metropolitan region to enlist volunteer student aides for use as guides during the AAAS Annual Meeting. After a short briefing on the exposition, volunteers from among these aides answered visitors' questions, and provided assistance with the distribution of registration forms and issue ballots. Several aides, also recruited through local high school science departments, received payment to assist organizations with moving and installing exhibits.

All student aides wore white laboratory coats with the words "Capital City Readout" printed in large letters on the back, which helped visitors to identify them as information sources. A similar technique is used by the Exploratorium in San Francisco, California, a science museum that encourages young visitors to explore exhibits freely; wearing bright red coats, students from the region are hired to act as "explainers," and can be approached by visitors seeking information on various exhibits. Exposition aides could simply have coats with the request "ask me" emblazoned on them. Five aides were on duty (at various stations in the hall) for the eight hours each day during which the exposition was open to the public.

COSTS AND FUNDING

Direct costs for Capital City Readout totaled approximately $22,000, exclusive of exhibit costs, for which most of the participating organizations paid (sixteen nonprofit and volunteer organizations received grants of $200 each from the AAAS for preparation of exhibits). All funds for the project were transferred from areas within the annual operating budget of the AAAS.

A schedule of estimated income and direct expense figures appears below. Costs not included in the total are for time of the editorial committee (estimated at about 500 man hours) and for office overhead.

Exhibit costs ranged from $200 for "home-made" types to over $200,000 for professionally designed and constructed displays. . . . Several exhibitors compensated for a relative lack of finances by investing great amounts of time in exhibits. Construction of

SALARIES (two full-time coordinators for eight months)		$10,752.96
OPERATIONS		
Clerical and miscellaneous services	$ 624.73	
Project coordinators' travel	510.33	
Supplies for exposition hall	4,132.11	
Printing (ballot book, posters, leaflets)	2,206.36	
Postage, telephone, books, and publications	861.72	
Grants to volunteer and nonprofit exhibitors	3,200.00	
		$11,535.25
TOTAL		$22,288.21

CO-OPT's puppet show, for example, required over 200 hours of one person's volunteer labor. Members of the League of Women Voters built their legislation game over the period of a week, in a volunteer's basement.

In most cases, grants of $200 barely covered expenses for exhibit materials, and they allowed absolutely no compensation for time spent by individuals; members of several volunteer and nonprofit organizations expressed dismay at this, and said they would be reluctant to participate in such an effort again unless more funds were made available to them.

PROBLEMS, PITFALLS, AND LESSONS LEARNED

Mainly due to amounts of time, money, and prior experience, there were some oversights and deficiencies in planning, which were reflected in the exposition.

- The exposition lacked complete cohesiveness and unification; exhibits did not present a synthesized view of local problems, options and opportunities.
- A few exhibits did not address local concerns; many did not identify and communicate salient aspects of community questions likely to be of significant interest to local residents.
- Citizen feedback on policy issues, and its potential uses, received insufficient attention during planning stages; thus,
- Issue ballots were difficult and time-consuming to count; several contained ambiguous and biased statements, which undermined their potential usefulness for accurately evaluating citizen attitudes.

In short, the exposition organizers lacked sufficient resources to identify properly, and plan for, potential hazards. They occasionally found it necessary to opt for the expedient in lieu of the preferred.

The opportunity to prepare this report has allowed us to study and reflect on many things that should have been considered and resolved before the exposition reached implementation stages. . . .

EVALUATION OF CAPITAL CITY READOUT

It is difficult to know how best to evaluate such a communication effort; many things can be measured easily, and some apparently cannot. Of course, readily measured aspects often provide only part of the information needed for evaluation. Museums, for example, generally measure various dimensions of attendance[3] —numbers of visitors and, where possible, their demographic characteristics and length of stay in the building—but few would defend such measures as trustworthy indicators of educational merit or amount of information communicated.

Psychologists, sociologists, and environmental designers have taken increasing interest in scientific exploration of human behavior in museum settings.[4] Literature in the field is vast.[5] It discusses the evaluation of exhibits from many viewpoints, ranging from surveys of viewer attitudes toward exhibits to unobtrusive observation of social interaction among museumgoers. Such research and evaluation techniques require substantial investments of money, apparatus, and labor in order to plan, measure, document, analyze, and report.

Few, if any, public information expositions have been designed or observed with scientific rigor, possibly because to do so has not been among the objectives, nor within the financial capabilities, of organizers, or because the pressures of a tight schedule make evaluation impossible. Aside from the intent to count ballots cast by citizens on a variety of local issues and to report results to interested organizations, evaluation of Capital City Readout received little consideration during planning stages because planners had insufficient time. Even so, Capital City Readout can be evaluated in several ways.

Perhaps the most important question to ask is: What (and how) did people learn from exhibits? While we cannot completely answer this question, we can show that citizens and organizations were interested, participated in, and benefited from the exposition. This section considers generally some evidence available from Capital City Readout, which can be used to evaluate and draw conclusions about the success of this endeavor and needs of similar future events.

Attendance

The figures below represent only those 4275 individuals who completed registration forms. Security guards and persons distributing registration forms at the entrance estimated a larger total of 5500 visitors (approximately 30% above the total registered).

Daily Registration

		Hours Open	Total	Regis-trants	Avg. per Hour	Per Cent Total
Tue.	26 Dec.	1 PM - 9 PM	8	1009	126	23.61
Wed.	27 Dec.	1 PM - 9 PM	8	856	107	20.02
Thu.	28 Dec.	1 PM - 9 PM	8	526	66	12.30
Fri.	29 Dec.	1 PM - 9 PM	8	622	78	14.55
Sat.	30 Dec.	10 AM - 6 PM	8	1262	158	29.52
TOTALS:		5	40	4275		100

Average per day - 855
Average per hour - 107

Naturally, the reliability of such estimates is open to question, due to the possibility of repeat visits and natural errors in casual counting procedures.

The figures reveal some useful information. A peak in attendance occurred on Saturday. Observers agree that the most "community visitors" (i.e., non-AAAS members) attended on the weekend; families were most in evidence at this time. Exposition staff and personnel manning the exhibits thought that daily attendance was highest about the noon hour on Saturday and about 1 PM on weekdays, and that numbers declined steadily toward evening. Total attendance was about one-half the anticipated number.[6] Despite rather extensive local promotion efforts, only slightly more than one-half the 5000 or so visitors were not affiliated with the AAAS and its annual meeting. The exposition was expected to reach substantial numbers of people; results were somewhat disappointing. There is some reason to expect that all attempts of this sort should expect such results.[7]

As is true in many institutional attempts to communicate with people outside the context of immediate personal needs and desires, the exposition organizers relied much on the use of advance publicity to persuade people to attend. Local media were contacted and most granted requests to devote space or air time to the exposition. "Community Bulletin Boards" in newspapers and on radio carried advance public notice of the exposition. The attendance figures show that it is probably unwise to place great faith in the power of mere announcements in the media to draw attendance. Most of those responsible for the project believe that low attendance figures resulted from three main factors:

- The exposition was inappropriately located in a hotel used primarily for conventions, rather than in a recognized community center.[8]
- The exposition was of relatively short duration.
- The promotional material did not sufficiently intrigue members of the population we sought most to attract, and did not adequately convey to people the expositions's objectives. . . .

Comments from Participants

Following the exposition, the authors mailed informal questionnaires to participating organizations, requesting that they:

- Assess the public educational value of their exhibits;
- Describe impacts (if any) of the exposition on their operations;
- Assess the value of the entire event in terms of public response and most effective exhibits; and
- Suggest ways to improve future "information exchanges."

Requests for follow-up comments from participants met with only scattered response. Significantly, most of those organizations who did reply were smaller ones, many of which were nonprofit and/or volunteer, and . . . had little or no prior experience with exhibits. Several governmental exhibitors also replied, most notably the Department of Health, Education and Welfare and the Bureau of Mines; but response from other larger organizations was weak. Exposition planning stages reflected a similar difference in the nature and number of responses from large as compared to small organizations; smaller organizations, without existing exhibit programs, were generally far more interested in meeting exhibit criteria . . . than were larger organizations, many of which already had exhibits and were convinced that these would suffice.

A selection of participants' comments is presented here, arranged according to the four areas mentioned above.

Discuss the educational content and value of your exhibit. Did visitors learn about programs and goals fostered by your organization? Could a member of the general public make intelligent issue decisions on the basis of information conveyed through your exhibits?

". . . most visitors to our booth spent most of their time asking questions and filling out information request cards. It seems likely that visitors would be likely to make more intelligent issue decisions after this contact."

—U.S. Department of Interior, Bureau of Mines

"We feel the general public could make intelligent decisions about planning their families, the best methods of contraception to use, and where to go for this service."

—Planned Parenthood of Metropolitan Washington, D.C.

"It was our intent that members of the public would be able to make an intelligent and thoughtful decision about abortion on the basis of information displayed. However, with abortion continuing to be such an especially sensitive issue, it may be more realistic for us to hope we were able to provoke, if not decision, at least more awareness."

—Preterm

Describe any ways in which participation in Capital City Readout has affected (or will affect) policies, programs, and/or operations in your organization. Has the exhibition had any impact of a lasting nature? Discuss any "peripheral" benefits the exposition may have brought to your organization.

"It was good for our organization to specifically list the problems in the District addressed by its centers, and to prepare maps. This has helped us to realize what our programs are providing, and many of the materials can be used in other exhibits. . . (and) will be useful for years to come"

—Planned Parenthood of Metropolitan Washington, D.C.

"(We know of) no effect other than several new contacts made with others in the recycling field, which have been followed up by visits or correspondence."

—U.S. Department of Interior, Bureau of Mines

"(We) want to explore the possibility of building on our experience with this display for a week-long open house exhibit in the HEW North Building. We also have to presume involvement by HEW during the 1976 Bicentennial Celebration, when 10 million visitors are expected in Washington. Almost certainly the D.C. Bicentennial Commission will ask major Federal departments to consider some kind of visitors' display. It is hard to justify the fact we are not doing something like this already."

—U.S. Department of Health, Education and Welfare

"In general, we can say that the Capital City Readout was a helpful forum for some of the ideas we have been pursuing with the Southeast Community. As a result of the exposure to the METRO planning effort, members of the District Government and the Government of Prince George's County were persuaded to request WMATA to study the Branch Avenue Line in the light of suggested alternatives. This is the first such study authorized by WMATA and while Capital City Readout may not have been the major contributing factor, it certainly played an important role in the accumulation of events leading up to the study authorization. Any exposure to our efforts and those of the Southeast Community are beneficial and we did receive inquiries from a number of people who wished to pursue further the details of our activities."

—George Washington University
Department of Urban and Regional Planning

Assess the value of this kind of exposition in terms of public response. What kinds of people seemed to visit your exhibit? How did they appear to react to what they were seeing?

"In general, people visiting our booths were those attending the meetings rather than the general public. This was something of a disappointment to us because we were particularly interested in reaching Washingtonians. It is also puzzling because publicity for the event was good."

—Planned Parenthood of Metropolitan Washington, D.C.

"Overall, attendance at the exhibit was less than we had anticipated. More advertising in the papers and on television might have attracted more local attendance."

—World Future Society

"Our visitors included all varieties—students, medical doctors, research workers and utility executives. In general, the people who viewed our exhibit were ordinary citizens; very few seemed to be scientists. Nearly all were very enthusiastic and asked many questions."

—U.S. Department of Interior, Bureau of Mines

"Scientists and science students."

—U.S. Department of Health, Education and Welfare

"Visitors to our exhibits were professionals in family planning, M.D.'s, nuclear physicists, young people, and anti-abortion activists."

—Preterm

Suggest ways in which your organization might be willing to participate in future public events like Capital City Readout. Define the problems you feel would have to be resolved in order to make the event more worthwhile.

". . . future exhibits of this type (should) have less participation by government agencies and institutional exhibitors . . . a smaller exhibit with some key points might be more effective. It may be useful to have nearby rooms available for small talks and discussions."

—World Future Society

"(We) would suggest more well-rounded involvement from the community."

—Preterm

". . . a snack bar at the exhibit would have been appreciated by many exhibitors and visitors."

—U.S. Department of Interior, Bureau of Mines

With a few exceptions, exhibitors enjoyed participating in Capital City Readout, and felt that the experience was worthwhile. Also apparently unanimous among exhibitors was the feeling that attendance should and could have been better. The seemingly contradictory comments about the nature of people who attended the exposition may be explained by the following:

- Different organizational representatives manned exhibits at different times of the day; those responding to the informal questionnaire were probably in the exposition hall at different times, and thus may have gained different impressions.

- Individual exhibits probably attracted individuals with advance interest in subject matter being presented; this process of audience selection may have led observers to decide (incorrectly) that those who visited their exhibits were representative of the entire audience.

An Opinion Survey by the Atomic Energy Commission

Out of interest in the exposition and concern for public response to its programs, the Atomic Energy Commission (which was one of the exhibitors) had twenty-three staff members attend Capital City Readout and subsequently asked them to answer a series of questions about the exposition. A member of the Atomic Energy Commission's staff prepared the following summary of results from the questionnaire. It is presented here as additional evaluative information.

Number Surveyed:	18 – From the Division of Biomedical and Environmental Research (DBER)
	1 – From the Oak Ridge Associated Universities (ORAU)
	4 – From the Office of Information Services (OIS)
Total:	23
Approach Used:	6 – Interviewed in person or by telephone.
	17 – Completed questionnaires.

Questions and responses:

1. What was the quality of the attendance at the exhibit? In other words, would you say there were many people from off the street visiting the exhibits, or would you say most of them seemed to be AAAS types?

The majority of those interviewed expressed the opinion that most of the attendees were AAAS types.

2. What use of the ballots did you observe? How seriously were the ballots being used?

Ten of those surveyed made no observation or had no comments. The majority of those responding to this question felt that there was no significant use being made of the ballots or that they were being used with little forethought. One person expressed the feeling that the ballots were being used, seemingly seriously, but were being cast before visiting the exhibit area. Another interviewee stated that there was much activity putting ballots in cans with little or no discussion of balloted issues. He stated that no questions or discussions were asked of him, or of PEPCO's staff, concerning the metropolitan siting proposition (Question No. 20 in the ballot book).

3. How many people attended? Do you feel the attendance was high or low?

The consensus of opinion was that attendance was low. Several compared the attendance to other meetings they had attended and a few others stated that they were there only in the evening or on the first day and therefore couldn't properly answer the question.

4. What type questions were asked at the Atomic Energy Commission exhibit? Generally, what was the reaction to our exhibit? Were most visitors friendly or antagonistic?

Most persons interviewed felt that the majority of the visitors to AEC's exhibits were friendly and interested. Persons visiting the "From Molecules to Man and His Environment" exhibit showed considerable interest in, and asked questions about, the cloud chamber and the radiation level display. Other questions concerned printed material. One person interviewed expressed his opinion that about 80% of the visitors were antagonistic. Two others indicated that they experienced antagonistic visitors and others that were skeptical.

5. What was the quality of other exhibits?

Generally, the opinion of those surveyed was that the quality of exhibits varied, from "excellent" to "poor," and that imagination and creativity were decisive factors, rather than money invested.

6. Do you have any other comments you would like to make?

Most of the additional comments concerned a feeling that attendance and effectiveness of AEC's exhibits were unfavorably affected by poor location in the exhibit hall.

Some Comparisons of Exhibits

Exhibits are usually evaluated by professionals from one or both of two perspectives:

- Influence on visitor behavior, as measured by the ability of exhibits to attract and hold attention, shape social groupings and interaction, and elicit other significant physical responses; and
- Influence on visitor attitudes, as measured by the amount and accuracy of information imparted, changes in visitor opinions and viewers' feelings toward "experiencing" exhibits.

The following list of criteria for evaluating exhibits, from a study of various Atomic Energy Commission exhibits[9] should indicate the range of areas currently considered by professionals.

Attractiveness of Exhibit
Ease of Comprehension

Unity within the Exhibit
Ability to Attract Attention
Ability to Hold Visitor Interest
Appropriateness of Exhibit Presentation
Accuracy of Information Presented
Location and Crowd Flow
Visitor Characteristics
Focus of Attention
Textual Material (labels, headings, etc.)
Relations of Exhibit to Surrounding Area and Other Exhibits
Design of Exhibit
- Size
- Physical layout
- Use of color
- Use of light

Exhibit Items
- Quantity
- Attractiveness
- Size

Communication Techniques
- Sound
- Motion
- Demonstrations
- Charts
- Film
- Models
- Auxiliary teaching techniques
- Audience participation

Because Capital City Readout was staged with insufficient funds and preparation time to permit rigorous evaluation of exhibits in the conventional sense, only observers' opinions are available to judge exhibit effectiveness.

A comment from one representative of the Atomic Energy Commission expresses well the sentiments of many who observed and contributed to the project:

"Some of the best exhibits and some of the worst were crude. Some offered much imagination and these were best attended and got their story across. There appeared to be no relationship between effectiveness and investment in the exhibit. What counted were the ideas, the imagination, and the people involved."

Among the relatively crude and inexpensive (less than $250 construction costs) exhibits, which attracted and sustained great visitor attention were:

- A legislative game devised by the League of Women Voters, in which visitors rolled dice and walked through a series of squares representative of stages in "getting your bill through Congress;"
- A display of contraceptive devices prepared by Planned Parenthood; and
- A puppet show titled "Don't Sleep While the Bull Dozes," produced by Co-Opt (Coalition on Optimum Growth), which dramatized the causes of urban congestion.

Certain intermediately elaborate exhibits (costing between $250 and $1000, exclusive of costs for equipment) also appealed to visitors strongly. These included:

Link Harper, Washington, D.C.

- A living-room mockup, prepared by the Mitre Corporation, in which visitors could sit on chairs or couches and experiment with interactive cable television equipment;
- A mockup of an abortion clinic reception room prepared by Preterm, in which visitors could study literature, photographs, and a short slide presentation about abortion procedures;
- A pair of air-inflated bags by Chrysalis Corporation containing rear projection screens, that resembled gigantic television screens and displayed movies on various topics; and
- A display sponsored by HEW, which permitted visitors to test equipment designed to aid physically handicapped persons.

Finally, several of the most elaborate exhibits (costing the organizations in some cases over $100,000) attracted and held viewer attention. Notable among them were:

- A temporary weather station, prepared by the National Oceanic and Atmospheric Administration, which displayed information about local weather conditions on machines linked to satellites; and
- A display and demonstration by NASA of equipment designed to monitor vital signs in space-bound astronauts.

Categorization and comparison of exhibits on the basis of cost alone is rather futile, since success in attracting and holding viewers seems dependent on other, more important factors. Indeed, several of the obviously very expensive exhibits appeared also to attract the least number of visitors. For example, the conventional "unpacked from a crate" displays offered by the Atomic Energy Commission, PEPCO, and several others, drew fewer visitors than the home-made variety, prepared by the League of Women Voters and others.

Location within the exhibit hall might be offered as a significant variable influencing exhibit attendance; however, observers agree that (1) a majority of people visited all areas, and (2) certain exhibits, remote from entries, exits, and major circulation paths, received substantial visitor attention.

Realizing their untested and unresearched nature, we, nonetheless, offer the following conclusions concerning exhibit effectiveness at Capital City Readout:

- The presence of human attendants added an important and useful educational dimension to exhibits since it allowed issue-oriented dialogue between visitors and representatives of organizations. Attendants also helped to assure that visitors understood the significance of exhibit materials in the context of the exposition's theme.
- Exhibits displaying objects of intrinsic interest—items that visitors could study, handle, or manipulate—were very attractive, but in some cases left the organizers with an impression that visitors were only preoccupied with equipment, and learned little about social issues or scientific principles. This effect was apparently diminished when exhibit attendants made efforts to explain the significance of interesting objects and equipment, as did personnel attending NASA's and HEW's exhibits.
- Exhibits with visual images and sounds discernible from remote locations drew the most visitors. Concise, unified presentations (for example, the sound and slide show offered by Preterm, and the Co-Opt puppet show) sustained visitor attention better than did longer or relatively undirected exhibits (for example, displays which required visitors to "browse").

Only a few exhibitors (notably Preterm, Planned Parenthood and Co-Opt) actually presented their ideas strongly in the local problem-and solution-oriented context suggested by publicity for Capital City Readout. Assuming the publicity worked, visitor expectations may not have been met by exhibits that failed to underscore issues.

A host of technical and logistical problems accompanied the exposition: orientation of exhibits to the hall entrance so visitors could, upon entering, have an overview of exhibits; noise interference among neighboring exhibits; layout of the exposition hall to meet space requirements of individual exhibitors; and crating, transport and storage of exhibit materials.

Also, visitors had no prescribed sequence in which to view exhibits and process issue ballots. One possible solution to this problem would have been to allow ballots to be cast directly in front of (or near to) exhibits concerned with the issue in question. No strong relationship existed between statements on the ballots and materials in exhibits. The content of each exhibit was not always clearly set forth, which could easily have been done with a large sign or placard identifying the exhibit's theme. This was an especially acute problem with continuous, unattended movies, where a sign would have helped visitors to identify films in progress. Finally, a map of the exposition hall, perhaps placed in the ballot book, would have provided visitors with an overview of exhibits and their sponsors.

Fundamental Problems with the Undertaking

Experiences with Capital City Readout and a review of comments from observers reveal several major areas that require consideration and improvement for future community information expositions:

- Synthesis and representation of a range of viewpoints may be difficult to achieve simultaneously. Synthesis requires strong coordination of planning and design, while presentation of balanced views may

Link Harper, Washington, D.C.

require participation by many interest groups. How can information be made comprehensive, understandable, and unbiased?

• Credibility and effectiveness are attributes that exposition sponsors and planners should possess if public trust is to be maintained in the honesty and worth of the effort. Strong affiliation with a single organization is likely to arouse public suspicion that motives behind exposition feedback activities are promotional, i.e., that information is unfairly slanted to "sell" attitudes toward problems and solutions favored by particular interest groups. Also, many persons believe (perhaps somewhat correctly) that most institutions are quite specialized, and might, therefore, have little ability or legitimate interest in affairs of the whole community. Appropriate and capable existing institutions, and ways to develop new ones, should be identified.

• Communication between organizations and citizens raises fundamental questions: what kinds of citizen feedback are useful to policy-making organizations, and to the citizens themselves; what kinds of information, and how much, do particular nonexpert audiences need on problem and policy topics; what are the best ways to present information on the complex and changing affairs of society? New methods of communication, taking advantage of advances in information technology, should be developed for use in these contexts.

• Maintenance of exhibits and citizen response-gathering programs over extended periods of time is costly. Every organization is constrained by limited human and financial resources. How can substantial costs be absorbed or offset?

• Location of exposition activities affects attendance characteristics. Centralization of exhibits creates geographic isolation; failure to locate expositions in recognized and highly trafficked civic areas leads to reliance on expensive and sometimes unreliable publicity devices. Citizen access to information is limited by the hours during which an exposition is "open" and persons who must be elsewhere at those times are excluded. Buildings carry with them cultural biases, to the extent that some people will avoid places they perceive as being "for someone else." Appropriate exposition locations, to reach intended audiences, should be identified.

• Linkage to similar or related activities in other regions of the nation would help to improve community information efforts. Information on events and methods used in other areas of the country is difficult to obtain and use effectively. How can reports on various experiences, and lessons learned from them, be disseminated to other communities?

• Evaluation should be made of potential impacts that innovative communication methods can have on public understanding of, and citizen participation in, community decision-making processes. Very little is known about applications of interactive display media to such processes.

We have identified these seven areas of concern for future popular expositions based on our direct personal experience with Capital City Readout, on comments provided by participants, on interviews with persons having related interests or abilities, and on research into literature on interactive display media. Each source of information posed substantial and important questions about ways to improve future exchanges of regional information, many of which are reflected in the concerns outlined above. . . .

Link Harper, Washington, D.C.

Summary of Evaluation

Out of six major goals for Capital City Readout, four were well satisfied, one was not, and evidence on another is not yet in . . .

- We demonstrated the viability of a new mode of communication on regional issues.
- Organizations—involved in identification and solution of problems—displayed information on a number of important regional problems. Citizens and organizational representatives discussed issues.
- Two thousand two-hundred citizens registered their views on local issues but these were not reported to organizations.
- We developed communication techniques and organizational procedures for possible use in other places.
- It remains to be seen whether the exposition stimulated similar subsequent efforts.

This chapter contains detailed analysis of attendance and exhibit effectiveness, and presents comments from participants. Salient comments and criticisms of the effort are summarized below.

Attendance at Capital City Readout totaled nearly 5500, which was slightly over one-half the anticipated number.[4] Promotion, while extensive, was not effective in persuading many members of low income and minority groups to attend. Location of the exposition in an expensive hotel probably discouraged attendance by local residents.

Smaller nonprofit and volunteer organizations were generally more responsive than large institutions to exhibit development guidelines, which requested displays focused on specific local problems and efforts being made to solve them. Most participating organizations expressed satisfaction with the exposition, although several cited low attendance as a disappointment. For several organizations, the exposition provided a valued first opportunity to build exhibits, some of which have received regular subsequent use.

Exhibit attendants added significantly to communication and education processes. They helped to create a dialogue between individual citizens and policy-making institutions. In exhibits, there appeared to be little relation between the amount of money invested and ability to attract and sustain visitor attention; imaginative design ideas were most important.

Large visual images, sounds, and objects of intrinsic interest attracted the most visitors to the exhibits. Static, undirected, two-dimensional displays did not attract and sustain peoples' attention.

A number of problems will confront similar future efforts. These include: synthesis of exhibit topics; credibility and effectiveness of organizers and sponsors; adequate planning and preparation time; forms and effective uses of citizen feedback; methods of communication through exhibit media; and planning and carrying through necessary evaluations of exposition effectiveness.

The exposition was successful in that it so far has satisfied four of the six major goals set for it. In addition, over 5000 people viewed the exhibits at an average rate of 1000 per day; local newspapers, television, and radio stations took considerable interest in the event; and many people and institutions have expressed interest in the concept.

NOTES AND REFERENCES

1. The editorial committee consisted of James Butler, director of the AAAS Communications Department; Philip Ritterbush, associate editor of the 1972 AAAS Annual Meeting and chairman of ORGANIZATION: RESPONSE; and Richard Scribner, director of the 1972 AAAS Annual Meeting and director of the AAAS Office of Science and Society.
2. The project coordinators were Joseph M. Valerio and Thomas V. Vonier of the School of Architecture of the University of Wisconsin.
3. The Chicago Museum of Science and Industry, for example, counts individuals who pass through museum doors and surveys license plates on cars parked outside for an indication of where visitors have come from. Length of visitors' average stay in the building is also computed. For the Museum of Science and Industry this figure is approximately 3 hours and 45 minutes—astonishing when compared to the Smithsonian Institution figure of 19 minutes (Museum of History and Technology), but modest in comparison to Disneyland's 8 hours.
4. The Smithsonian Institution Office of Museum Programs now has a Division of Psychological and Sociological Studies, which, among other things, sponsors projects and symposia involving interdisciplinary teams comprised of social scientists, architects, behavioral psychologists, and various members of the museum regular staff.
5. Comprehensive bibliographies on the subject have been compiled by Ross Loomis of the Smithsonian Institution and Harris Shettel of the American Institutes for Research.
6. An average daily attendance of 2000 was anticipated (10,000 total). Philip Ritterbush of the editorial committee made these estimates on the basis of his experience with attendance figures while on the staff of the Smithsonian Institution Office of Museum Programs.
7. In a forthcoming book, tentatively titled: *What's In A Node?,* Chandler H. Stevens of Rensselaer Polytechnic Institute suggests that all early experiments of this sort can expect disappointing results.
8. When asked if he were interested in attending the exposition, one local cabdriver confidentially assured the questioner that the event was "only for scientists attending the convention."
9. Shettel, Harris, *An Evaluation of Existing Criteria for Judging the Effectiveness of Science Exhibits* (submitted to the Atomic Energy Commission under purchase order number NY-65-366), American Institutes for Research, Pittsburgh, Pennsylvania, 1965, Appendix B.

13

WESSPAC Wesley Town Simulation Package

Luis H. Summers
Department of Architectural Engineering

Thomas Douglas
Division of Man-Environment Relations

Jack Nasar
Division of Man-Environment Relations

The Pennsylvania State University

INTRODUCTION

Facing the present technological and social flux, the planner and decision maker require a wide variety of predictive tools. As a holistic technique, simulation gaming can portray the movements of an otherwise complex and intangible process, but if gaming simulation is to be an accurate predictor, it must abstract from the empirical situation those factors that are most relevant to the process. From this perspective, simulation gaming is situation specific. A game developed for one situation and used for a similar problem requires considerable modification for its second application.

This game models a specific situation, Wesley Town. In order to model this situation accurately, the relevant interest groups were delineated. The task then became one of selecting from the wide range of existing games, those which best model the dynamics of the various groups involved. Each game required further modification to approximate more closely the specific situation and to make the game more playable.

It would be unrealistic to expect a single game to do more than generally depict a complex situation accurately. Games designed to model general aspects of role interactions sacrifice credibility while situation specific simulations lack general applicability. In order to take advantage of the strengths of both types, a hybrid simulation is proposed to model the dynamics involved in the planning of a new town.

Wesley Town is a new town to be built outside of Washington, Pennsylvania. In the initial stage of planning, the John Wesley Foundation hoped to build a remedial community for minority groups of five thousand people. The progression of this initial proposal to the final plan involved the interaction of various groups ranging from local residents to developers, media, to new town planners.

In WESSPAC three social systems were to be

modeled: Life, Career, and Decision. Life is modeled by the game WESSPOWER. Career is modeled by the games WESSPRO, WESSMEDIA, WESSDEV, and WESSWORK, and Decision is modeled by the game WESSPON. Depending on your goal these games can be played alone or in various combinations. When the full simulation package is used, the following steps are taken. Initially all individuals participate in the game of Life, WESSPOWER. Participants are then assigned to each of the four Career games, WESSPRO, WESSMED, WESSDEV, and WESSWORK. Players from each of the career games represent that group's interest and his/her own in the more encompassing Decision game, WESSPON. The overall structure of these games is shown in figure 13-1.

SOCIAL PROCESSES	GAME ANALOGS	
Life	WESSPOWER	
Career	WESSWORK WESSDEV	WESSMEDIA WESSPRO
Decision	WESSPON	

Figure 13-1
Social systems model.

WESSPOWER

Introduction

The game used in WESSPAC to simulate the different population groups encountered in Wesley Town is a modification of STARPOWER. STARPOWER, designed by Gary Shirts (1969), is a game in which citizens of a three-tiered society—squares, circles, and triangles—try to progress from one level to another through the accumulation of wealth. After a few rounds of play, once an upper wealthy class has been established, they are given a chance to make the rules for the game. No matter how well-intentioned the members of the wealthy class are, they invariably make rules that are considered by the other groups to be unfair, condescending, and often fascistic. Usually the game ends when members of the lower class refuse to obey the new rules, or stage a revolt against the upper class. Players in WESSPOWER are given entry roles for the career games.

Steps of Play

WESSPOWER, WESSPAC's version of STARPOWER, uses a monetary base for the accumulation of wealth, rather than chips. At the start of the game, participants are divided into three groups, squares, circles, and triangles. Each square, circle, and triangle is given $3000, $2000, and $1000 respectively. They are then told that they have a period of three minutes to trade economic tokens. Economic tokens are traded in order to maximize wealth. They are told that, in addition to the face value of money, the accumulation of certain denominations, will increase their wealth. If you have more than two bills of the same denomination, a bonus is added to your total. This bonus depends on the number of the bills you have and the denomination of the accumulated currency. Table 13-1 gives the various bonuses. Values in the Bonus Table are good for round one only. For subsequent rounds the operator will post the Bonus Table to be used in that round. Participants may also increase wealth by combinations of bills of different denominations (see table 13-2). Values in the Wealth Increase Table are good for the first round only. For subsequent rounds the operator will post the Wealth Increase Table to be used during that round. Participants are then asked to stand up and trade, no talking unless touching, and if no trade is consummated they must remain touching until the end of the trading period. At the end of the three minute trading period, participants total their scores and record them on the chalkboard next to their initials. Participants are next given two minutes to distribute a bonus gift of approximately 20% of the basic salary ($2500). This bonus gift is tied to the influence each group has in

TABLE 13-1
Bonus Table*

Number of Bills	Denomination *$100*	*$200*	*$300*	*$500*	*$1000*
3	20	30	40	50	100
4	30	60	60	100	200
5	40	90	120	150	300
6	50	120	160	200	400
7	60	150	200	250	500
8	70	180	240	300	600
9	80	210	280	350	700
10	90	240	320	400	800
11	100	270	360	450	900

*add these values to your score

TABLE 13-2
Wealth Increase Table**

Denomination Combinations *$100*	*$200*	*$300*	*$500*	*$1000*	*Bonus*
1	1	1	1	1	$1000
1	1	1	1		500
1	1	1			300
	1	1	1	1	900
		1	1	1	700

**add these values to your score

WESSPON. Distribution must be by group consensus. New scores are tabulated. Individuals with highest scores migrate to upper class, those with lower scores to lower class, and the rest remain in the middle class.

The last step of each round, after migration has occurred, consists of paying subsistence. Table 13-3 shows the costs related to subsistence for the first round. Due to inflationary tendencies and rising cost of living this value will change from round to round, and will be posted by the operator each round. This is the end of one round. At the beginning of the next round participants are given additional wealth, which models personal income and the inflationary trend of our economy. Income and inflation are not the same for all groups, and are a function of each group's level on the WESSPON board.

Game Operator Instructions

Dividing the participants

The participants are divided into three approximately equal groups named: squares, circles, and triangles. Have the groups sit in separate circles of chairs. Each person has a symbol representing his/her group, i.e. the squares are given a square symbol, the circles a circular symbol, and the triangles a triangular symbol. Each participant is given an envelope containing his/her income. Draw the player's symbol on the envelope. The game doesn't work very well if there are fewer than twenty-six participants, the ideal number is thirty, and the maximum is somewhere around forty.

Assigning initial income

Each square, circle, and triangle is given $3000, $2000, and $1000 respectively. One circle and one triangle should be given the same income as a square. Operator should arrange to have play money bundles in the denominations and quantitites indicated in table 13-4.

Assigning roles

On the players' envelopes, identify their group (square, circle, or triangle) by drawing the shape on the envelope. Roles for the career games (WESSWORK, WESSMED, WESSDEV, and WESSPRO) must now be assigned to each player. The minimum number of players is twenty-six, eight square, nine circles, and nine triangles. Table 13-5 indicates the game and the role to be given diverse players.

TABLE 13-3
Subsistence Table

Group:	*Squares*	*Circles*	*Triangles*
Cost:	$2000	$1500	$500

TABLE 13-4
Game Tokens Table/Player*

Denomination	*100*	*200*	*300*	*500*	*1000*
Number of tokens/player	8	6	4	2	2

*total number of tokens = no. of players X values in table

TABLE 13-5
Role Assignment Table

	Roles								
Game	*Squares*	*Min*	*Max*	*Circles*	*Min*	*Max*	*Triangles*	*Min*	*Max*
WESSPRO	W. Professional	1	1	P. Professional	3	6		1	3
WESSDEV	Housing	1		Housing	1		Housing	1	
	Business	1		Business	1		Government	1	
	Commerce	1		Commerce	1				
WESSMED	Conservative	2		Liberal	1		Liberal	1	
WESSWORK	Industrialist	2		Industrialist	2		Industrialist	5	

Example

Suppose you have twenty-eight participants. There are nine squares, nine circles, and ten triangles. All the squares, one circle, and one triangle receive $3000. The other circles receive $2000 and the remaining triangles receive $1000. Be sure players receive their income in a variety of denominations.

Explaining the rules

A. Tell the participants that this is a game that involves trading and bargaining and that the three persons with the highest scores will be declared the winners.
B. Read the Steps of Play, and emphasize scoring rules.

C. Explain the following rules of bargaining:
   1. They have ten minutes to improve their scores.
   2. They improve their scores by trading advantageously with other squares, circles, and triangles.
   3. Only "one for one" trades are legal. "Two for one" or any other are illegal.
   4. Persons must be holding hands to effect a trade.
   5. Once a participant touches the hand of another participant a token of unequal value or color must be traded. If a couple cannot consummate a trade, they have to hold hands for the entire ten minute trading session.
   6. There is no talking unless hands touch.
   7. Persons with folded arms do not have to trade with other persons.
   8. All tokens should be hidden. **This rule should be strictly enforced.**
   9. **Do not** reveal that the squares are given tokens of a higher value than the circles or triangles.
   10. Any other rules that you deem appropriate.

D. Starting the Trading Session
   1. After the rules have been explained, start the trading session. Tell them it will last ten minutes.
   2. During the trading session, your assistant should be putting each participant's initials on the chalkboard.
   3. After ten minutes of trading, have each group return to its circle of chairs.
   4. Have the participants compute their scores for the trading session, and record the scores on the chalkboard opposite their initials.

E. Starting the Bonus Points Session
   1. Announce that there will be a "bonus "points" bargaining session.

F. Explain the rules for the Bonus Points Bargaining Session. The rules are:
   1. Hold up a bonus chip (the special chips) and tell them that this is a bonus chip.
   2. Give each group three bonus point chips.
   3. Tell them that each bonus chip is worth $100.
   4. Their task during the session is to distribute the bonus chips to members of their group.
   5. The decision regarding the distribution of chips must be a unanimous vote.
   6. The chips must be distributed in multiples of 100.
   7. They have five minutes to distribute the bonus chips. If the groups have not distributed the chips at the end of the five minutes, the chips will be taken back by the director and no one will receive them.

G. Start the Bonus Points Bargaining Session. After about five or ten minutes, end the bonus bargaining session. Have those people who receive bonus points record them on the chalkboard opposite their initials. Collect all bonus chips.

H. Put those people with the highest scores in the square group. If there is a circle or a triangle who has a higher score than a square, have him/her trade groups. Any changes should be announced to the participants. It should be made known that so and so, who was a square, has become a circle, because he/she did not receive enough points, and so and so, who was a circle, is now a square because he/she received a higher number of points than a square. In any event, it is important that the participants know that the squares are made up of those people with the highest scores. Migrating players must leave their role envelopes on their chair. They may take their valuables with them.

I. Be sure to collect subsistence from each participant (see Subsistence Table). The operator may want to increase subsistence values with the passage of time.

J. Start the Second Round
   NOTE: Repeat this cycle—bargaining session, bonus session, reclassification, for one or two times or until the participants understand the process and the fact that the squares are high scores.

K. After about the second bonus session, announce that because the squares have worked hard they now have the authority to make the rules for the game. Although any group may suggest rules for the game, the squares will decide which rules will be implemented. You might suggest to the squares that they might want to make rules like: require triangles and circles to bargain with the squares even though they have arms folded; require triangles and circles to give squares the tokens they ask for regardless of whether the triangles and circles want to trade or not; or, redistribute the tokens on a more equal basis, and so on.

L. Announce any rules that the squares establish to all of the participants unless the squares want the rules kept secret.

M. From then on, play it by ear.

Encourage mobility

In order to encourage mobility the operator should change the values on the Bonus Table, and in the Wealth Increase Table. Another way to encourage mobility is to establish a lottery with prizes. Participants would all contribute to a pot, lots drawn, and the winner would be able to advance socially.

Conclusion

At the conclusion of two rounds, have players move to their assigned career roles.

WESSWORK

Introduction

WESSWORK is a game about work, production, concensus decisions, and market clearing. A basic material, three-product (objects, substances, and services) economy is simulated. Each player specializes in the processing of one to three product-specific raw material(s) from which a product is made. Economies of scale encourage the purchase of large quantities of raw materials, and the market encourages balanced productivity. Both objectives are possible only if production cartels (cooperative ventures) are formed.

WESSWORK's Goals, Players, Components, and Form

This game tries to simulate a simple economy where work, material, capital, specialization, and intraindustry cooperative cartels define productivity. Nine raw materials (A,B,C,D,E,F,G,H,I) are industrially processed and combined to form three products, ABC, DEF, and GHI.

Each player is the manager of an industry that is capable of processing raw materials for one of the three products. The player has the option to process one, two, or all three of the raw materials needed to assemble the product. Should the player process all three materials needed for his/her product in equal quantities, he/she can assemble the product and sell it in the general market (operator) at the going rate (a function determined by balanced productivity—equal numbers of products ABC, DEF, and GHI). Note, however, that the unit cost of raw materials is cheaper by bulk, and that productivity increases through specialization (economies of scale: see tables 13-6 through 13-9).

TABLE 13-6
Raw Material Price Chart

	Cost of Raw Materials Per Unit								
Quantity	*A*	*B*	*C*	*D*	*E*	*F*	*G*	*H*	*I*
1 to 20	1	1	1	1	1	1	1	1	1
21 to 50	.75	.75	.75	.75	.75	.75	.75	.75	.75
51 to 100	.50	.50	.50	.50	.50	.50	.50	.50	.50

TABLE 13-7
Specialization Chart

Number of Products Manufactured	*Specialization Index*
1	4
2	3
3	2

WESSWORK Rules

Initially each player declares, by writing on top of his/her form (see figure 13-2 on page 228), the product for which he/she is going to process materials.

Step A. With the capital assets from the last round totals, D2. (during the first round each player is given a starting capital) players make a decision to invest in the purchase of raw materials, and in the work needed for raw material processing. Work expenditure = Price of raw material (see table 13-6). Values in this chart should be used for round one only and are liable to change according to market fluctuations (see official chart posted by game operator).

Step B1. Now you are given a chance to work. You must complete, in a given time period, as many lines as possible of LIFT RIVET FIT ASSEMBLE if you process materials A, B, and/or C (objects); PRESS PUMP BUBBLE POUR if you process materials D, E, and/or F (substances); and TYPE FILE SHUFFLE SIGN if you process materials G, H, and/or I (services). The number of lines completed will determine your industry's WYI (Work Yield Inventory).

Step B2. Each industry may increase its productivity by specialization (see table 13-7).

Step B3. Each industry may increase its productivity by investing in technological innovation, industrialization (see tables 13-8 and 13-9).

TABLE 13-8
Industrialization Chart

Industrialization Level	*Industrialization Index*	*Cost WESSCREDS*
1	1	0
2	2	20
3	3	50
4	4	100
5	5	200
6	6	400

TABLE 13-9
Productivity Chart

WSI Index = Work Yield Inventory + Industrialization Index + Specialization Index	*Productivity Index = PI*
10	0.5
12	1
14	1.5
16	2
18	2.5

Step B4,5. Compute the WSI index by adding the work yield inventory (WYI) plus the specialization index (SI), plus the industrialization index (II). With the WSI index, enter the productivity chart and obtain the productivity index (PI). This index will be used as a multiplier of your industrial output.

Step C1,2. Units processed can now be computed by multiplying number of units purchased times the productivity index. These units are then added to the unsold units from the previous round, and the total number of units on hand can be determined.

Step C3. Players are given five minutes to establish product and balanced economy cartels. The conference can be used to allocate productivity, lend money to other players, or borrow money from other players. All transactions between players are private one, and the method of enforcement is up to the players. Should the role of broker be played in this game, they will act as the intermediaries between the national economy and the industries.

Step C4. During this step each industry determines the number of products sold. The market only recognizes assembled products ABC, DEF, and GHI. The operator will tabulate the total number of products made. The brokers or industry representatives will then receive income from the national economy according to the following formula:

TOTAL INDUSTRIAL = INCOME IN WESSCREDS	Equal number of products combination ABC, DEF, and GHI × X + remaining equal number of products combinations ABC, DEF, and/or ABC, GHI, and/or DEF, GHI. XY + remaining single products ABC, and/or DEF, and/or GHI × Z.

For round one, only $X = 15$, $Y = 9$, and $Z = 3$. Operator will announce other values during later rounds. A product may only be counted once when determining industrial income. The brokers or industry representatives will proceed to divide in an equitable way the overall industrial income among the various producers. The number of units sold can now be determined.

Step C5-8. The remaining inventory is determined by subtracting units sold from total units. The units sold times the price per unit determines the revenue for each product. These revenues can then be added, and the total revenue for the industry determined.

Step D1. It costs ½ a WESSCRED/ROUND to store one unit of inventory. Inventory storage cost is determined by multiplying remaining inventory times 0.5 WESSCREDs.

Step D2,3. WESSCREDs from the last round are obtained from step E of the previous round. During the first round this value is given (or entered on your sheet as step D2, Round 2) to you by operator. The material and labor costs are obtained from step A6.

Step D4. Enter industrial decision expenses

Step D5. The operator will distribute cards that will indicate fluctuations in the local economy due to global economy, weather, fire and theft, and other events not directly under the local controls.

Step D6. Any transactions incurred during step C3, cartel conference, should be entered here. Payments and loans are entered with a plus (+) sign.

Step E1. WESSCREDs on hand are found by adding, algebraically, all D's. This value should also be entered in line D2 of next round. Under no circumstances should this value be a negative one.

WESSWORK Game Operator Information

000. Read the *Player's Instruction Booklet.*

00. The distribution of WESSCREDs at the beginning of the game varies with the number of participants. Each participant is randomly given a player's tally sheet (figure 13-2) that contains either a 50, 100, 150, 250, 350, or 500 number on it. The distribution formula for this is found in figure 13-3. For instance if there are twenty players, two will receive Tally Sheets indicating they begin with 350 WESSCREDs, and two will receive Sheets indicating that they begin with 100 WESSCREDs. The remainder will be given 50 WESSCREDs each.

0. The steps of play for the game are contained on the Tally Sheet. Twenty-seven steps make one round of play. They are explained as follows:

Step A0 Initially each player must declare which product he/she is going to process materials for, objects (ABC), substances (DEF) and/or services (GHI).

Step A1 Player must indicated the raw material(s) he/she plans to process for the product declared in Step A0.

Step A2 Based on the quantity of units to be purchased each player must indicate the unit cost. Operator can at his/her discretion change the basic unit price each round by posting a new unit price matrix. Should the unit price matrix be changed, the multiplier values X, Y, and Z in the total Industrial Income (see Step C4 on this page) must also be changed.

Step A3 Based on the number of WESSCREDs on hand, each player must decode as many lines as possible of LIFT RIVET FIT ASSEMBLE if you process materials A, B, and/or C (objects); PRESS PUMP BUBBLE POUR if you process materials D, E, and/or F (substances); and TYPE FILE SHUFFLE SIGN if you process G, H, and/or I (services). Write these four words one line at a time. The number of lines completed in forty-five seconds is the WYI and should be indicated on the number of raw material units to be purchased. A player is not allowed to purchase more units than his/her capital assets permit. Each player may lend or borrow WESSCREDs to or from another player during Step C3.

Figure 13-2
WESSWORK tally sheet.

INDUSTRY NAME ______

	ROUND			
	COMPANY PRODUCT			
A1	Raw Material Label			
A2	Cost per Unit			
A3	#of Units Purchased			
A4	Cost			
A5	Total Material Cost = A4's			
A6	Total Material and Labor Costs = A5 X 2			
B1	Work Yield Inventory			
B2	Specialization			
B3	Industrialization			
B4	WSI Index = B1 + B2 + B3			
B5	Productivity			
C1	Units Processed A3 X B5			
C2	Total Units = C1 + Remaining Inventory from Last Round			
C3	Cartel Conference	//// //// ////	//// //// ////	//// //// ////
C4	Units Sold			
C5	Remaining Inventory			
C6	Price per Unit Sold			
C7	Revenue			
C8	Total Revenue C7's			
D1	Cost to Store Inventory			
D2	WESSCREDs from Last Round			
D3	Material and Labor Costs, A6			
D4	Industrialization Expenses			
D5	Random Natural Forces			
D6	Transactions			
E	WESSCREDs on Hand			

Figure 13-3
WESSCRED distribution formula.

No. Players	WESSCRED Starting Balance*	Total Cards	Status Quo	#	+	#	-	#	+ With TL 2 or - at TL 1
12-15	1-350 1-100 remainder 50	15	3	1 1 2	+60 +30 +20	2 3 1	-60 -30 -20	1 1	+20 -20 +20 -50
16-17	1-350 1-15-15 remainder -50	19	4	1 2 2	+60 +30 +20	2 3 3	-60 -30 -20		S
18-19	1-350 1-100 1-250 remainder -50	19	4	1 2 2	+60 +30 +20	2 3 3	-60 -30 -20		A
20-21	2-350 2-100 remainder 50	21	5	1 2 3	+60 +30 +20	2 3 3	-60 -30 -20		M
22-23	1-500 1-350 2-100 remainder -50	23	6	1 2 3	+60 +30 +20	2 4 3	-60 -30 -20		E
24-25	1-500 1-350 3-100 remainder 50	25	6	1 3 3	+60 +30 +20	3 1 4 2	-60 -50 -30 -20		
26-27	1-500 1-350 3-100 remainder 50	27	7	1 1 3 3	+60 +50 +30 +20	3 1 4 2	-60 -50 -30 -20		
28-29	1-500 1-350 4-100 remainder 50	29	7	1 1 4 3	+60 +50 +30 +20	3 1 4 3	-60 -50 -30 -20		
30-31	1-500 1-350 4-100 remainder 5	31	7	1 1 4 3	+60 +50 +30 +20	3 1 4 4	-60 -50 -30 -20	2 1	+20 -20 +20 -50
32-33	1-500 1-350 4-100 remainder 50	33	8	1 1 4 3	+60 +50 +30 +20	3 1 5 4	-60 -50 -30 -20	1 2	+20 -20 +20 -50
34-35	1-500 1-350 5-100 remainder 50	35	8	2 1 4 3	+60 +50 +30 +20	4 1 5 4	-60 -50 -30 -20	1 2	+20 -20 +20 -50

* Enter this value in player's information sheet
** Randomly distribute these cards each round during step D5.

Step A4 Cost of purchasing raw materials is obtained by multiplying cost per unit, A2, times the number of units purchased, A3.

Step A5 Total material cost is obtained by summing all A4's.

Step A6 Labor costs in WESSCRED are set equal to material costs. Therefore, total Material and Labor Costs are obtained by multiplying total Material Costs times two (2). The operator may at his/her discretion modify this formula by posting the appropriate ratios.

Step B1 This step is a fun gimmick where participants determine their Work Yield Inventory (WYI) for each round. Give participants forty-five seconds in which to write as many lines as possible of LIFT RIVET FIT ASSEMBLE if they process materials A, B, and/or C (objects); PRESS PUMP BUBBLE POUR if they process materials D, E, and/or F (substances); and TYPE FILE SHUFFLE SIGN if they process materials G, H, and/or I (services). Write these four words one line at a time. The number of lines completed in forty-five seconds is the WYI and should be indicated in the square of the Tally Sheet. Typically participants will complete between five and ten lines in this time period.

Step B2 Productivity is linked to specialization. Highest productivity is achieved by specializing in the process of one product. The manner in which specialization affects productivity is indicated in the Specialization Chart, and is called the Specialization Index (SI).

Step B3 Industrialization is a function of investment. The Industrialization Index (II) and its related cost is obtained from the Industrialization Chart.

Step B4 Compute the WSI index by adding the Work Yield Inventory (WYI) plus the Specialization Index (SI), plus the Industrialization Index (II).

Step B5 With the WSI Index, enter the Productivity Chart and obtain the Productivity Index (PI). This index will be used as a multiplier of your industrial output.

Step C1 The number of units processed is obtained by multiplying the Number of Units Purchased (A3) times the Productivity (B5).

Step C2 The total number of units is obtained by adding the newly processed units (C1) to the units in inventory (C5 from previous round).

Step C3 Cartel Conference. The game operator announces a set time for negotiations. Five minutes usually provides ample time. Ten minutes is a maximum. The game operator may wish to set a time limitation according to such factors as time available for play, intensity of negotiations, and/or interesting discussions. During this time participants may negotiate loans, establish manufacturing agreements (cartels) leading to the production of a product, and discuss means and ways of establishing balanced production. The conditions of the bargaining are entirely in the hands of the participants. Lending and borrowing rules are the responsibilities of the parties involved in the agreement, and enforcement is not the responsibility of the operator. Records of loans should be kept by players in the space allocated in D6.

Step C4 Each industry determines the number of products sold. The market only recognizes assembled products ABC, DEF, and GHI. The operator tabulates the total number of products made. The brokers or industry representatives receive income from the national economy according to the formula:

TOTAL INDUSTRIAL = INCOME IN WESSCREDS	Equal number of products combination ABC, DEF, and GHI $\times X$ + remaining equal number of products combinations ABC, DEF, and/or ABC, GHI, and/or DEF, GHI. XY + remaining single products ABC, and/or DEF and/or GHI $\times Z$.

For round one only, $X = 15$, $Y = 9$, and $Z = 3$. Operator will announce other values during later rounds. A product may only be counted once when determining industrial income. The brokers or industry representatives will proceed to divide in an equitable way the overall industrial income among the various producers. The number of units sold can now be determined. Should the operator decide to increase the cost of raw materials, equal proportional values in the coefficients X, Y, and Z must be made.

Step C5 Remaining inventory is obtained by subtracting units sold (C4) from Total Units (C2).

Step C6 Price per unit sold is obtained from Step C4.

Step C7 Revenue is then equal to Units sold (C4) times Price per Unit sold (C6).

Step C8 Total revenue is then obtained by adding individual unit revenues, summation of the C7's.

Step D1 It costs ½ a WESSCRED/ROUND to store one unit of inventory. Inventory Storage Cost is determined by multiplying remaining inventory times 0.5 WESSCREDs.

Step D2 Player should enter value from Step E of the previous round. During the first round each player will be given a starting number of WESSCREDs. See WESSCRED distribution formula.

Step D3 Total Material and Labor Costs are obtained from Step A6.

Step D4 At this point players may change industrialization levels by modernizing their factory. There are 6 levels of industrialization, each with its associated cost. To go from one level to another, pay the cost difference between levels.

Step D5 Let each participant randomly select from the NATURAL/SOCIAL FORCES CARDs. This is a chance factor indicating that weather conditions or social-political economic conditions have influenced your industry. The number and value of the cards should be controlled according to the DISTRIBUTION FORMULA. Be sure to collect all

NATURAL/SOCIAL FORCES CARDs and prepare them for the next round.

Step D6 Loans or repayments to other players are recorded as negative values, revenues as positive values.

Step E WESSCREDs on hand are equal to the algebraic summation of all Ds.

How to Modify the Game

WESSWORK is an adaptation of BALDICER—A Game on World Hunger and Population Control by Georgeann Wilcoxon (John Knox Press, Box 1176, Richmond, Virginia 23209). Although BALDICER was not conceived as a frame game (a game that can be easily modified to fit other circumstances), its model has all the characteristics of a frame game. The model used in WESSWORK should only be used to simulate situations that can be modeled by an accounting process, i.e. population models, municipal budgeting, budgeting in general, and other iterative situations. It should not be used, for example, as a space planning model, since location is not modeled by the WESSWORK structure. It could well be, however, that a WESSWORK type game would be included as a submodel of a land-use game.

It is therefore important to identify the type of situation, actors, and rules that are to be modeled. A causal diagram similar to the one used for WESSWORK (see figure 13-4) should then be constructed. This will help in identifying important simulation components and how they interact with each other. The causal diagram should then be followed by an economic flow diagram. The unit of trade and the corresponding exchanges should be determined. A form should then be made and a game should be played as soon as possible.

Calibration follows game play. Calibration insures that the game does not blow up economically, and enhances credibility of the simulation. It is important to remember that games should not be all strategy. Some fun is needed; therefore, the inclusion of gimmicks such as the task in WESSCRED where players write LIFT RIVET FIT ASSEMBLE is important. WESSCRED could just as easily have included a chart that would have determined the WYI in a probabilistic manner; however, it would not have had the same flavor as the WYI task.

Finally, it is important to keep some steps of the game a bit ambiguous, so that players can make their own interpretations from which healthy discussions and even arguments can arise. In WESSCRED this is done during the market clearing situation. The players must come out with a formula for partitioning the total industrial revenue.

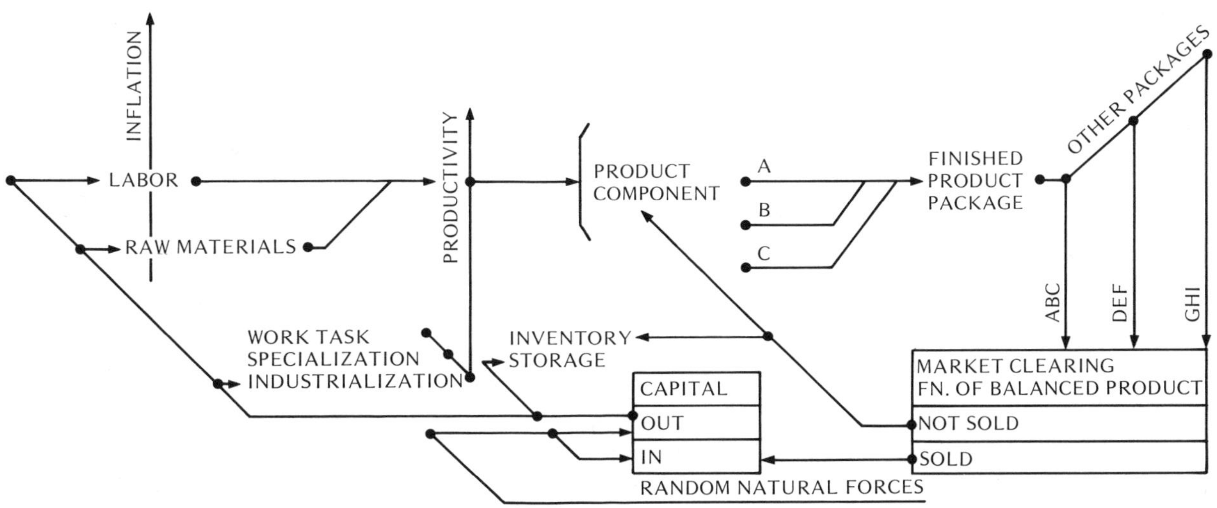

Figure 13-4
WESSCRED-game causal flow diagram.

WESSDEV

Introduction

This gaming module simulates the forces that affect the actions of different types of land developers during the development of a new town. Each team of developers begins the game with equal resources of $200,000 and a few plots of land. The object of the game is to maximize investment profits and to have the most money at the end of the game. People have the choice of investing their limited resources in either the construction of the community or land speculation. Players can choose to build a type of building that corresponds to their developer role and receive a fixed income in each subsequent round, or they may choose to purchase land at predevelopment bargain prices at the beginning of the game, selling the land later at inflated prices as the community is built. A GOVERNMENT DEVELOPMENT SECTOR is also simulated in this game. Unlike the other private development sectors, the GOVERNMENT SECTOR is primarily concerned with the quality of life of the new community during subsequent development.

The game operator represents Wesley Town Associates, a nonprofit corporation that plans to subcontract out most of the construction. The game begins with a general construction plan. As figure 13-5 illustrates, Wesley Town is to be developed within three phases. At the beginning of the game the various types of developers can buy options on any tract of land within the community, but can only build within phase one. After at least one half of the land within phase one is developed, land values double and construction in the phase two area can begin. After all of phase one tracts have been developed and at least one half of phase two tracts have been developed, land prices double again and construction can begin on phase three tracts.

In all there are 234 tracts of land that can be developed in this game. Each tract has been assigned an identification number that is composed of a letter that designates its vertical displacement and a number that designates its horizontal displacement. As stated before, these tracts are divided into three development phases. The development phase scheduled for any tract of land is noted within parentheses. Tract A1 for example is located in phase 1, while tract K5 is in phase 2, and Z9 is in phase 3.

There are four different types of developers in this game. These are 1) residential developers, 2) commercial developers, 3) service developers, and 4) governmental developers. The actual roles will be explained in the following sections but roughly the following interactions exist. The three teams of residential developers build housing within the community. After these developers build the community's housing, they sell the labor that housed there to the various community employers (Commercial, Services, and Government). These developers also buy commercial services from the Commercial sector and government services from the Government sector. The goal of the Housing development sector is to make as much money as possible. The second development sector, the Commercial sector, buys labor from the Housing sector and government services from the Government sector. It also in turn sells commercial services to the Housing sector. The third development sector, the Service developers, build offices and other white collar concerns that employ residents in the community. This sector essentially is involved in services that they sell to the outside world. This income is brought into the community and used to build additional service institutions, employ residents, and to purchase local government services. The fourth development sector, the Government sells government services to the other sectors and employs labor from within the community. Unlike the first three sectors

PHASE I

A1 (I)	A2 (I)	A3 (I)	A4 (I)	A5 (I)	A6 (I)	A7 (I)	A8 (I)	A9 (I)
B1 (I)	B2 (I)	B3 (I)	B4 (I)	B5 (I)	B6 (I)	B7 (I)	B8 (I)	B9 (I)
C1 (I)	C2 (I)	C3 (I)	C4 (I)	C5 (I)	C6 (I)	C7 (I)	C8 (I)	C9 (I)
D1 (I)	D2 (I)	D3 (I)	D4 (I)	D5 (I)	D6 (I)	D7 (I)	D8 (I)	D9 (I)
E1 (I)	E2 (I)	E3 (I)	E4 (I)	E5 (I)	E6 (I)	E7 (I)	E8 (I)	E9 (I)
F1 (I)	F2 (I)	F3 (I)	F4 (I)	F5 (I)	F6 (I)	F7 (I)	F8 (I)	F9 (I)
G1 (I)	G2 (I)	G3 (I)	G4 (I)	G5 (I)	G6 (I)	G7 (I)	G8 (I)	G9 (I)
H1 (I)	H2 (I)	H3 (I)	H4 (I)	H5 (I)	H6 (I)	H7 (I)	H8 (I)	H9 (I)
I1 (I)	I2 (I)	I3 (I)	I4 (I)	I5 (I)	I6 (I)	I7 (I)	I8 (I)	I9 (I)
J1 (I)	J2 (I)	J3 (I)	J4 (I)	J5 (I)	J6 (I)	J7 (I)	J8 (I)	J9 (I)

PHASE II

K1 (II)	K2 (II)	K3 (II)	K4 (II)	K5 (II)	K6 (II)	K7 (II)	K8 (II)	K9 (II)
L1 (II)	L2 (II)	L3 (II)	L4 (II)	L5 (II)	L6 (II)	L7 (II)	L8 (II)	L9 (II)
M1 (II)	M2 (II)	M3 (II)	M4 (II)	M5 (II)	M6 (II)	M7 (II)	M8 (II)	M9 (II)
N1 (II)	N2 (II)	N3 (II)	N4 (II)	N5 (II)	N6 (II)	N7 (II)	N8 (II)	N9 (II)
O1 (II)	O2 (II)	O3 (II)	O4 (II)	O5 (II)	O6 (II)	O7 (II)	O8 (II)	O9 (II)
P1 (II)	P2 (II)	P3 (II)	P4 (II)	P5 (II)	P6 (II)	P7 (II)	P8 (II)	P9 (II)
Q1 (II)	Q2 (II)	Q3 (II)	Q4 (II)	Q5 (II)	Q6 (II)	Q7 (II)	Q8 (II)	Q9 (II)
R1 (II)	R2 (II)	R3 (II)	R4 (II)	R5 (II)	R6 (II)	R7 (II)	R8 (II)	R9 (II)

Figure 13-5
WESSDEV game board.

described above, which attempt to maximize their capital gain, the Government sector attempts to maintain a high quality of life within the community by maintaining certain air and water quality and recreation standards.

PHASE III

S1 (III)	S2 (III)	S3 (III)	S4 (III)	S5 (III)	S6 (III)	S7 (III)	S8 (III)	S9 (III)
T1 (III)	T2 (III)	T3 (III)	T4 (III)	T5 (III)	T6 (III)	T7 (III)	T8 (III)	T9 (III)
U1 (III)	U2 (III)	U3 (III)	U4 (III)	U5 (III)	U6 (III)	U7 (III)	U8 (III)	U9 (III)
V1 (III)	V2 (III)	V3 (III)	V4 (III)	V5 (III)	V6 (III)	V7 (III)	V8 (III)	V9 (III)
W1 (III)	W2 (III)	W3 (III)	W4 (III)	W5 (III)	W6 (III)	W7 (III)	W8 (III)	W9 (III)
X1 (III)	X2 (III)	X3 (III)	X4 (III)	X5 (III)	X6 (III)	X7 (III)	X8 (III)	X9 (III)
Y1 (III)	Y2 (III)	Y3 (III)	Y4 (III)	Y5 (III)	Y6 (III)	Y7 (III)	Y8 (III)	Y9 (III)
Z1 (III)	Z2 (III)	Z3 (III)	Z4 (III)	Z5 (III)	Z6 (III)	Z7 (III)	Z8 (III)	Z9 (III)

Figure 13-5 continued.

Steps of Play

Step 1 The various teams bid on land by submitting closed bids to the bank.

Step 2 Teams develop the tracts of lands for that round. Teams build in the order of SERVICE, RESIDENTIAL, COMMERCIAL, and GOVERNMENT DEVELOPERS.

The following form (figure 13-6) is used by the various teams during the first two steps of play and handed into the game operator at the beginning of the third step of play.

Step 3 Teams receive income for that round: SERVICE developers receive $150,000 for each fully employed company from the game operator; RESIDENTIAL DEVELOPERS receive wages from the other developers; COMMERCIAL DEVELOPERS receive $4,000 from every labor unit that they supply commercial services; GOVERNMENTAL DEVELOPERS receive sewage fees from the other three developers. They also collect a recreation fee from the RESIDENTIAL DEVELOPERS. In addition to collecting money from the developers within the game, the GOVERNMENT DEVELOPERS also collect matching funds from the game operator. These monies represent matching state and federal funds.

Step 4 GOVERNMENT DEVELOPERS calculate the various quality of life indices for the entire system. Four quality of life indicators are used: 1) WATER QUALITY, 2) AIR QUALITY, 3) UNEMPLOYMENT, and 4) RECREATION STANDARDS. The community must meet certain standards for these measures if they are to receive matching state and federal funds for the next round.

The Business Service Sector

Two teams of two individuals each represent the BUSINESS SERVICE SECTOR in this game. These developers construct clean industries that employ residents within the community and sell their products to the outside world. The needs of these industries are described in table 13-10.

Each BUSINESS SERVICE company follows the steps of play outlined in step one. After bidding on land and purchasing their successful bids, the BUSINESS SERVICE DEVELOPER builds offices on land that he owns.

If there is a city utility service line running along the square where the developer is building, (s)he must link up to a city sewage system and pay a fee each round to the GOVERNMENT SECTOR. If the city utility lines do not run along an edge of the property, the developer must construct a septic tank system. Once a septic tank system is constructed the developer does not have to pay any additional fees. An office cannot receive any income unless it is linked up to the city utility system or has its own septic system.

After an office is built, the BUSINESS SERVICE DEVELOPER must find employees for the business. This labor is purchased from the RESIDENTIAL HOUSING DEVELOPER at a minimum labor fee of $1000 for each labor unit or $50,000 for 100 labor units. In order to receive income for any office complex a BUSINESS SERVICE DEVELOPER must have 100 labor units for each. No outside income will be paid to any office complex unless it is fully employed.

The Housing Sector

Three teams of two individuals each represent the RESIDENTIAL HOUSING DEVELOPMENT SECTOR in this game. Each team can develop three different types of housing units: low density housing (R1), medium density housing (R2), and high density housing (R3). Information necessary for this sector is included in table 13-11.

Figure 13-6
WESSDEV economic form.

ROUND ________

TEAM				
	HOUSING TEAM	1 ( )	2 ( )	3 ( )
	BUSINESS TEAM	1 ( )	2 ( )	
	COMMERCIAL TEAM	1 ( )	2 ( )	
	GOVERNMENT	( )		

PART I—Land bids

We would like to enter bids on the following tracts

TRACT ________	BID $________	ACCEPTED	YES ( ) NO ( )
TRACT ________	BID $________	ACCEPTED	YES ( ) NO ( )
TRACT ________	BID $________	ACCEPTED	YES ( ) NO ( )
TRACT ________	BID $________	ACCEPTED	YES ( ) NO ( )
TRACT ________	BID $________	ACCEPTED	YES ( ) NO ( )
TRACT ________	BID $________	ACCEPTED	YES ( ) NO ( )

AMOUNT OWED OPERATOR FOR LAND PURCHASES ——→ $________

PART II—Building

We would like to construct the following:

LAND USE ________	TRACT ________	SEPTIC TANK SYSTEM BUILT	YES ( )
LAND USE ________	TRACT ________	SEPTIC TANK SYSTEM BUILT	YES ( )
LAND USE ________	TRACT ________	SEPTIC TANK SYSTEM BUILT	YES ( )
LAND USE ________	TRACT ________	SEPTIC TANK SYSTEM BUILT	YES ( )
LAND USE ________	TRACT ________	SEPTIC TANK SYSTEM BUILT	YES ( )
LAND USE ________	TRACT ________	SEPTIC TANK SYSTEM BUILT	YES ( )

BUILDING COSTS OWED TO GAME OPERATOR ——→ $________

Each RESIDENTIAL DEVELOPER company follows the steps of play outlined in the previous section. After bidding on land and purchasing their successful bids, the RESIDENTIAL DEVELOPERS build housing on the land that they own.

If there is a city utility service line running along the square where the developer is building, (s)he must link up to a city sewage system and pay a fee each round to the GOVERNMENT SECTOR. If the city utility lines do not run along an edge of the property, the developer must construct a septic tank system. Once a septic tank system is constructed the developer does not have to pay any additional fees. A housing unit cannot receive any income unless it is either linked to a city utility system or has its own septic system. It is important to note that a high density housing unit (R3) can only be built within the areas served by the city sewage system.

After a housing unit is constructed, the various RESIDENTIAL DEVELOPERS attempt to sell the labor housed within their units to the other developers. The minimum wage suggested in this game is $1000 for each labor unit. The number of labor units that a developer receives for each housing type and their minimum income is described in table 13-11. Although a minimum income is set for this game there is no reason why the RESIDENTIAL DEVELOPERS cannot bargain with the other sectors for higher wages during the play of the game. The RESIDENTIAL DEVELOPERS are cautioned about this practice however. If wages were to increase, it is possible that inflation could begin within this economic system.

After the HOUSING SECTORS receive their incomes, they must pay a number of living costs. Each labor unit requires commercial services, which are purchased from the COMMERCIAL DEVELOPERS. These commercial services cost $4000 for each labor unit. The commercial service costs for a R1 for example are $20,000. The actual commercial service costs are described in table 13-10.

TABLE 13-10

Land Use	*Construction Cost*	*Needed Labor*	*Sewer Fees* *	*Construction Cost of a Septic System* **	*Outside Income Generated*
Office	$50,000	100	$1000	$5000	$150,000

* This amount must be paid each round to the GOVERNMENT SECTOR if the office is linked up to a sewer system.

** This amount is paid only once during the actual construction of the office.

TABLE 13-11

Land Use	*Construction Cost*	*Labor*	*Sewer Fees* *	*Construction of Septic System* **	*Commercial Costs*	*Recreation Fees*	*Income Produced*
R1	$60,000	50	$5000	$15,000	$20,000	$5000	$50,000
R2	$150,000	150	$15,000	$30,000	$60,000	$15,000	$150,000
R3	$280,000	300	$30,000	#	$120,000	$30,000	$300,000

\# must be constructed on a sewage system

* This amount must be paid each round to the GOVERNMENT SECTOR if the R1, R2, or R3 is linked up to a sewer system.

** This amount is paid only once during the actual construction of the R1, R2, or R3.

TABLE 13-12

Land Use	*Construction Cost*	*Needed Labor*	*Sewer Fees* *	*Construction of a Septic Tank System* **	*Population Served*	*Income Generated*
Store	$50,000	100	$1000	$5000	500	$200,000

* This amount must be paid each round to the GOVERNMENT SECTOR if the store is linked up to a sewer system.

** This amount is paid only once during the actual construction of the store.

In addition to paying the COMMERCIAL DEVELOPER, it is necessary that the RESIDENTIAL DEVELOPER pay the GOVERNMENT SECTOR various recreation and sewage fees if the housing units that they own are linked up to the city sewage system. If the housing units considered are not hooked to the city system but are linked up to their own septic system, the RESIDENTIAL DEVELOPER need only pay the GOVERNMENT SECTOR recreation fees. The recreation fees are used to construct recreation facilities within the community. The sewage fees are used to maintain existing sewage lines and to construct new utility lines.

The Commercial Sector

Two teams of two individuals each represent the COMMERCIAL SECTOR in this game. These COMMERCIAL DEVELOPERS construct commercial facilities that serve the community. The necessary information for this role is summarized in table 13-12.

Each COMMERCIAL DEVELOPER follows the steps of play outlined in section one. After bidding on land and purchasing their successful bids, the COMMERCIAL DEVELOPER builds stores on the land that (s)he owns.

If there is a city utility service line running along the square where the developer is building, (s)he must link up to a city sewage system and pay a fee each round to the GOVERNMENT SECTOR. If the city utility lines do not run along an edge of the property, the developer must construct a septic tank system. Once a septic tank system is constructed the developer does not have to pay any additional fees. A store cannot receive any income unless it is either linked to a city utility system or has its own septic system.

The COMMERCIAL SECTOR supplies commercial services to the local community. Before it can operate it must first employ 100 labor units for existing stores. The COMMERCIAL DEVELOPERS buy this labor from the RESIDENTIAL DEVELOPERS. If a store is fully employed it can sell commercial services to residents. Each store can supply 500 labor units. A store receives $4000 for each labor unit supplied: $20,000 for each R1, $60,000 for each R2, and $120,000 for each R3.

The Government Sector

The GOVERNMENT DEVELOPMENT SECTOR is represented by one team of three individuals in this game. This sector builds utility lines, utility facilities, and recreational facilities. One of the most important functions of this team is to calculate the quality of life index for this game.

The GOVERNMENT SECTOR collects fees from the other development sectors to finance its facility development. Sewer fees are collected from offices, stores, and housing units linked to the city utility system, and recreation fees are collected from each housing unit whether it is connected to the city utility system or not. In addition to these fees, the GOVERNMENT SECTOR also receives matching funds from the game operator. If for example the GOVERNMENT SECTOR collected $500,000 from the other sectors in fees, it would receive an additional $500,000 from the game operator. This matching funding is contingent on the community meeting the quality of life standards, which are given in table 13-13. The various fees and development costs associated with the GOVERNMENT SECTOR are given in tables 13-14 and 13-15.

As table 13-13 suggests, the community quality of life index is composed of four basic components: UNEMPLOYMENT, RECREATION QUALITY, WATER QUALITY, and AIR QUALITY. The calculation of these various indices is described below.

TABLE 13-13
Quality of Life Standards

Unemployment Standard	No more than 5% of the population. This is calculated in Table 9.
Recreation Standards	Adequate recreation facilities are available for everyone in this community. This is calculated in Table 11.
Water Quality Standards	No more than a water quality index of 500 as calculated in Table 12.
Air Quality Standards	No more than an air quality index of 500 as calculated in Table 10.

TABLE 13-14
Fee Rates

Land Use	*Sewer Fees if Connected to the City Utility System*	*Recreation Fees*	*Total Government Fees if Connected to the City System*	*Total Government Fees if Not Connected to the City System*
Office	$1000	*	$1000	0
Store	$1000	*	$1000	0
R1	$5000	$5000	$10,000	$5000
R2	$15,000	$15,000	$30,000	$15,000
R3	$30,000	$30,000	$60,000	$30,000

* does not apply

TABLE 13-15
Development Costs for the Government Sector

Project Name	*Development Costs*	*Space Needs*	*Labor Needs*	*Units Served*
Sewer/utility link (one square)	$10,000	*	*	2 tracts
Water treatment plant	$50,000	1 tract	50 L	50 WQ units
Recreation facility I	$50,000	1 tract	50 L	10 WQ units
Recreation facility II	$100,000	2 tracts	*	15 RQ units
Bus system	$50,000	1 tract	50 L	100 AQ units

* not significant

Unemployment The unemployment index is calculated using the following form (figure 13-7). The GOVERNMENT SECTOR first counts the housing units that have been developed within the community and then enumerates the employment potential. The difference between these totals is then used to calculate the unemployment index.

Air quality The air quality index is calculated using the form as shown in figure 13-8.

The GOVERNMENT SECTOR first enumerates the various land-use types developed and multiplies each total by a specific air pollution index. Another value is calculated for the amount of public transportation that is available to the community. The difference between the amount of air pollution that is produced by the various land uses and the amount saved through the use of a public transportation system determines the air quality index.

Recreation quality The recreation index is calculated using the form as shown in figure 13-9. The GOVERNMENT SECTOR first counts the number of the three housing types that are developed in the community. By multiplying these by a specific recreation index for each housing type a recreation need is established for the community. This is then compared to the recreation services provided. The difference of these defines the recreation index.

Water quality The water quality index is calculated using the form as shown in figure 13-10. The GOVERNMENT SECTOR first counts the number of developed land-use types that are connected to the city utility system. Next the number of land-use types that are connected to their own septic system are enumerated. Third, the number of water treat-

Figure 13-7
WESSDEV form used to calculate unemployment.

ROUND ______

STEP 1
FILL OUT THE FOLLOWING

land	#developed	employment potential	labor produced	employment product	labor product
R1	______	*	50	*	______
R2	______	*	100	*	______
R3	______	*	300	*	______
OFFICE	______	100	*	______	*
STORE	______	100	*	______	*
WATER TREATMENT PLANT	______	50	*	______	*
RECREATION FACILITY I	______	50	*	______	*
BUS SYSTEM UNIT	______	50	*	______	*

STEP 2
ADD THE PRODUCT COLUMNS

TOTAL	*	*	*	______	______

STEP 3
PUT THE EMPLOYMENT PRODUCT SUM HERE ——→ ______

STEP 4
SUBTRACT THESE SUMS AND PUT THE DIFFERENCE HERE ——→ ______

STEP 5
DIVIDE THE DIFFERENCE FOUND IN STEP 4 BY THE EMPLOYMENT PRODUCT SUM FOUND IN STEP 3

THIS ANSWER SHOULD NOT BE MORE THAN .05

Figure 13-8
WESSDEV form used to calculate community air quality.

STEP 1
FILL OUT THE FOLLOWING

land use	# developed	AQ index	PRODUCT = # times AQ index
R1	______	10	______
R2	______	30	______
R3	______	60	______
OFFICE	______	10	______
STORE	______	10	______
BUS SYSTEM UNITS	______	-100	______

STEP 2
ADD THE LAST COLUMN ⟶ ______
THIS FIGURE SHOULD NOT BE MORE THAN 500

Figure 13-9
WESSDEV form used to calculate community recreation standards.

STEP 1
FILL OUT THE FOLLOWING

land use	# developed	RQ index	product = # times RQ
R1	______	1	______
R2	______	3	______
R3	______	6	______
RECREATION FACILITY I	______	-10	______
RECREATION FACILITY II	______	-15	______

STEP 2
ADD THE LAST COLUMN ⟶ ______
THIS TOTAL CANNOT BE MORE THAN ZERO

ment plants that have been developed by the city are enumerated. These totals are then multiplied by a WQ index for each specific type. The summation of these products gives the water quality index for the community.

In order for the GOVERNMENT SECTOR to receive matching funds from the game operator, (s)he must submit these four forms to the Game Operator at the end of each round of play.

Figure 13-10
WESSDEV form used to calculate community water quality.

STEP 1
FILL OUT THE FOLLOWING

land use	# developed	WQ index	Product (# times WQ)
R1 connected to city system	______	10	______
R1 with own septic system	______	30	______
R2 connected to city system	______	30	______
R2 with own septic system	______	60	______
R3 connected to the city system	______	60	______
OFFICE connected to city system	______	1	______
OFFICE with own septic system	______	10	______
STORE connected to city system	______	1	______
STORE with own system	______	10	______
WATER TREATMENT PLANTS	______	-50	______

STEP 2
ADD THE THIRD COLUMN OF THIS TABLE ——————→ ______
THIS TOTAL CANNOT EXCEED 500

WESMED

Introduction

This gaming module simulates the forces that affect the actions of newscasters and other media personnel during the development of a new town. Two rival news programs are pitted against each other in a contest for audience rating. Each of the two teams portrayed have contrasting political views. One of the news programs, WHAT NEWS, is fairly conservative in its political outlook. The producers of this program see themselves as the defenders of the community's moral fiber and in general take a dim view of any government action that may upset this moral equilibrium. The media personnel at WHAT NEWS generally reflect this attitude. The other news program, WHYY NEWS, represents the other side of the political coin. The producers of WHYY NEWS see themselves as the defenders of the social underdog and generally oppose the conservative viewpoint championed by the WHAT NEWS team. The media personnel at WHYY NEWS are a pretty liberal bunch and generally agree with the liberal stand taken by the program's producers.

There are nine different roles used in this game. Each role, its game task, where it works, and the political viewpoint of the role is summarized in table 13-16 on pages 240 and 241.

Steps of Play

Step 1 NEWSCASTERS' MEETING—The WHAT NEWS team and the WHYY NEWS team meet separately to discuss their various strategies for the newscast. It is suggested that they meet on different sides of the room to allow a certain degree of competitive secrecy to develop. The AD team should get together at this time and trade ideas of how they might present their potential products. Examples of ads are given in APPENDIX II. This first step of play should last no longer than ten minutes. When the reporters leave their respective meetings they should have a good idea of what story they might write and should know what types of people they should interview.

Step 2 RESEARCH PERIOD—During this period the respective reporters should interview people for their story. Every news story should contain at least one quote in it. The anchor people at this time begin to organize their notes for the debate that follows the newscasts. The ad men can either walk around and ask people for ideas about products or commercials or they can begin writing their ads at this time. This stage should last no longer than fifteen minutes.

Step 3 WRITING PERIOD—During this step of play the various reporters should sit down and begin composing their stories. If they are having any trouble they should first check with their respective anchor person for help. Reporters should also at this time invent a stage name that they can use during the broadcast. The two anchor people should schedule their shows at this time (i.e., decide what order they are going to present the three stories and where they are going to place the two ads that must be included within their program). The ad people should be finishing up their ads. It is suggested that they rehearse them a number of times to be sure that the ads do not go beyond thirty seconds. This stage should not last longer than twenty minutes.

TABLE 13-16
Roles for WESSMED

Role	*Political Bias*	*Game Task*
WHAT NEWS ANCHOR PERSON	conservative	(S)he coordinates the WHAT NEWS program. (S)he makes sure that everyone understands their task and checks in on everyone on the WHAT NEWS staff to make sure that everything will be ready by the newscast. During the newscast the anchor person introduces the show and provides an introduction to each news story. In addition to this, the Anchor person must also debate his/her counterpart from WHYY NEWS during the POINT AND COUNTER POINT program that follows the news.
WHAT NEWS SPECIAL INTEREST REPORTER	conservative	The special interest reporter's assignment is to find a good special interest story that relates to the issue under consideration in the newscast. (S)he interviews people who are participating in the other CAREER game, and from these interviews (s)he attempts to write a short four minute story that (s)he will read during the WHAT NEWS PROGRAM. This story should reflect the political outlook of the producers of WHAT NEWS and advocate the "correct" side of the issue as a reporter from WHAT NEWS would see it.
WHAT NEWS POLITICAL REPORTER	conservative	The political news reporter's assignment is to find a story that has a political perspective and relates to the issue considered in the newscast. (S)he interviews people who are participating in the other CAREER game, and from these interviews (s)he attempts to write a short four minute story that (s)he will read during the WHAT NEWS PROGRAM. This story should reflect the political outlook of the producers of WHAT NEWS and advocate the "correct" side of the issue as a reporter from WHAT NEWS would see it.
WHAT NEWS ECONOMIC REPORTER	conservative	The economic news reporter's assignment is to find a story that has an economic perspective and relates to the issue considered in the newscast. (S)he interviews people who are participating in the other CAREER game, and from these interviews (s)he attempts to write a short four minute story that (s)he will read during the WHAT NEWS PROGRAM. This story should reflect the political outlook of the producers of WHAT NEWS and advocate the "correct" side of the issue as a reporter from WHAT NEWS would see it.
WHYY NEWS ANCHOR PERSON	liberal	(S)he coordinates the WHYY NEWS program. (S)he makes sure that everyone understands their task and checks in on everyone on the WHYY NEWS staff to make sure that everything will be ready by the newscast. During the newscast the anchor person introduces the show and provides an introduction to each news story. In addition to this, the Anchor person must also debate his/her counterpart from WHAT NEWS during the POINT AND COUNTER POINT program that follows the news.

TABLE 13-16 Continued

Role	*Political Bias*	*Game Task*
WHYY NEWS SPECIAL INTEREST REPORTER	liberal	The special interest reporter's assignment is to find a good special interest story that relates to the issue under consideration in the newscast. (S)he interviews people who are participating in the other CAREER game, and from these interviews (s)he attempts to write a short four minute story that (s)he will read during the WHYY NEWS PROGRAM. This story should reflect the political outlook of the producers of WHYY NEWS and advocate the "correct" side of the issue as a reporter from WHYY NEWS would see it.
WHYY NEWS POLITICAL REPORTER	liberal	The political news reporter's assignment is to find a story that has a political perspective and relates to the issue considered in the newscast. (S)he interviews people who are participating in the other CAREER game, and from these interviews (s)he attempts to write a short four minute story that (s)he will read during the WHYY NEWS PROGRAM. This story should reflect the political outlook of the producers of WHYY NEWS and advocate the "correct" side of the issue as a reporter from WHYY NEWS would see it.
WHYY NEWS ECONOMIC REPORTER	liberal	The economic news reporter's assignment is to find a story that has an economic perspective and relates to the issue considered in the newscast. (S)he interviews people who are participating in the other CAREER game, and from these interviews (s)he attempts to write a short four minute story that (s)he will read during the WHYY NEWS PROGRAM. This story should reflect the political outlook of the producers of WHYY NEWS and advocate the "correct" side of the issue as a reporter from WHYY NEWS would see it.
ADD PEOPLE	none	The advertizing people in this simulation do not have anything to do with the various news programs and as such do not have any political biases that interfere with their work. These two to three people are responsible for writing four thirty-second commercials, which are shown during the two news programs. These commercials can be about imaginary products or they can be paid political announcements sponsored by the various interest groups involved within the planning of WESLEY TOWN.

Step 4 FIRST NEWSCAST—The two anchor people should flip a coin to see who presents first. A possible time scheduling is given in APPENDIX I. It is important to note that this newscast cannot last longer than fifteen minutes. This will be timed by the game operator. While one news team is presenting, the other news team should take notes that might be helpful during the POINT AND COUNTER-POINT DEBATE that follows the two news programs.

Step 5 SECOND NEWSCAST—same as above.

Figure 13-11
WESMED rating form.

WHAT CAREER GAME DID YOU PLAY?
WESSDEV ( )
WESPRO ( )
WESCRED ( )

WHICH WAS THE BEST NEWS CAST?
WHAT NEWS ( ) conservative news
WHYY NEWS ( ) liberal news

WHICH NETWORK WON THE DEBATE?
WHAT NEWS ( )
WHYY NEWS ( )

WHICH ADVERTISED PRODUCT DID YOU LIKE BEST?

Step 6 POINT AND COUNTER-POINT DEBATE—After the newscasts, the two anchor people will debate the issue that will be considered during the WESSPON game. The format will be as follows: First debater will be given three minutes for his or her opening comment. The other anchor person will then be given three minutes to counter this presentation. The first debater will then have one minute for rebuttal as will the second debater.

Step 7 RATINGS—Before the newscasts it is suggested that a rating form be distributed among the various career players. These people will be asked to rate the newscasts and the ads and decide who has won the debate. The suggested form is shown above (see figure 13-11).

Appendix I

A possible scheduling form:

1. INTRODUCTION BY THE ANCHOR PERSON: Hello, I am NOSMO KING, and this is WHAT . . .
2. THIRTY-SECOND AD
3. POLITICAL NEWS
4. ECONOMIC AND SOCIAL NEWS
5. THIRTY-SECOND AD
6. SPECIAL INTEREST NEWS
7. THE REMAINING TIME CAN BE USED BY THE ANCHOR PERSON TO SUM UP THE NEWS FOR TONIGHT

Appendix II

Ideas for commercials:

1. Procter and Gamble's new dessert topping—TWO IN ONE—TWO IN ONE can be used as a dessert topping as well as a floor wax—Stops yellow build up on your teeth and your kitchen floor.

2. DANTE's new trash compactor and incinerator —This new product turns kitchen scraps, paper, small dogs, and even tin cans into ashes. . . . Save on your trash bills, help the environment, and rid the world of unsightly trash. Battery operated by 30 D batteries. Batteries are not included.

3. America is going through a new craze—KIWI BERRIES—Yes KIWI BERRIES! This tropical fruit imported from the tropical paradise of Tierra del Fuego—is shaped like a small lemon—but is a dull brown. It's covered by strands of delicate brown hair. But when you open them up, there is where the fun begins. KIWI BERRIES are bright green inside and have numerous tiny black spots that are rumored to be seeds. Surprise your friends. Watch the surprised look on their faces when they bite into a ripe KIWI BERRY. They make great presents! Buy them in any one of your many local stores.

4. BASS-O-MATIC—Tired of scaling and gutting the fish that you had so much fun catching yourself? Well try this new product by AVOCO—BASS-O-MATIC—no more scaling, no more unsightly gutting, just drop your fish into this powerful blender and turn it on. Why pour those vitamins down the drain? Just blend and drink. THAT'S SOME BASS.

For your BASS-O-MATIC just send $29.99 to Post Office Box 1976, SHOPPING MALL, NEW JERSEY. Hurry and send off today, we have only 6 million of these little babies and they are selling like hotcakes.

WESPRO

Player Instructions

This game attempts to model the effects of two forces acting on the professional's job role—the desire for professional respect (PROFESSIONALISM) and the desire for monetary gain (DOLLARS). In each round, each player chooses his or her own goal: Each can choose to either maximize PROFESSIONALISM, or to maximize DOLLARS. At the end, the game has two winners—the player with the highest PROFESSIONALISM score, and the player with the highest DOLLAR score.

Various positive and negative PROFESSIONALISM and DOLLAR points are associated with each set of objectives available to each player (see figure 13-12).

In each round, the player or team is to develop a project that consists of one to five of those objectives. In selecting objectives for a project, the player should try to select those that (s)he thinks will maximize that player's particular goal (PROFESSIONALISM or DOLLAR). However, players should be aware that a project maximizing PROFESSIONALISM can receive too low a DOLLAR score, and similarly a project maximizing DOLLAR score can have too low a PROFESSIONALISM score. In both cases the project is not implemented, and the player receives penalty points.

In each round of play except the first, players can invest up to 10% of their DOLLARS or PROFESSIONALISM score in status. Each 1000 DOLLAR status investment translates to 1 PROFESSIONAL CREDIT. Each 1 PROFESSIONALISM investment translates to 1000 DOLLARS CREDIT. Part or all of the CREDITS can be used in future rounds to offset a potentially low PROFESSIONAL or DOLLAR score.

Figure 13-12
WESPRO objectives list.

Profitable Objectives (Dollars)

1. Maximize the number of dwelling units per land area.
2. Level the land before development.
3. Use standardized housing types.
4. The site plan is guided by attempts to minimize sewage line cost.
5. Drain storm water into city system.
6. Clear out trees and shrubs before development.
7. Use inexpensive materials.
8. Build small units.
9. Provide cheaply built plumbing.
10. Minimize investment in sound, smell, and heat insulation.

Professional Objectives (Professionalism)

A. The number of units should be based upon a desirable growth rate for the area.
B. The development should respond to, not alter, the site.
C. A variety of housing types should be provided.
D. The development should provide its own storm drainage.
E. Trees and natural elements should be sensitively added.
F. Long lasting materials should be used.
G. Buildings should express their use.
H. Energy-efficient heating systems should be used.
J. The development should function efficiently.
K. The quality of the designs for the units should be regulated.

Profitable and Professional Objectives

I. Include a mix of income and racial groups.
II. Laundry rooms should be provided within the development.
III. Commercial areas should be developed to serve the residences.
IV. The development should be built such that it can be easily policed by local authorities.

Steps of Play

Step 1 From the OBJECTIVES LIST, choose one to five objectives to make up your first project and write them on the SCORE SHEET at 1 (see figure 13-13 on page 244). Indicate WESPOWER group.

Step 2 Hand sheets to game operator, who will assess and note your project scores in the spaces marked 2. Then mark your total at 3.

Step 3 You can invest up to 10% of your total score in status at 4. Insert new total at 5.

Step 4 Transfer that investment to 6 at the rate of \$1000 invested = 1 PROFESSIONAL CREDIT, and 1 PROFESSIONAL investment = \$1000 CREDIT.

Step 5 Note your total CREDIT accrued at 7.

Step 6 Note how much of that CREDIT you wish to invest for your next project in the three places marked 8.

Step 7 Record your total CREDIT at 9.

Step 8 Reiterate steps 1 through 9.

Game Operator Instructions

This game can be played with any number of players or teams of players. However, it is recommended that, unless you computerize the game operator's assessment of project scores, you have one scorer for each five to ten players or teams. In this way, the play of the game will not be unnecessarily delayed.

The game operator should run the game as follows:

1. Hand out one OBJECTIVE LISTS, one SCORES SHEET, and one PLAYER INSTRUCTION sheet to each player or team.
2. Read the introduction to the PLAYER INSTRUCTION sheet aloud.
3. Tell players to take a quick look at the OBJECTIVES LIST.
4. Initiate the first round of play by reading each of the steps of play, and allowing players to complete each task.
5. The game should be played for at least four rounds, but can go on for longer.
6. On completion of the game, open the floor to discussion.

Your project scoring chart is shown below. Only the game operator and his scorers should have access to this sheet. Associated with each objective are two scores—a DOLLAR and a PROFESSIONALISM score. Make sure to calculate credit invested in reaching the final score.

	DOLLAR	PROFESSIONALISM
1.	3000	-1
2.	3000	-3
3.	2000	0
4.	3000	-2
5.	2000	-2
6.	1000	-2
7.	2000	-2
8.	3000	-3
9.	1000	-3
10.	2000	-3
	DOLLAR	PROFESSIONALISM
A.	-1000	+3
B.	-1000	+3
C.	-1000	+3
D.	-1000	+2
E.	-2000	+2
F.	-1000	+1
G.	-3000	+2
H.	-2000	+2
J.	-3000	+3
K.	-2000	+3
	DOLLAR	PROFESSIONALISM
I.	+1000	+1
II.	+1000	+1
III.	+1000	+1
IV.	+1000	+1

If a player scores less than -4000 DOLLARS or less than -4 PROFESSIONALISM points his project is

Figure 13-13
WESPRO score sheet.

ROUND		*SCORES* PRO	*SCORES* DOLLAR	*CREDITS*	PRO	DOLLAR
CREDIT ______	BONUS	______	______			
PROJECT	PROJECT SCORE	② ______	② ______			
	TOTAL	③ ______	③ ______	PREVIOUS TOTAL	______	______
①	STATUS INVEST-MENT	④ ______	④ ______	CREDIT TRANSFER	⑥ ______	⑦ ______
	TOTAL	⑤ ______	⑤ ______	TOTAL	⑦ ______	⑦ ______
CREDIT ⑧ ______	BONUS	______	______	CREDIT INVESTED	⑧ ______	⑧ ______
PROJECT	PROJECT SCORE	______	______			
	TOTAL	______	______	TOTAL	⑨ ______	⑨ ______
	STATUS INVEST-MENT	______	______	CREDIT TRANSFER	______	______
	TOTAL	______	______	TOTAL	______	______
CREDIT ______	BONUS	______	______	CREDIT INVESTED	______	______
PROJECT	PROJECT SCORE	______	______			
	TOTAL	______	______	TOTAL	______	______
	STATUS INVEST-MENT	______	______	CREDIT TRANSFER	______	______
	TOTAL	______	______	TOTAL	______	______
CREDIT ______	BONUS	______	______	CREDIT INVESTED	______	______
PROJECT	PROJECT SCORE	______	______			
	TOTAL	______	______	TOTAL	______	______
	STATUS INVEST-MENT	______	______	CREDIT TRANSFER	______	______
	TOTAL	______	______	TOTAL	______	______

not approved and instead of his actual point tally he is given a penalty score as follows:

DOLLARS CALCU-LATED	DOLLAR	PROFESSION-ALISM
-4000	-4000	-1
-5000	-5000	-2
-6000	-6000	-3
-7000	-7000	-4
-8000	-8000	-5
-9000	-9000	-6
-10000	-10000	-7
-11000	-11000	-8
-12000	-12000	-9
-13000	-13000	-10
-14000	-14000	-11
-15000	-15000	-12
PROFESSION-ALISM CALCULATED	**DOLLAR**	**PROFESSION-ALISM**
-4	-1000	-4
-5	-2000	-5
-6	-3000	-6
-7	-4000	-7
-8	-5000	-8
-9	-6000	-9
-10	-7000	-10
-11	-8000	-11
-12	-9000	-12
-13	-10000	-13
-14	-11000	-14
-15	-12000	-15

If this game is played in connection with WESPOWER, then the squares, triangles, and circles start with different bonus points after WESPOWER. This should be announced to the players in the round immediately following WESPOWER.

Any square receives both a 2 point PROFESSIONAL BONUS and a $2000 DOLLAR BONUS.

Any triangle can choose either a 1 point PROFESSIONAL BONUS, or a $1000 DOLLAR BONUS.

Circles receive nothing.

Modifications for Specific Purposes

This game represents a frame game that a game designer can adapt to fit certain specific purposes. To adapt the game to your specific purposes, you should take the following steps:

1. Identify the individual, group, role, or object that has two or more opposing forces acting upon it, or guiding its actions.
2. Identify those forces.
3. Identify a set of resultant actions or objectives that relate to these forces. To each action or objective, attach values based upon your estimate or an informed estimate of how each action or objective benefits or hurts each force.
4. Play the game several times using the above forces and values, and successively calibrate the set of actions and their values to better fit the reality that you are attempting to model.

WESPON

Introduction

WESPON (WESLEY TOWN POLICY NEGOTIATIONS) is an adaptation of the basic Policy Negotiations frame model. People representing various interest groups attempt to have issues passed that would maximize both their influence in the community as well as their voting strength within the public forum. These interest groups include professionals, developers, and workers. The winning members of the games representing each group represent their group in the WESPON game.

Player and Game Operator Instructions

Each subgroup within each career game (WESPRO—PROFESSIONALS, and DOLLAR; DEVELOPER—GOVERNMENT, and DEVELOPERS; WESWORK CITIZENS—squares, circles, and triangles) sends its winner to the WESPON game as a representative. Each representative initially receives ten influence chips. Each representative receives a copy of the game board for WESPON (shown in figure 13-14), and places the board in front of his/herself. Each number on the game board represents an issue of the same number from the list of issues (figure 13-15), which each representative also receives.

While the other career game participants do not sit at the WESPON game table, they play an important role in trying to influence both their own representative and the other representatives.

Finally, there is a group of three JUDGES (a CONSERVATIVE, a LIBERAL, and an ADVERTISER from the MEDIA game). The JUDGES each have a copy of the judging form shown in figure 13-16 on page 248.

Figure 13-14
WESPON game board.

1	YES	NO	2	YES	NO	3	YES	NO
4	YES	NO	5	YES	NO	6	YES	NO
7	YES	NO	8	YES	NO	9	YES	NO
10	YES	NO	11	YES	NO	12	YES	NO
13	YES	NO		YES	NO		YES	NO
			SAVINGS IN CONSERVATIVE MEDIA			SAVINGS IN LIBERAL MEDIA		

Figure 13-15
WESPON issues list.

1. The worker's input on physical elements will be given prime consideration in the final building plan. If this issue fails, professionals will have the final say.

2. There will be no noticeable differences in the facades of garden apartments within the planned community. Although rents may vary, they will be determined by interior facilities rather than upon the elaborate nature of the exterior design.

3. A high diversity of housing types, for each worker type, will be constructed. This proposal represents an additional cost to the developer. If this proposal fails, the developer may build at his own discretion.

4. There will be a separation of pedestrian and vehicular traffic incorporated with the plan. A local transit system will eliminate the need for a car within the development.

5. An evaluative study by the media of the worker's satisfaction with the housing will receive the full cooperation of the developer. This study, to be conducted at five, ten, and 25 year periods, has received financial support from the federal government.

6. The community should be economically integrated. If this issue passes, 25% of the community will be low income, 50% middle income, and 35% high income. If the issue fails, 10% will be low income, 50% middle, and 40% high.

7. No large scale development should be allowed in this area.

8. Wesley Town should be a remedial community where disenfranchised poor could be trained for work and life in the surrounding communities.

9. During the initial planning phases, the plan was seen as being as socially integrated as possible, including elderly and minority housing. If yes, 50% minorities, 12% elderly, and 38% other. If no, 12% minorities, 12% elderly, and 76% other.

10. It has been suggested by developers and professionals alike that a new PRD ordinance be instituted. Such an ordinance would break the standard lot and allow more freedom in the development and planning.

11. The local school board suggests that any new development of 160 units or more supply an elementary school for residents. The costs of this school would be absorbed by the developer in building costs.

12. It was suggested at a local country club that architects involved within the planning of Wesley Town develop a code of aesthetics for the development. This may make houses and condominiums more expensive.

13. Local union leaders have suggested that construction within the development should provide work for the local poor and minority residents. This may raise building costs and subsequently lower profits if any type of on-the-job training is instituted.

Steps of Play

Step 1 The first step of play consists of two brief news reports presented by the CONSERVATIVE and LIBERAL media. The time allocated to each report is to be a function of the number of chips stored in that MEDIA by the players in step 2, *allocation to savings.*

Step 2 The second step of play has six activities taking place *simultaneously:* Agenda setting; Policy vote; Allocation to savings; Advertising; Bargaining between representatives; Team influence on representatives; Judges' feedback. These are described in detail below.

Agenda setting The representatives place their influence chips on the YES or NO for issues on the issue board on which they wish to have a vote, or no vote, accordingly. The issue with the highest total score wil be placed on the agenda for a vote next round. The total score is calculated by adding all of the YES votes, and then by subtracting all of the NO votes from that total.

Policy vote The representatives vote on the issue placed on the agenda in the previous round. The policy option currently on the agenda may be passed or defeated by a 2 to 1 margin. If a 2 to 1 margin is not attained, the motion is tabled and returned to a 0 value on the Agenda board.

In voting, each player allocates part or all of their chips to YES or NO for the issues under concern.

Allocation to savings Players can also allocate chips to savings by placing them in either the CONSERVATIVE or LIBERAL media box. Chips assigned to savings can be withdrawn at a future round for a more important vote or agenda item. The chips in savings are counted each round to determine how much time each media will have to present their stories. Thus players should save in the media that best reflects their own interests. Players may attempt to bargain with the media to give them air time in exchange for saving chips.

Purchasing advertising Players can buy advertising with the media, but these chips are considered lost. Each media station determines the amount of chips required for a specific amount of advertising time. The station keeps these chips on their own chart, and while the player cannot get them back, the chips are reflected in the total chip count for that station.

Bargaining between representatives Players in this game are encouraged to interact with one another and to set up bargains and tradeoffs with other players to get issues PASSED, DEFEATED, or on the agenda.

Team influence on representatives A representative can attempt to use his career game constituents to assist him in his bargaining by sending them to team members from other representatives with specific deals. The constituents can attempt to influence their representative's selection of issues, and votes, by indicating how much bargaining support or resistance they would give for the various issues. Thus, if for instance, team members disapprove of their representative's choice, they could try to counter his bargaining efforts by bargaining with other players to stop that choice. The constituents from one career game cannot talk directly with the representative from another career game, they can only talk directly with their own representative and with other constituents from their own game and from other games.

Judging feedback After each vote, depending on the favorableness of the outcome of the vote, the various groups can gain or lose influence chips to be received in the next round. The JUDGES will be responsible for assessing the consequences. Using the JUDGES' form (figure 13-16), they are to assess the possible consequences of each PASSED, DEFEATED, or TABLED issue brought to a vote on the groups. This assessment must be completed before the vote is tallied.

Step 3 In this step of play, the game operator tallies the agenda item votes, the vote on the issues, the media votes, and announces whether the issue is PASSED or DEFEATED, indicates the agenda item for the next round, and indicates the media time for broadcast at the start of the next round of play.

The game operator indicates the total agenda tally for each issue on a chalkboard visible to all players.

The JUDGES announce the shifts in influence for the next round, and briefly describe their reasoning. As the judges announce the shifts, the game operator distributes chips accordingly.

Step 4 Reiterate steps 1 through 3.

Adaptations for Specific Purposes

1. Identify the problem of interest.
2. Identify specific issues relating to the problem, and identify any possible outside actions or events that might affect the situation.
3. Identify the groups or individuals who influence and are affected by the various issues, and possibly set guidelines for how these influences work.
4. Insert these roles, influences (in terms of initial number of chips), and issues into the framework used here. In this framework, you can omit or keep the news media role depending on its influence on your particular problem. The judges may play the role of assessing or indicating any outside forces that may crop up as a function of time or specific decisions, and assessing the impact of these forces on future decisions.

Figure 13-16
WESPON judge's form.

ROUND NUMBER	POLICY OPTION ON AGENDA	GROUP	EFFECT OF PASSAGE (plus or minus influence chips)	EFFECT OF DEFEAT (plus or minus influence chips)	EFFECT OF TABLING (plus or minus influence chips)
		1 PRO			
		2 DOL			
		3 GOV			
		4 DEV			
		5 SQ.(hi)			
		6 CIR(mid)			
		7 TRI(lo)			
		1 PRO			
		2 DOL			
		3 GOV			
		4 DEV			
		5 SQ(hi)			
		6 CIR(mid)			
		7 TRI(lo)			
		1 PRO			
		2 DOL			
		3 GOV			
		4 DEV			
		5 SQ(hi)			
		6 CIR(mid)			
		7 TRI(lo)			
		1 PRO			
		2 DOL			
		3 GOV			
		4 DEV			
		5 SQ(hi)			
		6 CIR(mid)			
		7 TRI(lo)			

Bibliography

INTRODUCTION

Arnstein, Sherry. 1969. A Ladder of Citizen Participation. *Journal of the American Institute of Planners* **35**(4):216, 224.

Davidoff, Paul. 1965. Advocacy and Pluralism in Planning. *Journal of the American Institute of Planners* **31**(4):331.

Dean, Andrea O. 1976. Community Design Centers: Practicing "Social Architecture." *Journal of the American Institute of Architects* **65**(1):38.

Delbecq, Andre L., and Van de Ven, Andrew H. 1971. A Group Process Model for Identification and Program Planning. *Journal of Applied and Behavioral Sciences* 7(July/Aug.):4.

Delbecq, Andre L.; Van de Ven, Andrew H.; and Gustafson, David H. 1975. *Group Techniques for Program Planning: A Guide to Nominal Group and Delphi Processes.* Glenview, Illinois: Scott Foresman.

Duke, Richard. 1974. *Gaming, the Future's Language.* Beverley Hills, California: Sage Publications.

Friedmann, John. 1973. *Retracking America: A Theory of Transactive Planning.* New York: Anchor Press.

Goodman, Robert. 1973. *After the Planners.* New York: Simon and Schuster.

Katz, Daniel, and Kahn, Robert. 1966. *The Social Psychology of Organizations.* New York: John Wiley and Sons.

Linstone, H., and Turoff, M. 1975. *The Delphi Method: Technique and Applications.* Reading, Mass.: Addison-Wesley.

Michael, Donald N. 1973. *On Learning to Plan and Planning to Learn.* San Francisco: Jossey-Bass.

Peattie, Lisa. 1968. Reflections on Advocacy Planning. *Journal of the American Institute of Planners* **34**(2):80–88.

Rogers, Everett, with Shoemaker, F. 1971. *Diffusion of Innovations.* New York: The Free Press.

Tester, David, ed. 1974. *Designing the Method.* North Carolina Student Publication of the School of Design, Vol. 23.

HISTORIC DISTRICT

Berk, Emanuel. 1974. *Downtown Improvement Manual.* Chicago: American Society of Planning Officials Press.

Blake, Peter. 1964. *God's Own Junkyard: The Planned Deterioration of America's Landscape.* New York: Holt, Rinehart and Winston.

Costonis, John J. 1974. *Space Adrift.* Urbana: University of Illinois Press.

Insall, Donald. 1974. *The Care of Old Buildings Today: A Practical Guide.* New York: Watson-Guptill Publication.

McNulty, Robert, and Kliment, Steven A. 1976. *Neighborhood Conservation: A Handbook of Methods and Techniques.* New York: Whitney Library of Design.

Rudofsky, Bernard. 1969. *Streets for People: A Primer for Americans.* Garden City, New York: Doubleday and Co., Inc.

Wurman, Richard Soul. 1971. *Making the City Observable.* Cambridge: MIT Press.

ENVIRONMENTAL AWARENESS

Clay, Grady. 1974. *Close-Up: How to Read the American City.* New York: Praeger Publishers.

Group for Environmental Education. 1971. *Our Man-Made Environment.* Cambridge: MIT Press.

Lynch, Kevin. 1960. *The Image of the City.* Cambridge: MIT Press.

McCoy, Michael. 1976. *Problem Solving in the Man-Made Environment.* Michigan: Cranbrook Academy of Art.

Nelson, Doreen. 1974. *City Building Educational Program.* Los Angeles: CBEP.

Rasmussen, Steen Eiler. 1962. *Experiencing Architecture.* Cambridge: MIT Press.

Sanoff, Henry. 1975. *Seeing the Environment.* North Carolina: Learning Environments.

Wurman, Richard S. 1972. *Yellow Pages of Learning Resources.* Cambridge: MIT Press.

Index